THE DESIGN OF SYNTHETIC INHIBITORS OF THROMBIN

ADVANCES IN EXPERIMENTAL MEDICINE AND BIOLOGY

Recent Volumes in this Series

Volume 338
CHEMISTRY AND BIOLOGY OF PTERIDINES AND FOLATES
Edited by June E. Ayling, M. Gopal Nair, and Charles M. Baugh

Volume 339
NOVEL APPROACHES TO SELECTIVE TREATMENTS OF HUMAN SOLID TUMORS: Laboratory and Clinical Correlation
Edited by Youcef M. Rustum

Volume 340
THE DESIGN OF SYNTHETIC INHIBITORS OF THROMBIN
Edited by Goran Claeson, Michael F. Scully, Vijay V. Kakkar, and John Deadman

Volume 341
CIRRHOSIS, HYPERAMMONEMIA, AND HEPATIC ENCEPHALOPATHY
Edited by Santiago Grisolía and Vicente Felipo

Volume 342
CORONAVIRUSES: Molecular Biology and Virus–Host Interactions
Edited by Hubert Laude and Jean-François Vautherot

Volume 343
CURRENT DIRECTIONS IN INSULIN-LIKE GROWTH FACTOR RESEARCH
Edited by Derek LeRoith and Mohan K. Raizada

Volume 344
MECHANISMS OF PLATELET ACTIVATION AND CONTROL
Edited by Kalwant S. Authi, Steve P. Watson, and Vijay V. Kakkar

Volume 345
OXYGEN TRANSPORT TO TISSUE XV
Edited by Peter Vaupel, Rolf Zander, and Duane F. Bruley

Volume 346
INTERACTIVE PHENOMENA IN THE CARDIAC SYSTEM
Edited by Samuel Sideman and Rafael Beyar

A Continuation Order Plan is available for this series. A continuation order will bring delivery of each new volume immediately upon publication. Volumes are billed only upon actual shipment. For further information please contact the publisher.

THE DESIGN OF SYNTHETIC INHIBITORS OF THROMBIN

Edited by

Goran Claeson
Michael F. Scully
Vijay V. Kakkar

and

John Deadman

Thrombosis Research Institute
London, United Kingdom

PLENUM PRESS • NEW YORK AND LONDON

Library of Congress Cataloging-in-Publication Data

The Design of synthetic inhibitors of thrombin / edited by Goran Claeson ... [et al.].
p. cm. -- (Advances in experimental medicine and biology ; v. 340)
"Proceedings of an International Symposium on the Design of Synthetic Inhibitors of Thrombin, held July 8-9, 1991, in London, United Kingdom"--T.p. verso.
Includes bibliographical references and index.
ISBN 0-306-44593-X
1. Antithrombins--Congresses. I. Claeson, Göran.
II. International Symposium on the Design of Synthetic Inhibitors of Thrombin (1991 : London, England) III. Series.
[DNLM: 1. Thrombin--antagonists & inhibitors--congresses. 2. Drug Design--congresses. 3. Hirudin--pharmacology--congresses. W1 AD559 v. 340 1993 / QV 193 D457 1991]
QP93.7.A58D47 1993
615'.718--dc20
DNLM/DLC
for Library of Congress 93-45361
CIP

Proceedings of an International Symposium on The Design of Synthetic Inhibitors of Thrombin, held July 8–9, 1991, in London, United Kingdom

ISBN 0-306-44593-X

A Division of Plenum Publishing Corporation
233 Spring Street, New York, N.Y. 10013

Printed in the United States of America

FOREWORD

In one generation, the numerous factors involved in blood coagulation have become real protein entities, isolated in pure form, expressed by recombinant DNA techniques, and subjected to structure elucidation by the modern methods of physical chemistry, viz., X-ray diffraction, and NMR, ESR and fluorescence spectroscopy. The major milestone in this field was the breakthrough achieved by W. Bode, R. Huber and their colleagues in 1989 in determining the crystal structure of human α-thrombin, inhibited with D-Phe-Pro-Arg chloromethyl ketone. The availability of this structure will greatly facilitate the interpretation of experiments designed to gain an understanding of the interatomic interactions between this enzyme and fibrinogen and its other substrates. At the same time, it provides a rational basis for the design and synthesis of inhibitors of thrombin, the subject of this symposium.

The symposium was organized in four sessions: (1) Structural features of the interaction of thrombin with substrates and inhibitors, (2) Synthetic inhibitors, (3) Hirudin and its analogues, and (4) Pharmacological and clinical considerations. This book contains summaries of most of the papers presented, and takes its righful place among two others that provide a comprehensive picture of our current knowledge about thrombin, viz. the 1977 volume entitled "Chemistry and Biology of Thrombin", edited by R.L. Lundblad, J.W. Fenton II, and K.G. Mann, and the 1992 volume entitled "Thrombin: Structure and Function", edited by L.J. Berliner.

The X-ray and spectroscopic work presented in this symposium has elucidated the structures of complexes of thrombin with fibrinopeptides, with peptides derived from hirudin, and with other inhibitors. These structures, together with the known chemistry of action of the enzyme, are guideposts for designing both peptides and non-peptide thrombin inhibitors, allowing the medicinal chemist and enzyme kineticist to explore many possibilities, taking advantage of various target sites on the surface of the thrombin molecule. One of the most potent inhibitors of thrombin is hirudin, and the chemistry of this compound and analogs is receiving considerable attention for anticipated clinical use as antithrombotic agents. The work described here offers the hope of increasing the arsenal of thrombin inhibitors that can function at various stages of the blood clotting process and in situations where natural inhibitors such as antithrombin III become ineffective. It is fair to say that the symposium has documented much progress in the field of thrombin inhibitors, and points the way to further breakthroughs in the future. In the words of one of the participants, quoting Churchill, "This is not the end, nor even the beginning of the end, but it is the end of the beginning".

Harold A. Scheraga
Todd Professor of Chemistry
Cornell University
Ithaca, New York

ACKNOWLEDGEMENTS

The organizers of this symposium are grateful to the following pharmacological companies for financial support.

Abbott Laboratories (USA)

Biogen (USA)

Ciba-Geigy

I.C.I. Pharmaceuticals

Immuno Ltd.

Lilly Industries Ltd.

Mitsubishi Chemical Industrial Ltd.

Organon International Ltd.

Sandoz (Nurenberg)

The Wellcome Research Laboratories

PREFACE

In current clinical practice the prevention and control of thrombosis is primarily dependent on two drugs, heparin and coumarin both developed in the first half of this century after serendipitous observations made with natural products. In modern times pharmaceutical developments may still arise from the natural source but also drugs can be designed with respect to known properties of target proteins or enzymes in purified systems. Within the thrombosis field, the development of thrombin inhibitors constitute the first designer drugs - an orally active drug being a key aim. From the early work of the Blombäck group upon the features of the primary structure of fibrinopeptide which favoured cleavage by thrombin, the design of specific chromogenic substrates was developed based upon the tight binding peptide D-Phe-Pro-Arg, similar to the structure of a number of the specific inhibitors of thrombin which have since been developed. In the mid-eighties with the advent of recombinant technology, hirudin became available and the elucidation of the structural features contributing its properties - the most potent and specific of all thrombin inhibitors - became possible. This information will undoubtedly lead to further advance in the design of synthetic thrombin inhibitors. With the publication of the crystal structures of thrombin it is an exciting period in this area of research especially as thrombin inhibitors are now becoming available as pharmaceuticals for assessment in the clinic. For this reason a meeting was organized at the Thrombosis Research Institute in London on July 8th and 9th 1991 to consider all aspects of the design of thrombin inhibitors. The meeting was attended by over 120 international delegates, the pharmaceuticals industry being particularly well represented. As organizers we were fortunate to enlist the interest of the eminent researchers in this field and are grateful for the quality of oral presentations which engendered most stimulating discussion throughout the meeting. Their written contributions attest to the quality of work which was presented. We are very grateful to Professor H.A. Scheraga, one of the pioneers in the area of thrombin specificity, for his support in providing a foreward for this work.

The editors are very grateful for the financial support from the pharmaceutical industry lists on the facing page. Our thanks to Mrs. S. Vost, our administrative secretary, for the meeting. We are particularly thankful to Mrs. Eileen Bayford for the energy, dedication and skill she has shown in completing the typescript for this book.

G. Claeson
M.F. Scully
V.V. Kakkar
J. Deadman

London, April 1993

CONTENTS

THE RATIONAL DESIGN OF THROMBIN-DIRECTED ANTITHROMBOTICS

J.W. Fenton II[1-3], F. Ni[4], J.I. Witting[1], D.V. Brezniak[1], T.T. Andersen[2], and A.B. Malik[3]

[1]New York State Department of Health, Wadsworth Center for Laboratories and Research, P.0. Box 509, Albany, New York 12201-0509; [2]Department of Biochemistry and Molecular Biology and [3]Department of Physiology and Cell Biology, The Albany Medical College of Union University, New Scotland Avenue, Albany, New York 12208, USA; and [4]Molecular Biology Section, National Research Council Canada, Biotechnology Research Institute, 6100 Avenue Royalmont, Montreal, Quebec H4P 2R2, Canada

There are few original ideas in Science and for that matter in life in general. Most often, such ideas are the "logical extension" of existing knowledge, practice, or state of the art. Originality might be considered as a form of pathology. That is, the incorrect recall or substitution of an idea or act. But it is the ability to recognize new associations which becomes creativity. In these regards, the imperfect mind creates new ideas and a creative mind must be able to sort out what is new or novel. The term "novel", however, in Science has acquired the connotation of "subject to consideration of patenting". Most often, novel ideas constituting inventions, are conceived when the time is correct or the circumstances are right and occur independently to more than one person. Hence, the old adage "necessity is the mother of invention". If such ideas are good enough, those who have conceived them are frequently overshadowed by those who promote them, and originality becomes the victim of promotion or marketing. Good Science is like any other business or human undertaking; the marketing, promotion, and/or propagation of an idea and its application to a useful purpose are all essential components in the inventive process. Otherwise, the best of ideas are soon forgotten.

The concept of bridge-binding double-ligand inhibitors of thrombin (Figure l) is an example of the case in point in the rational design of novel therapeutics. The general structure-function relationships of the blood clotting enzyme, thrombin[1-10] were known (Figure 2). Such relationships were furthermore known for the leech-derived thrombin inhibitor, hirudin[11-14], and how it interacted with thrombin had been reviewed[15,16]. However, prior to the knowledge of the crystallographic structures of human α-thrombin[17] or its hirudin complexes[18,19], the bridged double-ligand concept was independently thought of within about two years by three persons. These persons were the first author in Albany, New York[20], the second author in Montreal, Quebec[21], and Dr. V.S. Chauhan in

The Design of Synthetic Inhibitors of Thrombin
Edited by G. Claeson, *et al.*, Plenum Press, New York, 1993

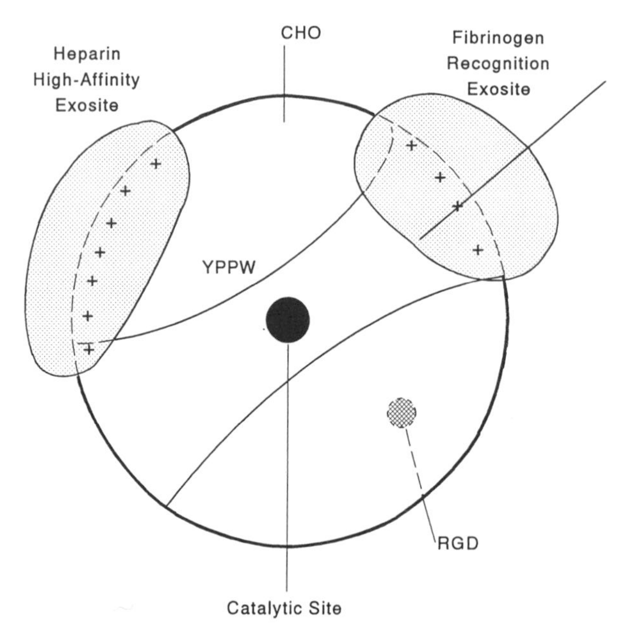

Figure 1 A diagramatic representation of the bridge-binding double ligand concept leading to the development of thrombin-directed inhibitors, termed "hirulogs".

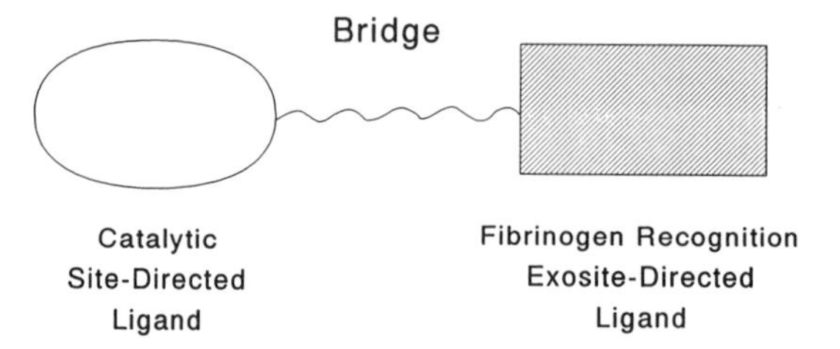

Figure 2 A conceptional drawing of the general structure of α-thrombin. The catalytic site, heparin high-affinity exosite, and fibrinogen recognition exosite are identified. The abbreviation CHO indicates the carbohydrate attachment, YPPW the Tyr-Pro-Pro-Trp60D tetrapeptide, and RGD the Arg-Gly-Asp189 tripeptide. The A chain resides on the backside and is not shown.

New Delhi, India[22]. In the first author's case the idea came from the underlying observation that hirudin binding blocked the fibrin(ogen) recognition exosite in addition to the catalytic site in thrombin[23] and from two minor studies carried out in collaboration with Dr. Jawed Fareed in Chicago, Illinois. The first confirmed[24-27] that peptides corresponding to the hirudin tail were inhibitors of the fibrinogen clotting activity of thrombin and speculated that these peptides blocked the fibrin(ogen) recognition exocite for α-thrombin[27]. The second demonstrated that leupeptin was an effective thrombin inhibitor although three orders of magnitude less potent than D-Phe-Pro-Arg-H[28]. Hence, associating the D-Phe-Pro-Arg ligand, which binds in the catalytic-site region[29,30], with the exosite-binding hirudin tail peptides was perhaps a rational thing to do.

Having worked on other affinity-labelling systems[31,32] and having completed with Dr. John M. Maraganore in Cambridge, Massachusetts a paper on affinity-labelling the fibrin(ogen) recognition exosite with a hirudin tail peptide[33], the first author was asked by Maraganore if the Cambridge group could make any peptides of interest. After thinking about the offer, the peptide of D-Phe-Pro-Arg-Pro-$(Gly)_x$-hirudin tail peptide was suggested. Here, the second Pro was included, since the Pro following Lys47 in hirudin had been thought to prevent cleavage[6,11], although the Lys-Pro48 bond had been shown to be inaccessible for cleavage[34]. The length of the $(Gly)_x$-bridge was estimated from the preceding affinity-labelling study and was subsequently experimentally determined[20]. For the second ligand, the desulfohirudin[53-64] or desulfohirugen (Figure 3) was used because of the experience of the Cambridge group with such peptides[26]. Owing to an oversight, the first peptide synthesized was D-Phe-Pro-Arg-$(Gly)_5$-desulfo-hirugen lacking the second Pro. As could be predicted, this peptide was essentially no better an inhibitor of fibrinogen clotting activity than hirugen, suggesting that its Arg-Gly bond was readily cleaved by thrombin. Upon hearing of this result, the Cambridge group was informed that they should have put in the second Pro as originally proposed to prevent cleavage. A few weeks later, a call came from Maraganore in Atlanta airport stating that the correct peptide (hirulog-l in Figure 3) had been a potent inhibitor of thrombin, approximately three orders of magnitude more so than hirugen with a K_i in the lower nanomolar range[20]. This potency was subsequently confirmed in Albany (Table l), but the inhibitor was found to slowly lose its potency upon incubation with thrombin[35,36]. This finding suggested that thrombin could slowly cleave the Arg-Pro bond, which was confirmed in Cambridge by peptide separation and sequencing techniques[37]. This prompted the preparation of analogs with a nonscissional bond with ß-homo Arg but these had slightly to appreciably reduced potencies and offered little or no advantages over hirulog-l[38]. Other investigations in Cambridge led to an analog with greater potency by substituting D-cyclohexylAla for the first residue[37]. Employing tight-binding kinetic techniques[36], this peptide was confirmed in Albany to be more potent than the original (Table l).

The independent conception of double-ligand thrombin inhibitors in Montreal, originates from some of the first author's writings[5,6,9] and the second author's NMR studies on thrombin interactions with fibrinopeptides[39-42] and a hirudin tail fragment[43]. NMR studies with Dr. Harold A. Scheraga in Ithaca, New York demonstrated that human fibrinopeptide A can still bind to thrombin after being released from the fibrinogen Aα chain[40]. In the thrombin-bound state, the aromatic side chain of residue Phe8 of fibrinopeptide A bends back toward the residue Gly14 in the peptide segment Gly-Val-Arg16 that precedes the cleaved Arg-Gly17 peptide bond. This observation provided an explanation for the fact the D-Phe-Pro-Arg ligand binds to the active site of thrombin[17]. The side chain of the D-Phe residue in the tripeptide mimics the contribution of the Phe8 in the natural substrate.

Hirugen

H_3N^{+}-Asn-Gly-Asp(−)-Phe-Glu(−)-Glu(−)-Ile-Pro-Glu(−)-Glu(−)-Tyr(SO_3^{-})-Leu-CO_2^{-}

Hirulog-1

H_3N^{+}-(D-Phe)-Pro-Arg(+)-Pro-Gly-Gly-Gly-Gly-Asn-Gly-Asp(−)-Phe-Glu(−)-Glu(−)-Ile-Pro-Glu(−)-Glu(−)-Tyr-Leu-CO_2^{-}

HR1

H_3N^{+}-Asn-Asp(−)-Gly-Asp(−)-Phe-Glu(−)-Glu(−)-Ile-Pro-Glu(−)-Glu(−)-Tyr-Leu-Gln-CO_2^{-}

P53

Ac-(D-Phe)-Pro-Arg(+)-Pro-Gln-Ser-His-Asn-Asp(−)-Gly-Asp(−)-Phe-Glu(−)-Glu(−)-Ile-Pro-Glu(−)-Glu(−)-Tyr-Leu-Gln-CO_2^{-}

Figure 3 Structures of hirugen[26], hirulog-1[21], and peptides HR1[43] and P53[21] portrayed in ionized states. Peptide HR1 corresponds to residues 52 through 65 of hirudin HV1 without the sulfonated Tyr63.

Table 1 Inhibition constants for hirudin-derived bridge-binding peptides with human α-thrombin.

Origin	Peptide	K_i(t=0) pM	k min^{-1}
Albany*	hirulog-1	2,560 ± 350	0.33 ± 0.08
	hirulog-B2	77.1 ± 0.4	0.20 ± 0.04
Montreal#	P53	2,800 ± 900	not determined
	P79	370 ± 30	not cleaved

* Values were determined at pH7.4 and -23°C by tight binding methods. The K_i is that for zero time binding, whereas k is the cleavage rate. The structure of hirulog-1 is in Figure 3 and that of hirulog-B2 differs by D-cyclohexyl Ala substituted in the first position.

Values were determined at pH7.8 and 25°C by conventional kinetic methods. The structure of P53 is in Figure 3 while that of P79 has the corresponding Arg ketomethylene bond and the Pro residue after Arg is replaced by Gly (44).

During the NMR studies of fibrinopeptide A, this author also became interested in the thrombin-hirudin interactions[5,6,9] and was perplexed by the suggestion that fibrinopeptide A may share part of the binding site of the acidic hirudin C-terminal tail on thrombin[13]. In an attempt to resolve this issue, we suggested to our collaborator, Dr. Yasuo Konishi, then with Monsanto at St. Louis, Missouri, to synthesize a peptide fragment corresponding to residues Gly42 to Gln65 of the hirudin tail. Owing to a problem with peptide synthesis, we decided to obtain only a shorter peptide comprising residues Asn52 to Gln65 (HRl in Fig.3). This peptide was found to bind to bovine α-thrombin by an analysis of NMR transferred NOEs[43]. Furthermore, the binding was not affected by the D-Phe-Pro-Arg ligand, which irreversibly blocks the catalytic site and adjacent regions of thrombin[29,30]. Thus D-Phe-Pro-Arg and the hirudin tail peptide represent two classes of ligands that bind to thrombin independently. At this point, we were convinced that fibrinopeptide A cannot occupy the same site as the hirudin tail peptide and were talking about fusing the two peptides together to see the enhanced binding. The D-Phe-Pro-Arg ligand was substituted for fibrinopeptide A and linked it to the hirudin tail peptide through residue Pro48, assuming that the Lys-Pro48 bond in hirudin corresponded to the Arg-Gly17 scissional bond in the Aα chain. The first peptide made included a D-Pro at position 48 of hirudin since it was thought to prevent cleavage. This peptide, however, showed very little enhanced binding[21]. This Pro was subsequently replaced by L-Pro in hirudin to give peptide P53 (Figure 3) with enhanced activity similar to those of the preceding independent groups (Table l).

With this chimeric peptide available, the second author went ahead to investigate the binding to thrombin by NMR spectroscopy and found out that there was a time-dependent change in the NMR spectra of the peptide in the presence of thrombin. Since there is only one Arg in the entire peptide (Figure 3), this NMR finding suggested that the peptide was cleaved at the Arg-Pro peptide bond by thrombin, which was later confirmed by peptide separation and sequencing analysis[21]. This observation prompted the synthesis of noncleavable analogs at the scissional Arg-Pro peptide bond[44]. Reasoning by analogy with human fibrinogen, which incorporates a glycine residue after Arg16, we have successfully incorporated a ketomethylene surrogate into P53 with enhanced inhibitory activity toward human thrombin (Table 1).

The independent origin in New Delhi comes from Chauhan's studies on analogues of hirudin tail fragments. Some of these peptides were examined in Albany because he could not send them through the mail and we could not safely ship thrombin on dry ice to India. In a letter to the first author, he suggested making a double ligand inhibitor, and the first author informed him that we had made such inhibitors. He subsequently informed us of the potency of his inhibitors but has not, to our knowledge, published his data[23].

The term "hirulog" comes from one coined by the Cambridge group when they were in search of a hirudin tail analogue more potent than hirugen. The term thought of in Albany was "thromboblock", which was particularly conceived on the notion that blocking thrombin conversion of fibrinogen into fibrin was the primary target for antithrombotic intervention. Although this is an important target in preventing thrombus formation, the underlying target is that of thrombin activation of factors V, VIII, and XI in amplification of thrombin generation[45-47]. Of note, γ-thrombin essentially lacks fibrinogen procoagulant activity, as the consequence of disruption of its fibrin(ogen) recognition exosite, but is capable of amplifying thrombin generation[48]. Furthermore, the hirudin tail fragment analogue, hirugen, can only particularly block thrombin generation and does so in a nonsaturable manner[49], suggesting that the fibrin(ogen) recognition exosite is not essential for amplification. On the other hand, hirugen blocks thrombin activation of platelets[50] and particularly certain endothelial cell responses[51]. Hirulog-l is not an inhibitor of trypsin, factor Xa, nor γ-thrombin[20,35,35] but specifically inhibits α- and ζ-thrombin which have similar fibrinogen procoagulant activities[36]. Thus, the design of thrombin-directed antithrombotics may well be directed toward differentiating between thrombin functions.

The fact remains that hirulog is an effective antithrombotic and has passed phase I clinical trials without exhibiting indications of either toxicity or immunogenicity[52]. In contrast, proteins of non-host physiological origins are prone to be immunogenic, whereas synthetic compounds of unnatural origins are most likely to be toxic. Herein lie two major problems in drug design and discovery.

One of the disadvantages of peptide therapeutics is that they usually have relatively short half lives and may require continual administration to maintain therapeutic levels. On the other hand, this may be an advantage in that antidote may not be necessary, since relatively rapid clearance should occur upon discontinuing administration. Another disadvantage of oligopeptides of the size of hirulog-1 is they are not normally intestinally absorbed and require injection or other routes of administration. In the case of an antithrombotic for acute situations, however, this is not a disadvantage, since the long standing agent, heparin, and its various derivatives, are administrated by injection under hospital settings.

The attractiveness of tripeptide and other short peptides for antithrombotics is that of possible oral administration, potentially fewer synthetic steps, and reduced manufacturing costs. Because of their small size, these agents are solely directed to regions at or near the catalytic site and are affinity-labelling reagents[29,53], transition-state analogues[28,54,55], or contain unnatural constituents[55-57], where the potential for toxicity is fairly high. As compared with those of other serine proteinases, the catalytic site of α-thrombin is situated in a very deep cleft or canyon[17]. Notably, the canyon walls are very hydrophobic, and the highly unusual Tyr-Pro-Pro-Trp 60D (chymotrypsin numbering) tetrapeptide is above the catalytic site (Figure 2). The topography formed by this tetrapeptide constitutes a unique subsite for D-Phe[17] and presumably similar residues accounting for the specificity of thrombin-directed tripeptide ligands[30,58].

Moreover, this subsite has been implicated in binding of hirudin[18] and fibrinopeptide A[39-42] and may comprise as essential component of the thrombin domain for leukocyte chemotaxis[59]. As such, it is one of the few structural features around which thrombin-specific short peptides may be designed. However, the potential toxicities of reactive or other groupings necessary for an inhibitor pose serious limitations in therapeutic development.

On the other hand, the fibrinogen recognition exosite is characterized by a shallow surface cavity with hydrophobic residues surrounded by positively-charged Arg or Lys residues[17,60] Attempts have been made to design shorter peptide mimics of the hirudin tail in the hope that such peptides may still be thrombin-specific inhibitors targeted toward this

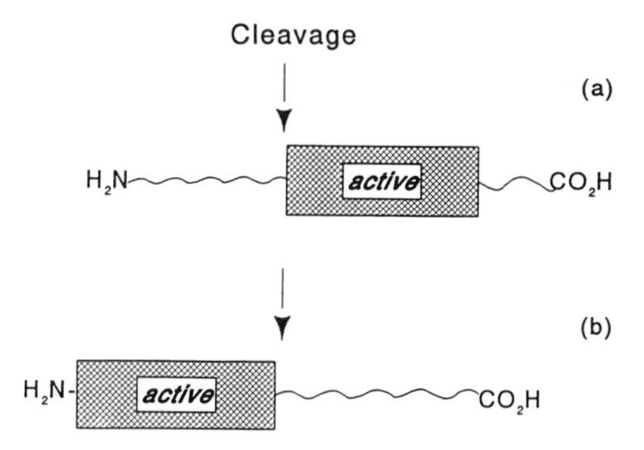

Figure 4 A schematic drawing where a proteinase activates a peptide product on the CO_2H-terminal side (a) or the NH_2-terminal side (b) of a preactivated peptide.

exosite. It has been demonstrated that the bulky 3_{10}-helix (residues Glu61 to Gln65) in the hirudin tail peptides can be substituted by a single hydrophobic residue with minimal loss of antithrombin activity[60]. Further truncation of these peptides to 4 residues resulted in a great loss of activity and specificity[61], and such tetrapeptides might be modified for other structural features of the exosite[62]. Such subtle structural features may further be modelled employing computer-aided molecular design that are subject of recent studies[63]. Thrombin has also an exosite for high-affinity binding of heparin (Figure 2). Other than heparin binding[64], the function of this exosite is unknown, although it may participate in factor V and VIII activation[65]. The design of double-ligand bridge-building inhibitors for this exosite and the catalytic site is conceivable but to our knowledge has not been attempted.

Additional unique structural features of thrombin include: i) the A chain attachment site, ii) the carbohydrate attachment site, and iii) the Arg-Gly-Asp189 tripeptide (Figure 2). The A-chain is a 36-residue remnant activation peptide attached by a disulfide bridge. It resides on the back side of thrombin and appears to be of little importance[17], although it may contribute some conformational stability to thrombin[3,6]. The carbohydrate attachment site is located within the major insertion in thrombin, as is also the Tyr-Pro-Pro-Trp60D tetrapeptide, but is removed from the catalytic site[17]. The carbohydrate side chain has no influence on thrombin activities[66] nor is it masked by complexing with hirudin[67], suggesting that it is of little interest in drug design. The Arg-Gly-Asp189 tripeptide, however, is of interest because of receptor binding properties of such peptides. In α-thrombin, this tripeptide appears to be buried[17] and requires protein denaturation for exposure in promoting cell spreading[68]. Synthetic peptides corresponding to the sequence containing the tripeptide have been observed to enhance wound healing[69].

Recent cloning and sequencing of the genes for thrombin receptors have opened a new avenue for drug design[70,71]. The thrombin receptors are seven-transmembrane proteins with a thrombin susceptible Arg-Ser bond followed by a conserved Ser-Phe-Phe/Leu-Leu-Arg-Asn-Pro- - - sequence, a spacing sequence, and then a negatively charged sequence similar to those of the fibrinogen Aα chain, hirudin tail fragments, and other proteins that interact with the thrombin exosite for fibrinogen recognition[70]. The synthetic 14-residue peptide from the scissional site causes platelet aggregation[71] and stimulates endothelial cell activation analogous to that induced by thrombin[72,73]. Thus, thrombin cleaves the receptor protein, creating a tethered biologically active amino-terminal peptide, which can be mimicked by synthetic analogues. If such biologically active peptides can be synthesized, then conceivably a variety of antagonist peptides can be prepared by appropriate amino acid substitutions. Taking the concept another step further then, a thrombin-directed ligand, such as D-Phe-Pro-Arg, could be added, where the peptide was biologically inactive until cleaved by thrombin. Thus, thrombin can potentially generate platelet and endothelial cell agonists, perfecting injury-directed antithrombotic therapy. This is perhaps a new concept in antithrombotic drug design, and the reverse activation mechanism can also be envisioned (Figure 4).

To understand thrombin is to understand haemostasis and *vice versa*[74]. The multiplicity of thrombin functions in haemostasis makes it a primary target for therapeutic intervention. The possibility of differentiating thrombin functions from injury through healing exist and provide unique challenges in drug design. Thrombin has distinct structural features that have so far permitted two inhibitor classes: thrombin-directed tripeptides and double-ligand bridge-binding oligopeptides. The former class takes advantage of unique subsites neighbouring the catalytic site but are not absolutely thrombin-specific and have potentially toxic groupings. The latter class utilizes the thrombin exosite for fibrin(ogen) recognition, as well as the preceding subsites. Such inhibitors are specific for thrombin forms with fibrinogen clotting activities and exhibit no known toxicity or immunogenicity. An additional class of thrombin-directed antithrombotics might consist of platelet and/or endothelial cell receptor antagonists generated from peptides with a thrombin scissional bond permitting injury-directed therapy.

ACKNOWLEDGEMENTS

We wish to thank Debra VonZwehl for secretarial assistance. This work was supported in part by NIH grants HL13160 to JWF and HL27016, HL32418, and HL45638 to ABM and funds to FN from the National Research Council of Canada.

REFERENCES

1. S. Magnusson, Thrombin and prothrombin, in: P.D. Boyer, Ed: The Enzymes, 3rd ed, Vol 3, Academic Press, New York, pp. 277 (1971).
2. S. Magnusson, T.E. Petersen, L. Sottrup-Jensen, and H. Claeys, Complete primary structure of prothrombin: isolation, structure, and reactivity of ten carboxylated glutamic acid residues and regulation of prothrombin activation by thrombin, in: E. Reich, D.B. Rifkin, and E. Shaw, Eds, Proteases and Biological Control, Cold Spring Harbor Laboratory, Cold Spring Harbor, NY, pp. 123 (1975).
3. J.W. Fenton II, B.H. Landis, D.H. Bing, R.D. Feinmann, M.P. Zabinski, S.A. Sonder, L.J. Berliner, and J.S. Finlayson, Human Thrombin: preparative evaluation, structural properties and enzymic specificity, in, D.H. Bing, Ed. The Chemistry and Physiology of the Human Plasma Proteins, Pergamon, New York, NY, pp. 151 (1979).
4. D.H. Bing, R. Laura, D.J. Robison, B. Furie, B.C. Furie, and R.J. Feldmann, A computer-generated three-dimensional model of the B chain of bovine α-thrombin, Ann. NY Acad. Sci, 370:496 (1981).
5. J.W. Fenton II, Thrombin specificity, Ann. NY Acad. Sci, 370:468 (1981).
6. J.W. Fenton II, and D.H. Bing, Thrombin active-site regions, Semin. Thromb. Hemost, 12:200 (1986).
7. B. Furie, D.H. Bing, R.J. Feldmann, D.J. Robison, J.P. Burnier, and B.C. Furie, Computer-generated models of blood coagulation factor Xa, factor IXa, and thrombin based upon structural homology with other serine proteinases, J. Biol. Chem, 57:3875 (1982).
8. D.H. Bing, R.J. Feldmann, and J.W. Fenton II, Structure-function relationships of thrombin based on the computer generated three-dimensional model of the B chain of bovine thrombin, Ann. NY Acad. Sci, 485:104 (1986).
9. J.W. Fenton II, Thrombin, Ann. NY Acad. Sci, 485:5 (1986).
10. J.W. Fenton II, Regulation of thrombin generation and functions, Semin. Thromb. Hemost, 14:234 (1988).
11. D. Bagdy, E. Barabas, L. Graf, T.E. Petersen, and S. Magnusson, Hirudin, Methods Enzymol. 45:669 (1976).
12. J. Dodt, H.P. Muller, V. Seemuller, and J.Y. Chang, The complete amino acid sequence of hirudin, a thrombin specific inhibitor, FEBS Lett, 165:180 (1984).
13. J.Y. Chang, The functional domain of hirudin, a thrombin-specific inhibitor, FEBS Lett, 164:307 (1984).
14. J. Dodt, V. Seemuller, R. Maschler, and H. Fritz, The complete covalent structure of hirudin, Localization of the disulfide bonds, Biol. Chem. Hoppe-Seyler, 366:379 (1985).
15. J.W. Fenton II, Thrombin interactions with hirudin, Semin. Thromb. Hemost. 15:265 (1989).
16. P.H. Johnson, P. Sze, R. Winant, P.W. Payne, and J.B. Lazar, Biochemistry and genetic engineering of hirudin, Semin. Thromb. Hemost. 15:302 (1989).
17. W. Bode, I. Mayr, V. Baumann, R. Huber, S.R. Stone, and J. Hofsteenge, The refined 1.9 A crystal structure of human α-thrombin: interaction with D-Phe-Pro-Arg chloromethyl-methyl ketone and significance of the Tyr-Pro-Pro-Trp insertion segment, EMBO J. 8:3467 (1989).
18. T.J. Rydel, K.G. Ravichandraw, A. Tulinsky, W. Bode, R. Huber, C. Roitsch, and J.W. Fenton II, The structure of a complex of recombinant hirudin and human α-thrombin, Science, 249:277 (1990).

19. M.G. Gruther, J.P. Priestle, J. Rahuel, H. Grossenbacher, W. Bode, J. Hofsteenge, and S.R. Stone, Crystal structure of the thrombin-hirudin complex: a novel mode of serine protease inhibition, EMBO J. 9:2361 (1990).
20. J.M. Maraganore, P. Bowidon, J. Jablonski, K.L. Ramachandran, and J.W. Fenton II, Design and characterization of hirulogs: a novel class of bivalent peptide inhibitors of thrombin, Biochemistry, 29:7095 (1990).
21. J. DiMaio, B. Gibbs, D. Munn, J. Lefebvre, F. Ni, and Y. Konishi, Bifunctional thrombin inhibitors based on the sequence of hirudin45-65, J. Biol. Chem. 265:21698 (1990).
22. V.S. Chauhan, Personal communication, 1990.
23. J.W. Fenton II, T.A. Olson, M.P. Zabinski, and D.G. Wilner, Anion-binding exosite of human α-thrombin and fibrin(ogen) recognition, Biochemistry, 27:7106 (1988).
24. J.L. Krstenansky, and S.J. Mao, Antithrombin properties of C-terminus of hirudin using synthetic unsulfated Nα-acetyl-hirudin45-65, FEBS Lett. 211:10 (1987).
25. S.J.T. Mao, M.T. Yates, T.J. Owen, and J.L. Krstenansky, Interaction of hirudin with thrombin: identification of a minimal binding domain of hirudin that inhibits clotting activity, Biochemistry 27:8170 (1988).
26. J.M. Maraganore, B. Chao, M.L. Joseph, J. Jablonski, and K.L. Ramachandran, Anticoagulant activity of synthetic hirudin peptides, J. Biol. Chem. 264:8692 (1989).
27. J.W. Fenton II, J.I. Witting, C. Pouliott, and J. Fareed, Thrombin anion-binding exosite interactions with heparin and various polyanions, Ann. NY Acad. Sci. 556:158 (1989).
28. J.I. Witting, J.L. Pouliott, J.L. Catalfamo, J. Fareed, and J.W. Fenton II, Thrombin inhibition with dipeptidyl argininals, Thromb. Res. 50:461 (1988).
29. C. Kettner, and E. Shaw, D-Phe-Pro-ArgCH_2Cl - a selective affinity label for thrombin, Thromb. Res. 14:969 (1979).
30. S.A. Sonder, and J.W. Fenton II, Proflavin binding within the fibrinopeptide groove adjacent to the catalytic site of human α-thrombin, Biochemistry 23:1818 (1984).
31. J.W. Fenton II, and S.J. Singer, Affinity labelling of antibodies to the p-azophenyltrimethyl-ammonium hapten and a comparison of affinity-labelled antibodies of two different specificities, Biochemistry 10:1429 (1971).
32. D.H. Bing, M. Cory, and J.W. Fenton II, Exosite affinity labelling of human thrombins. Similar labelling on the A chain and B chain/fragments of clotting α- and nonclotting ß-thrombins, J. Biol. Chem. 252:3587 (1977).
33. P. Bourdon, J.W. Fenton II, and J.M. Maraganore, Affinity labelling of lysine-149 in the anion binding exosite of human α-thrombin with a N^{α}-dinitro fluorobenzyl-hirudin C-terminal peptide, Biochemistry 29:6379 (1990).
34. J. Dodt, S. Kohler, T. Schmitz, and B. Wilhelm, Distinct binding sites of Ala48-hirudin1-47 and Ala48-hirudin48-65 on α-thrombin, J.Biol. Chem. 265:713 (1990).
35. J.W. Fenton II, J.I. Witting, P. Bourdon, and J.M. Maraganore, Thrombin-specific inhibition by a novel hirudin analog, Circulation 82:III-659, (abstract) (1990).
36. J.I. Witting, P. Bourdon, J.M. Maraganore, and J.W. Fenton II, Thrombin-specific inhibition by hirulog-1 and cleavage of its arginyl-3-propyl-4 bond, Thromb. Haemost. 65:829 (abstract), (1991).
37. J.M. Maraganore J.W. Fenton II, T. Kline, P. Bourdon, J. Witting, J.A. Jablonski and C. Hammond, Modifications in hirulog peptides yielding improved antithrombin activities, Thromb. Haemost. 65:830 (Abstract).

38. T. Kline, C. Hammond, P. Bourdon, and J.M. Maraganore, Hirulog peptides with scissile bond replacements resistant to thrombin cleavage, Biochem. Biophys. Res. Commun. 177:1049 (1991).
39. F. Ni, H.A. Scheraga, and S.T. Lord, High-resolution NMR studies of fibrinogen-like peptides in solution: resonance assignments and conformational analysis of residues 1-23 of the Aα chain of human fibrinogen, Biochemistry 27:4481 (1988).
40. F. Ni, Y. Konishi, R.B. Frazier, H.A. Scheraga, and S.T. Lord, High resolution NMR studies of fibrinogen-like peptides in solution: interaction of thrombin with residues 1-23 of the Aα chain of human fibrinogen, Biochemistry 28:3082 (1989).
41. F. Ni, Y.C. Meinwald, M. Vasquez, and H.A. Scheraga, High-resolution NMR0 studies of fibrinogen-like peptides in solution: structure of a thrombin-bound peptide corresponding to residues 7-16 of the Aα chain of human fibrinogen, Biochemistry 28:3094 (1989).
42. F. Ni, Y. Konishi, L.D. Bullock, M.N. Rivetna, and H.A. Scheraga, High-resolution NMR studies of fibrinogen-like peptides in solution: structural basis for thebleeding disorder caused by a single mutation of Gly_{12} to Val_{12} in the Aα chain of human fibrinogen Rouen, Biochemistry 28:3106 (1989).
43. F. Ni, Y. Konishi, and H.A. Scheraga, Thrombin-bound conformation of the C-terminal fragments of hirudin determined by transferred nuclear Overhauser effects, Biochemistry 29:4479 (1990).
44. J. DiMaio, F. Ni, B. Gibbs, and Y. Konishi, A new class of potent thrombin inhibitors that incorporates a scissile pseudopeptide bond, FEBS Lett. 282:47 (1991).
45. T. Lindhout, R.Blezer, and H.C. Hemker, The anticoagulant mechanism of action of recombinant hirudin (CGP 39393) in plasma, Thromb. Haemost. 64:464 (1990).
46. X.J. Yang, M.A. Blajchman, S. Graven, L.M. Smith, N. Anvari, and F.A. Ofosu, Activation of factor V during intrinsic and extrinsic coagulation. Inhibition by heparin, hirudin and D-Phe-Pro-Arg-CH_2Cl, Biochem. J. 272:399 (1990).
47. D. Gailani, and G.J. Broze Jr, Factor XI activation in a revised model of blood coagulation, Science 253:909 (1991).
48. J.W. Fenton II, G.B. Villaneuva, F.A. Ofosu, and J.M. Maraganore, Thrombin inhibition by hirudin: how hirudin inhibits thrombin, Haemostasis 21 (suppl 1):27 (1991).
49. F.A. Ofosu, and J.W. Fenton II, Unpublished data.
50. J.A. Jakubowski, and J.M. Maraganore, Inhibition of coagulation and thrombin-induced platelet activities by a synthetic dodecapeptide modelled on the carboxy-terminus of hirudin, Blood 75:399 (1990).
51. M.A.A. DeMichele, D.G. Moon, J.W. Fenton II, and F.L. Minnear, Thrombin's enzymatic activity increases permeability of endothelial cell monolayers, J. Appl. Physiol. 69:1599 (1990).
52. A. Dawson, P. Loynds, K. Findlen, E. Levin, T. Mant, J. Maraganore, D. Hanson, J. Wagner, and I.Fox, Hirulog-1: a bivalent thrombin inhibitor with potent anticoagulant properties in humans, Thromb. Haemostas. 65:830 (abstract), (1991).
53. C. Kettner, L. Mersinger, and R. Knabb R, The selective inhibition of thrombin by peptides of boroarginine, J. Biol. Chem. 265:18289 (1990).
54. S. Bajusz, E. Barabas, P. Tolnay, E. Szell, and D. Bagdy, Inhibition of thrombin and trypsin by tripeptide aldehydes, Int. J. Pept. Prot. Res. 12:217 (1978).

55. J. Hauptmann, and F. Markwardt, Pharmakologie synthetischer thrombin inhibitoren, Beitrag zur Wirkstoffor-Schung, 26:1 (1986).
56. R. Kikumoto, Y. Tamao, T. Tezuka, S. Tonomura, M. Mara, K. Ninomiya, A. Hijikata, and S. Okamoto S, Selective inhibition of thrombin by (2R, 4R)-4-methyl-1-[N^2-[(3-methyl1,2,3,4-tetrahydro-8-quinolinyl) sulfonyl)-L-arginine)]2-piperidine-carboxylic acid, Biochemistry 23:85 (1984).
57. D. Turk, J. Stürzebecher, and W. Bode, Geometry of binding of the N_α tosylated piperidides of m-amidino-, p-amidino- and p-guanidino phenylalanine to thrombin and trypsin. X-ray crystal structures of their trypsin complexes and modelling of their thrombin complexes, FEBS Lett. 287:133 (1991).
58. S.A. Sonder, and J.W. Fenton II, Thrombin specificity with tripeptide chromogenic substrates: Comparison of human and bovine thrombins with and without fibrinogen clotting activities, Clin. Chem. 32:934 (1986).
59. R. Bar-Shavit, A. Kahn, M.S. Mudd, G.D. Wilner, K.G. Mann, and J.W. Fenton II, Localization of a chemotactic domain in human thrombin, Biochemistry 23:397 (1984).
60. S.Y. Yue, J. DiMaio, Z. Szewczuk, E.O. Purisima, F. Ni, and Y. Konishi, Characterization of the interactions of a bifunctionalinhibitor with α-thrombin by molecular modelling and peptide synthesis, Protein Engineering 5:77 (1992).
61. Z. Szewczuk, D. Leonard, D. Munn, J. DiMaio, F. Ni, S.Y. Yue, and Y. Konishi, Design of thrombin "exo-site" inhibitors based on structure-activity study, Proc. 12th American Peptide Symposium (abstr.) In press.
62. F. Ni, D.R. Ripoli, and E.O. Purisima, Conformational stability of a thrombin-binding peptide derived from the hirudin C-terminus, Biochemistry 31:2445 (1992).
63. N.C. Cohen, J.M. Blaney, C. Humblett, P. Gund, and D.C. Barry, Molecular modelling software and methods for medicinal chemistry, J. Med. Chem. 33:883 (1990).
64. F.C. Church, C.W. Pratt, C.N. Noyes, T. Kalayanamit, G.B. Sherill, R.B. Tobin, and J.B. Meade, Structural and functional properties of human α-thrombin, phosphorylated α-thrombin, and γ_T-thrombin. Identification of lysyl residues in α-thrombin that are critical for heparin and fibrin(ogen) interactions, J. Biol. Chem. 264:18419 (1989).
65. G.L. Hortin, Sulfation of tyrosine residues in coagulation factor V, Blood 76:946 (1990).
66. McD. Horne III, and H.R. Gralnick, The oligosaccharide of human thrombin investigations of functional significance, Blood 63:188 (1984).
67. T.A. Olson, S.A. Sonder, G.D. Wilner, and J.W. Fenton II, Heparin binding in proximity to the catalytic site of human α-thrombin, Ann. NY Acad. Sci. 485:96 (1986).
68. R. Bar-Shavit, V. Sabbah, M.G. Lampugnani, P.C. Marchisio, J.W. Fenton II, I. Vlodavsky, and E. Dejana, An Arg-Gly-Asp sequence within thrombin promotes endothelial cell adhesion, J. Cell. Biol. 112:335 (1991).
69. K.C. Glenn, G.H. Frost, J.S. Bergman, and D.H. Carney, Synthetic peptides bind to high-affinity thrombin receptors and modulate thrombin mitogenesis, Peptide Res. 1:65 (1988).
70. V.B. Rasmussen, V. Vouret-Craviai, S. Jallat, Y. Schlesinger, G. Pates, A. Pavirani, J.P. Lecoca, J. Pouyssegur, and E. Van Obberghen-Schilling, cDNA cloning and expression of a hamster α-thrombin receptor coupled to Ca^{2+} mobilization, FEBS Lett. 288:123 (1991).

71. T.K.H. Vu, D.T. Hung, V.I. Wheaton, and S.R. Coughlin, Molecular cloning of functional thrombin receptor reveals a novel proteolytic mechanism of receptor activation, Cell 64:1057 (1991).
72. J.R. Ngaiza, and E.A. Jaffe, A 14 amino acid peptide derived from the amino terminus of the cleaved thrombin receptor elevates intracellular calcium and stimulates prostacyclin production in human endothelial cells. Biochem. Biophys. Res. Commun. 179:1661 (1991).
73. A.B. Malik, and J.W. Fenton II, Thrombin-mediated increase in vascular endothelial permeability, Semin. Thromb. Hemost. 18:193 (1992).
74. J.W. Fenton II, F.A. Ofosu, D.G. Moon, and J.M. Maraganore, Thrombin structure and function: why thrombin is the primary target for antithrombotics, Blood Coagul. Fibrinol. 2:69 (1991).

X-RAY CRYSTAL STRUCTURES OF THROMBIN IN COMPLEX WITH D-PHE-PRO-ARG AND WITH SMALL BENZAMIDINE- AND ARGININE-BASED "NON-PEPTIDIC" INHIBITORS

Wolfram Bode

Max-Planck-Institut für Biochemie
D-8033 Martinsried, Germany

Thrombin plays a key role in thrombosis and haemostasis. Amongst other functions, thrombin effects thrombus formation through conversion of fibrinogen into fibrin and induction of platelet aggregation. Thrombin has therefore been implicated in various disease processes such as myocardial infarction, stroke or pulmonary embolism. In such cases, the quick administration of selective thrombin inhibitors might offer an attractive means of antithrombotic therapy[1]. Apart from a few natural protein inhibitors such as antithrombin III and hirudin, a large number of synthetic inhibitors have been found[2-9]. Some of these inhibitors (in particular MQPA = MD805, see Figure 1) are already in clinical trials. The recent availability of experimental thrombin structures[10,11] allows rationalization of these screening results with respect to substitution effects, conformation and shape[19,23]. In conjunction with interactive graphics and new computational methods, such experimental structures now offer the unique possibility of tailoring existing drugs, or of designing new compounds and elaborating them into drugs.

We have established the 1.9 Å crystal structure of human α-thrombin in complex with D-Phe-Pro-Arg CH_2Cl (PPACK-thrombin) and have introduced a new sequence numbering system for thrombin based on the topological equivalence with chymotrypsinogen and trypsin[10-12]. Recently, we have completed the crystallographic refinement of this PPACK-thrombin (R-value of 0.156) and have discussed the thrombin structure in detail including its active site cleft and interaction with PPACK[11,12]. These and other structural aspects of thrombin (such as its interesting electrostatic properties and its probable interactions with various ligands have been further outlined elsewhere[12,16,17,24,25]. The refined PPACK-thrombin model has further served to solve the X-ray crystal structures of human α-thrombin complexes with hirudin[13,14], hirugen and a fibrinogen-derived peptide[15-17], as well as those of bovine thrombin complexes with a fibrinopeptide A-related peptide or with benzamidine[18]. Up to now, we have only been able to obtain relatively poorly diffracting crystals of human thrombin in complex with noncovalently binding synthetic inhibitors (Bode, unpublished data). We therefore explored[19,20] the

The Design of Synthetic Inhibitors of Thrombin
Edited by G. Claeson, *et al.*, Plenum Press, New York, 1993

mode of interaction with thrombin of several arginine-, benzamidine- and isocoumarine-based inhibitors through modelling them to thrombin assuming homologous interactions as to bovine trypsin[19-22]. These "modelled" complexes allowed some plausible explanations for the selective and tight binding towards thrombin of most of these inhibitors[19,20], but required experimental verification and higher precision.

We are now able to grow reproducibly well-diffracting crystals of bovine thrombin in complex with various thrombin-specific small inhibitors. The refined 2.3 Å crystal structures of complexes with the lead inhibitors NAPAP and MQPA, and with the related

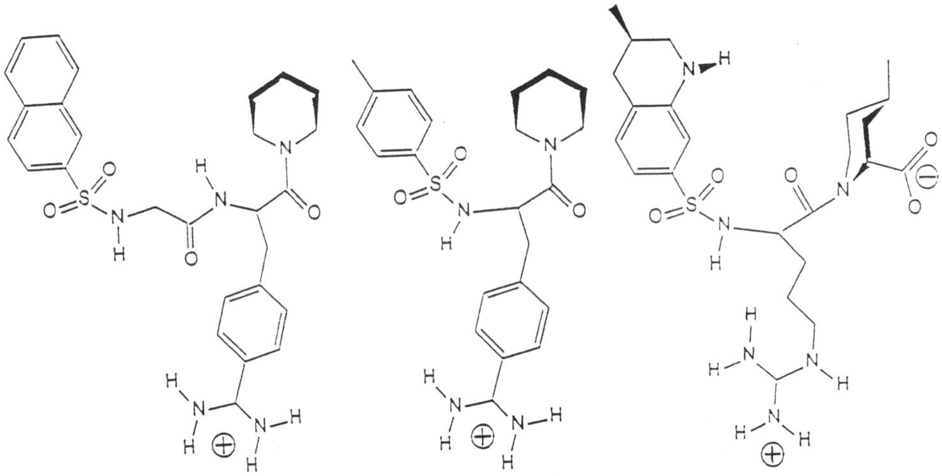

NAPAP 4-TAPAP MQPA

Figure 1 Chemical formulas of NAPAP (Nα-(2-naphthyl-sulphonyl-glycyl)-DL-p-amidino-phenylalanine-piperidine); 4-TAPAP (Nα-(4-toluene-sulphonyl)-DL-p-amidino-phenylalanine-piperidine);MQPA((2R,4R)-4-methyl-1-[Nα-((RS)-3-methyl-1,2,3,4-tetrahydro-8-quinolenesulphonyl)-L-arginyl]-2-piperidine carboxylic acid).

4-TAPAP, are being communicated elsewhere[23,24]. The topologies of the "non-peptidic" inhibitors investigated, together with their chemical names, are given in Figure 1. From our various thrombin structures we know that human and bovine thrombin exhibit similar active-site geometries; these structural results are therefore equally valid for both thrombin species[5,18,23].

Figure 2 shows the deep narrow "canyon"-like active-site cleft of human α-thrombin, the most remarkable feature of the thrombin surface[10-12]. This cleft is bordered and shaped by four extending polypeptide segments on one side (forming the "north rim"[10]), and two loop segments on the opposite side (the "south rim"). Most important for

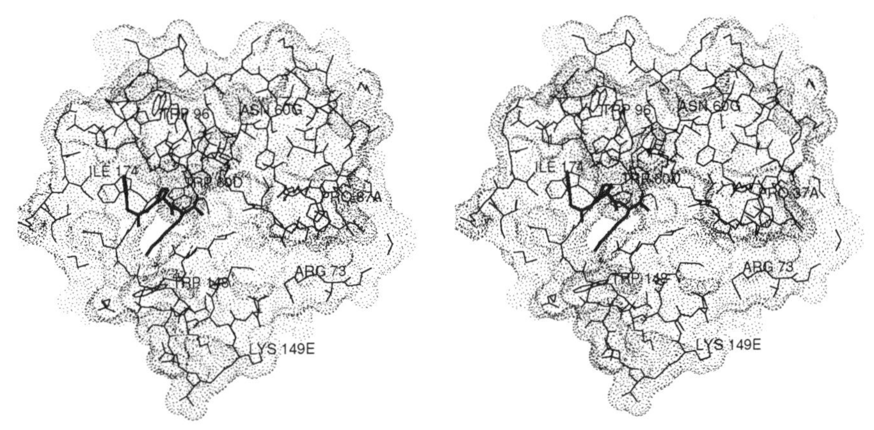

Figure 2 "Standard view" towards the thrombin molecule, displayed together with the bound PPACK molecule and the Connolly dot surface of thrombin. The active-site cleft runs from left to right across the molecular surface; the bound PPACK molecule is emphasized by bold connections. Only the front parts of the thrombin molecule and of its molecular surface are displayed. The surface "hole" close to the centre corrsponds to the entrance to the specificity pocket. The "60-insertion loop", in particular Trp60D, is partially occluding the active site (from[12]).

the narrow specificity of thrombin towards macromolecular substrates and inhibitors is the characteristic "60-insertion loop", which protrudes particularly far into the cleft and partially occludes access to the active site residues and to the specificity pocket.

Figure 3 shows the PPACK molecule as it is bound to the active-site cleft of thrombin. Its peptidyl moiety juxtaposes the extended thrombin segment Ser214-Trp215-Gly216 in a twisted antiparallel manner. The Arg3I side chain extends into the thrombin S1-pocket in a zig-zag-like conformation, very similar to that found in arginine-P1-protein inhibitor-trypsin complexes[26,27]. The guanidyl group opposes the carboxylate group of Asp189 at the bottom of the pocket, with each of the two terminal nitrogens placed in favourable and equal hydrogen/ionic bond distance to each carboxylate oxygen allowing optimal charge compensation (see Table 1); the third guanidyl nitrogen (NE) hydrogen bonds to Gly219 O. Ala190 of the thrombin pocket is almost perpendicular to the guanidyl group of Arg3I, and the buried solvent molecule 305 is located within its plane (see Figure 3).

The P3-(D-Phe1I) and P2-residue (Pro2I) of PPACK nestle tightly into the extended, quite hydrophobic depression near the entrance to the thrombin specificity pocket (lined by Ile174, Trp215, segment 97-99, His57, Tyr60A and Trp60D, see Figures 2 and 3), which is most certainly the suggested "apolar binding site"[28]. Its "back" part is made up of the same structural elements as in trypsin[29]; in thrombin, however, it is further screened off from bulk water by the mainly hydrophobic Tyr60A-Trp60D loop on one side and Ile174 on the other side. In the PPACK-thrombin complex, the pyrrolidine ring of Pro2I is almost fully shielded from bulk water; it has a completely hydrophobic environment, and seems to be perfectly accommodated through its polyproline II-conformation.

The side chain of the first PPACK-residue, D-Phe1I, fits into the notched, mainly hydrophobic cleft made by Ile174, Trp215, segment 97-99 and Tyr60A (Figure 3). Not only these favourable hydrophobic contacts, but also the perpendicular aryl-aryl ("edge-on") arrangement and the aryl-carbonyl contact with the carbonyl group of Glu97A, seem

Table 1 Some hydrogen bonding distances (in Å) for the experimental thrombin inhibitor complexes[11,23,24].

Thrombin residue	atom	NAPAP residue	atom		4-TAPAP residue	atom		MQPA residue	atom		PPACK residue	atom	
Asp189	OD1	PapaI3	NG2	2.9	PappI2	NG1	2.7	ArgI2	NH1	3.1	Arm3I	NH2	2.7
	OD2	PapaI3	NG1	3.0		NG1	3.0		NH1	2.8	Arm3I	NH1	2.8
Ser195	OG							MCPII3	OP2	3.0		N	3.1
Gly216	N	GlypI2	O	3.1		O	3.3	ArgI2	O	3.0	Phe1I	O	3.1
	O	GlypI2	N	2.7		N	2.9		N	2.8		N	2.7
Gly219	O	PapaI3	NG1	2.8	-	-			NE	2.7	Arm3I	NH2	2.8

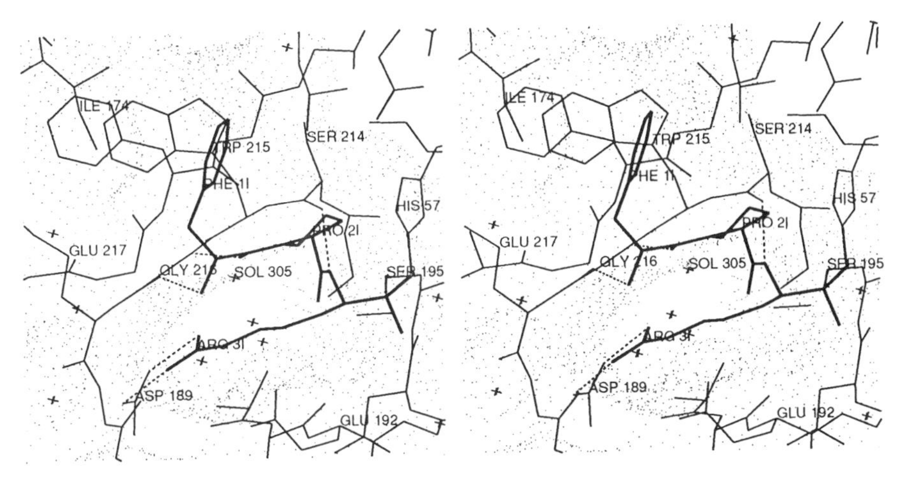

Figure 3 Interaction of the PPACK molecule (thick connections) with the thrombin active site (thin connections), displayed together with the Connooly surface of thrombin and some localized solvent molecules (crosses). The "60-insertion loop" is cut off to allow site (thin connections), displayed together with the Connolly surface of thrombin and an unobscured view into the S2-cavity. Most of the hydrogen bonds formed between PPACK and thrombin are shown in addition (dashed lines). The Arg31 carbonyl group forms a tetrahedral hemiketal structure covalently linked with Ser195 OG and (via the methylene group) with His57 NE2. Standard view as in Figure 2 (from[11]).

to contribute significantly to binding strength (see 11,19). The ammonium and the carbonyl group of D-PheII make favourable hydrogen-bond contacts with G1y216 (see Figure 3 and Table 1). An L-configured diastereomer PheII-residue at P3 would be unable to interact with thrombin in such a favourable manner (in particular when linked to a proline; see Figure 3); simultaneous hydrogen bond formation through its amino group and favourable aryl interaction would be mutually exclusive. The naphthyl, the tosyl and the chinolyl moieties of certain benzamidine- and arginine-derived synthetic inhibitors[11,19-21] (see Figure 7), as well as the phenolic side chain of Tyr3HI of hirudin[13] and the phenyl moiety of the fibrinogen Aa-chain Phe8[16-18] interact in a similar favourable manner with this cleft[23]. We have therefore proposed the designation "aryl binding site" for this cleft[11,19], distinguishing it from the other part of the "apolar binding site" (the "S2-cavity").

Figure 4 displays a section of the final electron density of the refined 2.3 Å NAPAP-thrombin complex[23] around the active site together with the corresponding model. NAPAP binds in a compact form to thrombin (very similar as to trypsin, see[19,29]), with its "main chain" juxtaposing thrombin segment Ser2l4-Gly2l6 in an antiparallel, slightly twisted manner; the amino and carbonyl groups of its glycine residue form favourable hydrogen bonds with Gly2l6 O and N of thrombin, respectively (see Table 1 and Figure 6). As in the trypsin complex[19], only the D-(or R-)-stereoisomer of the p-amidino-phenylalanine is bound; its side chain is sandwiched between main chain segments Alal90-

Figure 4 Active-site region of the complex formed between NAPAP (thick lines) and bovine thrombin (thin lines) superimposed with the final 2Fobs-Fcalc electron density. The view is towards the thrombin surface; the active-site residues are to the right, the specificity pocket is at the back. Contour surface is at 0.9 δ (from[23,24]).

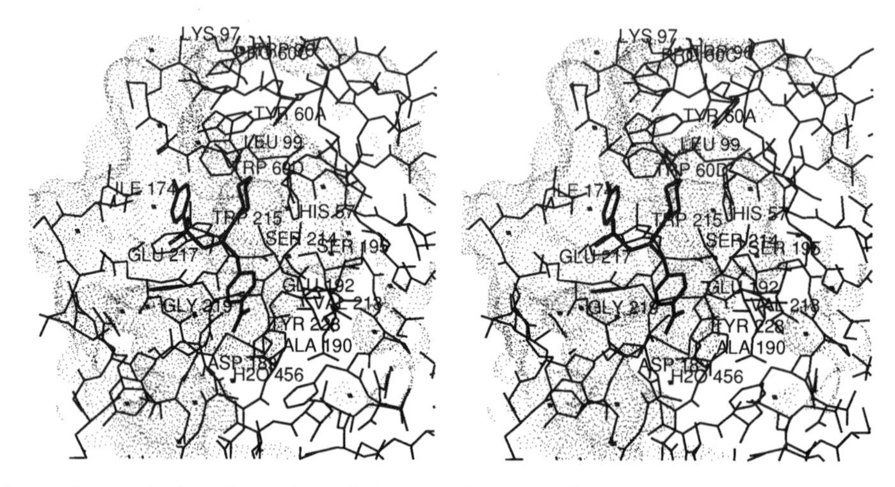

Figure 5 Active-site region of the complex formed between 4-TAPAP (thick lines) and bovine thrombin (thin lines) superimposed with the Connolly surface (from[23,24]).

Glu192 and Trp215-Gly216 of thrombin, and its two distal amidino nitrogens just juxtapose the two carboxylate oxygens of Asp189 (see Table 1). Unlike in the PPACK-human thrombin complex and the "free" bovine thrombin[7], there is no electron density for a buried solvent molecule (such as Sol305 in PPACK-thrombin (see Figure 3) or Sol416 in NAPAP-trypsin[19]) fixed between the amidino group and the "back" of the pocket made by Phe227, Tyr228 and Val213 (see Figure 7).

The ß-naphthyl group of NAPAP is almost perpendicular to the indole moiety of Trp215, and nestles in the notched, characteristic "aryl binding site" of thrombin[19]. Both oxygen atoms of the sulphonyl group are directed away from the thrombin surface and are in contact with bulk water. The carboxy-terminal piperidine ring of NAPAP, which is clearly confined to one chair conformation (see Figure 4), is squeezed between the naphthyl moiety and the His57 imidazole side chain of thrombin, completely filling the characteristic hydrophobic S2-cavity of thrombin[10,11]. The carbonyl group of the amidinophenylalanine "residue" is remote from the OG atom of the reactive Ser195 of thrombin and, contrary to the Arg3I carbonyl group of PPACK, does not point into the oxyanion-hole between Gly193 N and Ser195 N (see Figure 7).

Figure 5 shows part of the 2.5 Å 4-TAPAP-thrombin structure[23] (see Figure 1). Similar to the 3-TAPAP analogue[20] (with the amidino group in m-position), the 4-TAPAP molecule is positioned alongside Ser214-Gly216 in such a manner that the amino and the carbonyl group of its central amidinophenylalanyl "residue" make hydrogen bonds with Gly216 N and O of thrombin (see Table 1). Consequently, insertion into the specificity pocket of its side chain is likewise only possible for the L stereoisomer. In contrast to 3-TAP AP[20], however, its amidino group is not directed towards H189, but more towards the "back" of the pocket, where it can interact with the Asp189 carboxylate group only via one of its nitrogen atoms, i.e. through "lateral" salt bridges/hydrogen bonds. The other nitrogen is merely in van der Waals contact with Va1213, Ala190, and Phe227 O (4.7 Å). The toluene group of 4-TAPAP binds towards the aryl binding site of thrombin but does not occupy it completely (see Figure 7). The piperidine ring is positioned between the toluene ring and the His57 imidazole ring, but does not fill the hydrophobic S2-cavity to the same extent as observed for NAPAP (see Figure 7), thus leaving some space that could be accommodated by additional ring substituents (see the discussion in reference[19]).

Figure 6 shows the final electron density and the corresponding model of the refined 2.3 Å MQPA-thrombin around the active-site region[23]. Similar to its docking geometry in trypsin[21] and to the 4-TAPAP-thrombin complex[23], the MQPA molecule is positioned alongside thrombin segment Ser214-G1y216, with the amino and carbonyl groups of the central arginine residue hydrogen bonding to Gly216 O and N of thrombin, respectively (see Table l). As a consequence, the slightly bent L-arginyl side chain adopts a less frequently observed conformation (i.e. with CG trans to HA) in order to insert into the S1-pocket. Similar to the amidino group of 4-TAPAP (see Figure 5) the distal guanidyl group is placed in such a manner that it forms a hydrogen bond to Gly219 O via NE, and hydrogen/ionic bonds to both Asp189 carboxylate oxygens through only one of its terminal nitrogens (see Table 1). The other terminal guanidyl nitrogen is hydrogen bond-connected via two buried solvent molecules to Phe217 0 (HSol 561) and Ser214 0 (HSol 616, see Figure 6). Both of these latter solvent molecules are key donors/ acceptors in a hydrogen bond network extending between Asp189, Phe227, Ser214, and the 2R-carboxylate and the arginine side chain of MQPA.

The quinoline moiety of MQPA rests with its long axis perpendicular to the indole moiety of Trp215, nestling with its hydrogenated ring in the "aryl binding cleft" of thrombin. Both diastereomers (with respect to the 3-methyl group) are in agreement with the density. The (2R,4R) bisubstituted piperidine exhibits a distinct chair conformation (see Figure 6) inserting into the characteristic S2-cavity of thrombin. The equatorial (4R)-

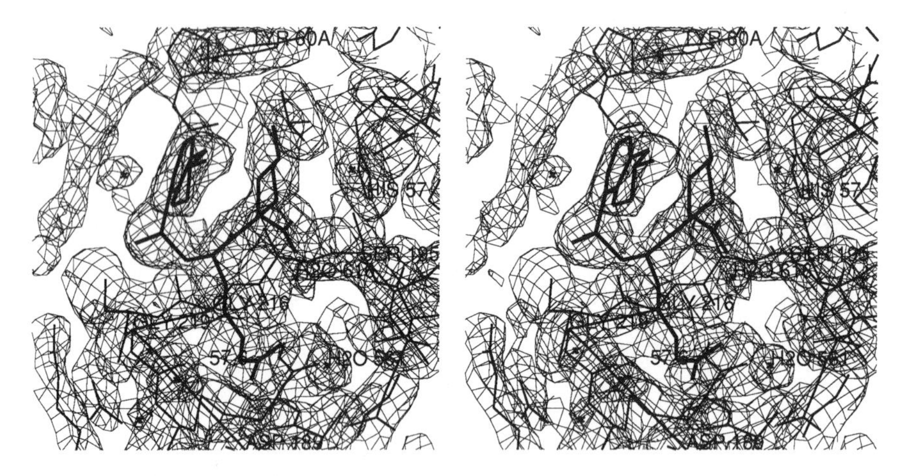

Figure 6 Active-site region of the complex formed between MQPA (thick lines) and bovine thrombin (thin lines) superimposed with the final 2Fobs-Fcalc electron density. The view is towards the thrombin surface; the active-site residues are to the right, the specificity pocket is at the back. Contour surface is at 0.9 δ (from[23,24]).

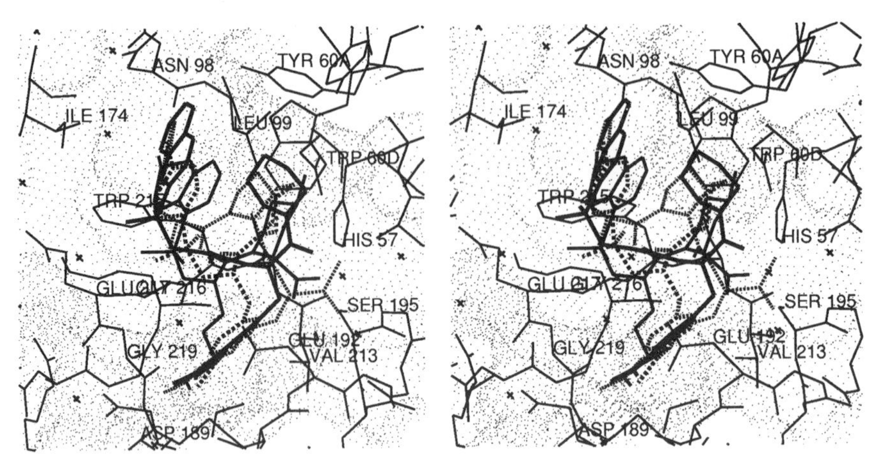

Figure 7 Active-site region of bovine thrombin (intermediate connections) displayed with the (NAPAP)-thrombin Connolly surface, superimposed with the experimentally determined inhibitors NAPAP (dashed thick lines), 4-TAPAP (thick connections), MQPA (dashed intermediate connections), and PPACK (intermediate connections). View as in Figure 2 (from[23,24]).

methyl group just fills the space in the cavity formed by Tyr60A and the imidazole ring of His 57, while the axial (2R)-carboxylate group points through the solvent accessible "channel"towards the oxyanion hole; one carboxylate oxygen (02) forms a hydrogen bond to Ser196 OG of thrombin. MQPA-analogues with different configurations at the piperidine moiety or with bulky substituents at the 4R-position would clearly make a significant worse fit to the S2-cavity in agreement with experimental results[6].

CONCLUSION

As shown in Figure 7, the four small inhibitors presented above interact with thrombin through various docking geometries. All four bind on the "fibrinopeptide" side forming twisted, two-stranded antiparallel ß-pleated sheet structures with thrombin segment Ser214-Gly216. All four are wedged into the furrowed thrombin surface adjacent to the active site; the favourable sterical fit in general explains their significantly higher affinity towards thrombin compared with trypsin (see[4,5,6,19,23,24]).

In detail, however, there are large differences between the peptidic and the "non-peptidic" compounds. PPACK binds in an essentially extended conformation (probably like "real" peptide substrates[16-18]), with its positively charged P1-side chain penetrating into the S1-pocket with favourable ion pair formation. The medium-sized P2-residue packs into the hydrophobic S2-cavity, and its hydrophobic P3-side chain snuggles up into the notched "aryl binding site" (in normal substrates, L-amino acid side chains at one P3-position would rather point away from the thrombin surface; instead, the P4-side chain would be oriented towards this hydrophobic cleft).

The "non-peptidic" benzamidine- and arginine-based compounds, however, bind in a more compact, "U-shape" conformation (which might be close to their "inherently" most stable state in case of high-affinity compounds, see[21]), so that the central basic residue is sandwiched in the S1-pocket, with both terminal, largely hydrophobic groups folded together and nestling in the adjacent hydrophobic S2-cavity and the aryl binding site. The docking geometry of the three "non-peptidic" inhibitors is remarkably different (Figure 7). From the crystal structures of their thrombin complexes (see Figure 7), however, some general rules clearly must be considered for antithrombotic design; their intermolecular interactions seem to be governed primarily by[19,20,23].

i) hydrogen bonds from the inhibitor "backbone" to Gly216 0 and N,

ii) tight insertion of the N-terminal aromatic group into the notched hydrophobic aryl binding site,

iii) squeezing of the carboxy-terminal piperidine(-like) moiety between the N-terminal aromatic group and the His57 imidazole ring,

iv) sandwiching of the amidinophenyl or the arginine moiety and some sort of ionic/hydrogen bonding interaction of the distal amidino (guanidino) group with the Asp189 carboxylate group.

It is noteworthy[23,24], however, that the (presumably most optimal) "opposing" arrangement of the distal basic group and the Asp189 carboxylate as observed for PPACK-thrombin and for NAPAP-thrombin is not always realized, and that this interaction does not primarily govern the overall docking geometry. Obviously, electrostatic interactions via "lateral" hydrogen bond/salt bridge contacts are still strong enough to compensate for the unfavourable ("non-saturating") arrangement of the second terminal amidino/guanidino nitrogen in the "back" of the pocket. In the MQPA-thrombin complex, this "back" space seems to be sufficiently well filled by two buried solvent molecules which establish an extended hydrogen bonding network including the arginine side chain. It is remarkable, however, that equivalent solvent molecules are not observed in the NAPAP- or in the 4-TAPAP complex, in spite of the similar presence of charged (amidino) groups and of almost as much empty space in the "back" side of the specificity pocket. The partially hydrophobic character of this pocket[10,11,24] might render thrombin more able to accommodate less charged groups in its S1-pocket. Antithrombotics lacking positively charged groups would be superior to charged drugs, due to facilitated intestine resorption (important for oral application) and diminished toxicity. The design of such non-charged drugs with high thrombin affinity and selectivity would therefore appear to be one of the great challenges of contemporary antithrombotic drug development. Crystallographic investigations with such compounds are currently underway in our laboratory.

Nevertheless, tight binding of such non-charged compounds would require particularly favourable interactions (i.e. excellent sterical fit) at other sites. The close hydrophobic interactions with the aryl binding site and with the S2-pocket of thrombin seem to be of particular importance. In all inhibitor-thrombin complexes, the hydrophobic surfaces of both inhibitor and thrombin are considerably reduced upon complex formation; the corresponding surface reduction is much smaller for trypsin complexes (see[19]). The much better fit to the thrombin surface of the tosyl/naphthyl moieties and of the piperidine rings are in qualitatively good agreement with the (50 to 100 fold[4-6,19,23,24]) stronger binding of NAPAP, MQPA and 4-TAPAP to thrombin compared with trypsin. In the case of 3-TAPAP[20] and 4-TAPAP[23], which show a relatively bad sterical fit to the S2-cavity, piperidyl substituents with appropriate stereo-configuration (such as those exemplified by the 2R-carboxylate and the 4R-methyl groups in MQPA, see also the discussion in[19]) should clearly improve the binding properties[23,24].

Comparison of the various thrombin crystal structures shows that the thrombin surface regions around the non-primed subsites (in particular the exposed Tyr60A-Trp60D loop, see Figure 2) is not particularly affected by binding of quite different ligands; its overall structure is essentially maintained[23]. Thus, the PPACK-human thrombin model[10,11] (being the most accurate thrombin structure elucidated so far) is a particularly good candidate for computer graphics-aided molecular design. Obviously, problems in predicting the correct binding geometry for an existing drug (see the quite unsuccessful prediction of the binding of MQPA[21,30]) and one presence and sites of buried solvent molecules, mean that accurate experimental thrombin-inhibitor structures (probably to at least 2.5 Å resolution) such as shown here are indispensable as reliable starting points for successful drug improvement. Thrombin, due to its furrowed surface, seems to be particularly well suited for such attempts.

ACKNOWLEDGEMENTS

The constant support by Professor R. Huber, the experienced help of H. Brandstetter, Dr. M. Bauer, D. Turk, Mrs. I. Mayr and Mrs. D. Grosse, the cooperation with Dr. S.R. Stone and Dr. J. Stürzebecher, and the careful proof reading by Dr. M.T. Stubbs are gratefully acknowledged. This work has been supported by the SFB207 of the Universität München and by the Fonds der Chemischen Industrie.

REFERENCES

1. J.W. Fenton II, F.A. Ofosu, D.G. Moon, and J.M. Maraganore, Thrombin structure and function: why thrombin is the primary target for antithrombotics, Blood Coag. Fibrinol. 2:69 (1991).

2. S. Bajusz, E. Barabas, P. Tolnay, E. Szell, and D. Bagdy, Inhibition of thrombin and trypsin by tripeptide aldehydes, Int. J. Peptide Prot.Res. 123:217 (1978).
3. C. Kettner, and E. Shaw, Inactivation of trypsin-like enzymes with peptides of arginine chloromethyl ketone, Method Enzymol. 80:826 (1981).
4. J. Stürzebecher, F. Markwardt, B. Voigt, G. Wagner, and P. Walsmann, Cyclic amides of Nα-arylsulfonylaminoacylated 4-amidinophenyl-alanine - tight binding of thrombin, Thromb. Res. 29:635 (1983).
5. J. Stürzebecher, P. Walsmann, B. Voigt, and G. Wagner, Inhibition of bovine and human thrombins by derivatives of benzamidine, Thromb. Res. 36:457 (1984).
6. R. Kikumoto, Y. Tamao, T. Tezuka, S. Tonomura, H. Hara, K. Ninomiya, A. Hijikata, and S. Okamoto, Selective inhibition of thrombin by (2R,4R)-4-ethyl-1-[N2-3-methyl-1,2,3,4-tetrahydro-8-quinolinyl)sulphonyl)-L-arginyl]-2-piperidinecarboxylic acid, Biochemistry 23:85 (1984).
7. K. Cho, T. Tanaka, R.R. Cook, W. Kisiel, K. Fujikawa, K. Kurachi, and J.C. Powers, Active-site mapping of bovine and human blood coagulation serine proteases using synthetic peptide 4-nitroanilide and thio ester substrates, Biochemistry 23:644 (1984).
8. C.M. Kam, W. Fujikawa, and J.C. Powers, Mechanism-based isocoumarin inhibitors for trypsin and blood coagulation serine proteases: new anticoagulants, Biochemistry 27:2547 (1988).
9. C. Kettner, L. Mersinger, and R. Knabb, The selective inhibition of thrombin by peptides of boroarginine, J. Biol. Chem. 265:18289 (1991).
10. W. Bode, I. Mayr, U. Baumann, R. Huber, S.R. Stone, and J. Hofsteenge, The refined 1.9 Å crystal structure of human α-thrombin: Interaction with D-Phe-Pro-Arg chloromethyl-ketone and significance of the Tyr-Pro-Pro-Trp insertion segment, EMBO J. 8:3467 (1989).
11. W. Bode, D. Turk, and A. Xarshikov, The refined 1.9 Å crystal structure of D-Phe-Pro-Arg chloromethylketone inhibited human α-thrombin. Structure analysis, overall structure, electrostatic properties, detailed active-site geometry, structure-function relationships, Protein Sci. 426 (1992).
12. W. Bode, M. and Stubbs, The spatial structure of thrombin as a guide to its multiple sites of interaction, Sem. Thromb. Hemost. 426:(1992).
13. T.J. Rydel, A. Tulinsky, W. Bode, R. and Huber, Refined structure of the hirudin-thrombin complex, J. Mol. Biol. 221: 583 (1991).
14. M.G. Grütter, U.P. Priestle, J. Rahuel, H. Grossenbacher, W. Bode, J. Hofsteenge, and Stone SR, Crystal structure of the thrombin-hirudin complex: A novel mode of serine protease inhibition, EMBO J. 9:2361 (1990).
15. E. Skrzykczak-Jankun, V.E. Carperos, K.G. Ravichandran, A. Tulinsky, M. Westbrook, and J.M. Maraganore, Structure of the hirugen and hirulog 1 complexes of α-thrombin, J. Mol. Biol. 221:1379 (1991).
16. M. Stubbs, H. Oschkinat, I. Mayr, R. Huber, H. Angliker, S.R. Stone, and W. Bode, The interaction of thrombin with fibrinogen - a structural basis for its specificity, Eur. J. Biochem. 206:187 (1992).
17. M. Stubbs, and W. Bode, A model for the specificity of fibrinogen cleavage by thrombin, Semin. Thromb. Hemost. In press (1992).
18. P.D. Martin, W. Robertson, D. Turk, R. Huber, W. Bode, and B.F.P. Edwards, The structure of residues 7-16 of the Aα-chain of human fibrinogen bound to bovine thrombin at 2.3 Å resolution, J. Biol. Chem. 267:7911 (1992).
19. W. Bode, D. Turk, and J. Stürzebecher, Geometry of binding of the benzamidine- and arginine-based inhibitors NAPAP and MQPA to human α-thrombin. X-ray crystallographic determination of the NAPAP-trypsin complex and modelling of NAPAP-thrombin and MQPA-thrombin, Eur. J. Biochem. 193:175 (1990).

20. D. Turk, J. Stürzebecher, and W. Bode, Geometry of binding of α-tosylated piperidides of m-amidino, p-amidino- and p-guanidino phenylalanine to thrombin and trypsin. X-ray crystal structures of their trypsin complexes and modelling of their thrombin complexes, FEBS. Lett. 287:133 (1991).
21. T. Matsuzaki, C. Sasaki, C. Okumura. and H. Umeyama, X-ray analysis of a thrombin inhibitor-trypsin complex, J. Biochem. (Tokyo) 105:949 (1989).
22. M.M. Chow, E.F. Meyer Jr, W. Bode, C-M. Kam, R. Radhakrishnan, J. Vijayalakshmi, and J.C. Powers, X-ray crystal structure of the complex of trypsin inhibited by 4-chloro-3-ethoxy-7-guanidinocoumarin: A proposed model of the thrombin-inhibitor complex, J. Am. Chem. Soc. 112:7783 (1990).
23. H. Brandstetter, D. Turk, W. Hoeffken, D. Grosse, J. Stürzebecher, P.D. Martin, B.F.P. Edwards, and W. Bode, The 2.2 Å resolution X-ray crystal structure of the complex of trypsin inhibited by 4-chloro-3-ethoxy-7-guanidinocoumarin: A proposed model of the thrombin-inhibitor complex, Mol. Biol. 226:1985 (1992).
24. W. Bode, H. Brandstetter, D. Turk, M. Bauer, and J. Stürzebecher, Crystallographic determination of thrombin complexes with small synthetic inhibitors as a starting point for the receptor-based design of antithrombotics, Semin. Thromb. Hemost. In press (1992).
25. W. Bode, and A. Karshikov, The electrostatic properties of thrombin: Importance for structural stabilization and for ligand binding, Semin. Thromb. Hemost. In press (1992).
26. W. Bode, J. Walter, R. Huber, H.R. Wenzel, and H. Tschesche, The refined 2.2 Å (0.22mm) X-ray crystal structure of the ternary complex formed by bovine trypsinogen, valine-valine, and the Arg15 analogue of bovine pancreatic trypsin inhibitor, Eur. J. Biochem. 144:185 (1984).
27. W. Bode, H.J. Greyling, R. Huber, J. Otlewski, and T. Wilusz, The refined 2.0 Å X-ray crystal structure of the complex formed between bovine ß-trypsin and CMTI-I, a trypsin inhibitor from squash seeds (*Cucurbita maxima*). Topological similarity of the squash inhibitors with the carboxypeptidase A inhibitor from potatoes, FEBS. Lett. 242:285 (1989).
28. L.J. Berliner, and Y.Y.L. Shen, Physical evidence for an apolar binding site near the catalytic center of human α-thrombin, Biochemistry 16:4622 (1977).
29. W. Bode, and P. Schwager, The refined crystal structure of bovine ß-trypsin at 1.8 Å resolution. II. Crystallographic refinement, calcium binding site, benzamidine binding site and active site at pH7, J. Mol. Biol. 98:693 (1975).
30. T. Matsuzaki, C. Sasaki, and H. Umeyama, A predicted tertiary structure of a thrombin inhibitor-trypsin complex explains the mechanisms of the selective inhibition of thrombin, factor Xa, plasmin and trypsin, J. Biochem. 103:537 (1988).

INHIBITOR BINDING TO THROMBIN : X-RAY CRYSTALLOGRAPHIC STUDIES

David W. Banner and Paul Hadváry

Pharmaceutical Research Departments
F. Hoffmann-La Roche Ltd.
Grenzacherstrasse 124
CH-4002 Basel, Switzerland

INTRODUCTION TO OUR CRYSTALLOGRAPHIC PROJECTS

It is a great pleasure to see all the excellent structures presented by Wolfram Bode. I am grateful to him for agreeing to leave me something new to present, namely our work on the binding of the Mitsubishi compound MD-8O5 (Argatroban) to human thrombin. Before I proceed to the details of that, I would first like to give a short overview of our work at Roche, Basle in this field. Our interest in thrombin inhibitors as clinical candidates led us to purify human thrombin and set about crystallization trials. As others had had poor success with uninhibited thrombin, (presumably because of autolysis), the first efforts were concentrated on the binary complex of thrombin with hirudin extracted from whole leeches. When I joined the project five years ago Fritz Winkler and Allan D'Arcy had found a hexagonal crystal form HEX1 (see Table l) which, however, only diffracted to about 3.5Å. We managed to grow some large crystals of this complex but, for some unknown reason, could never collect good native data and were also unable to find heavy atom derivatives. We looked for alternatives, and reading the work of Stuart Stone and Jan Hofsteenge decided to try the ternary complex of PPACK-thrombin - hirudin, hoping (Figure 1) to gain information about binding in the P1′, P2′, etc. sites as well as the active site and the anion binding exosite. We found after some time an orthorhombic crystal form (ORl, Table l) which was difficult to reproduce and optimized instead a second orthorhombic form, OR2, which was used for heavy atom trials with no success. We later found one very large crystal of form OR2 and collected from it data to 2.35Å. As a third attempt we decided to use, instead of hirudin, just a hirudin C terminal peptide, hoping to prevent autolysis and to give us different crystals. This worked well and we solved the structure of the human thrombin-PPACK- hirudin peptide complex in the tetragonal Tl form using the heavy-atom method. The two hirudin containing forms were then solved by molecular replacement (with the assistance of Wolfgang Janes and Christian Oefner, whom we thank). To our disappointment we have not been able to interpret the weak electron density for the hirudin N-terminal domain in either crystal form, and have concluded that in the presence of PPACK this domain is bound unspecifically. The well ordered hirudin density corresponds to residues 55-65, i.e. the same peptide used to obtain the Tl form, and both structures turn out to be similar. There is, however, a difference in that the natural hirudin is sulphated on Tyr 63. The S03-group makes two very good hydrogen bonds to thrombin, consistent with the tighter binding of sulphated peptides. The hirudin complexes both present Trpl48 towards the active site, although the two conformations of the loops (Lysl45 - Lysl49E, chymotrypsin numbering) are not strictly identical. In contrast, this loop swings away from the active site in the Tl crystal form and is stabilised by a crystal contact. Otherwise the three PPACK ternary complexes are remarkably similar, and so the refined 2.35Å hirudin complex was most useful in helping to give a good model for complexes in the Tl crystal form, where it is not possible to measure data beyond about 3Å resolution. As far as we can tell, our structures, determined independently, are entirely consistent with those described at this meeting and the preceding Amsterdam meeting by Wolfram Bode and Al Tulinsky.

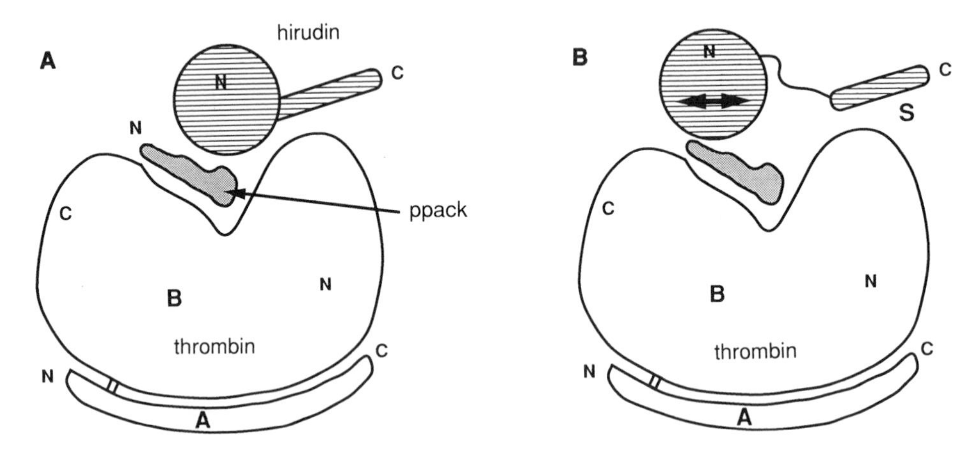

Figure 1A – Scheme of expected hirudin binding to PPACK-thrombin.

B – Scheme of observed hirudin binding to PPACK-thrombin.

Active Site-Directed Inhibitors - General Comments

It has been possible to grow crystals of the Tl form with a number of active site directed inhibitors other than PPACK. The structures of four of these (?PPACK, NAPAP, MD-805 and benzamidine) have been accepted for publication (J. Biol. Chem. Vol. 266 No.30 Oct 25, 1991 pp2OO85-20093). We have very similar experience to Wolfram Bode with NAPAP analogues and have also a very recent fibrinopeptide structure. In our case we have the sequence Gly-Asp-Phe-Leu-Ala-Glu-Gly-Gly-Gly-Val-Arg*Gly-Pro-Arg corresponding to residues 6-19 of the fibrinogen Aα chain. To prevent cleavage, the -NH of glycine 17 has been replaced by -CH_2. Again, our structure seems identical to that just presented by Wolfram Bode, but with a C-terminal extension, which turns out towards solvent rather than finding any tight binding in possible S2′, S3′ sites. This raises a

Table 1 Thrombin crystal data.

Crystal forms

Form	+Hirudin	+Hirudin peptide	+PPACK	Best resolution
HEX1	√			3.5Å
OR1	√		√	2.9Å
OR2	√		√	2.35Å
T1		√	√	2.9Å

Crystallizing conditions

Form	PEG4000	Salt	Buffer	pH
HEX1	20%	300mM $CaCl_2$	100mM Glycine	9.5
OR1	18%	300mM $CaCl_2$	10mM Tris pH7.5	sank to ≈3.8
OR2	25%	150mM $LiSO_4$	100mM HEPES	6.5
T1	20%	200mM $CaCl_2$	100mM HEPES	7.5

Note: All the above conditions should be taken as indicative rather than definitive e.g. if seeding is used it is found that optimum growth conditions are significantly different from the best conditions without seeding.

Unit Cell Data

Form	Space group	Cell (Å)
HEX1	$P6_1/P6_5$	a=b=125.7; c=130.6
OR1	$P2_12_12$	a=108.2; b=80.2; c=45.8
OR2	$P2_12_12_1$	a=108.9; b=68.4; c=57.6
T1	$P4_32_12$	a=b=90.8; c=132.5

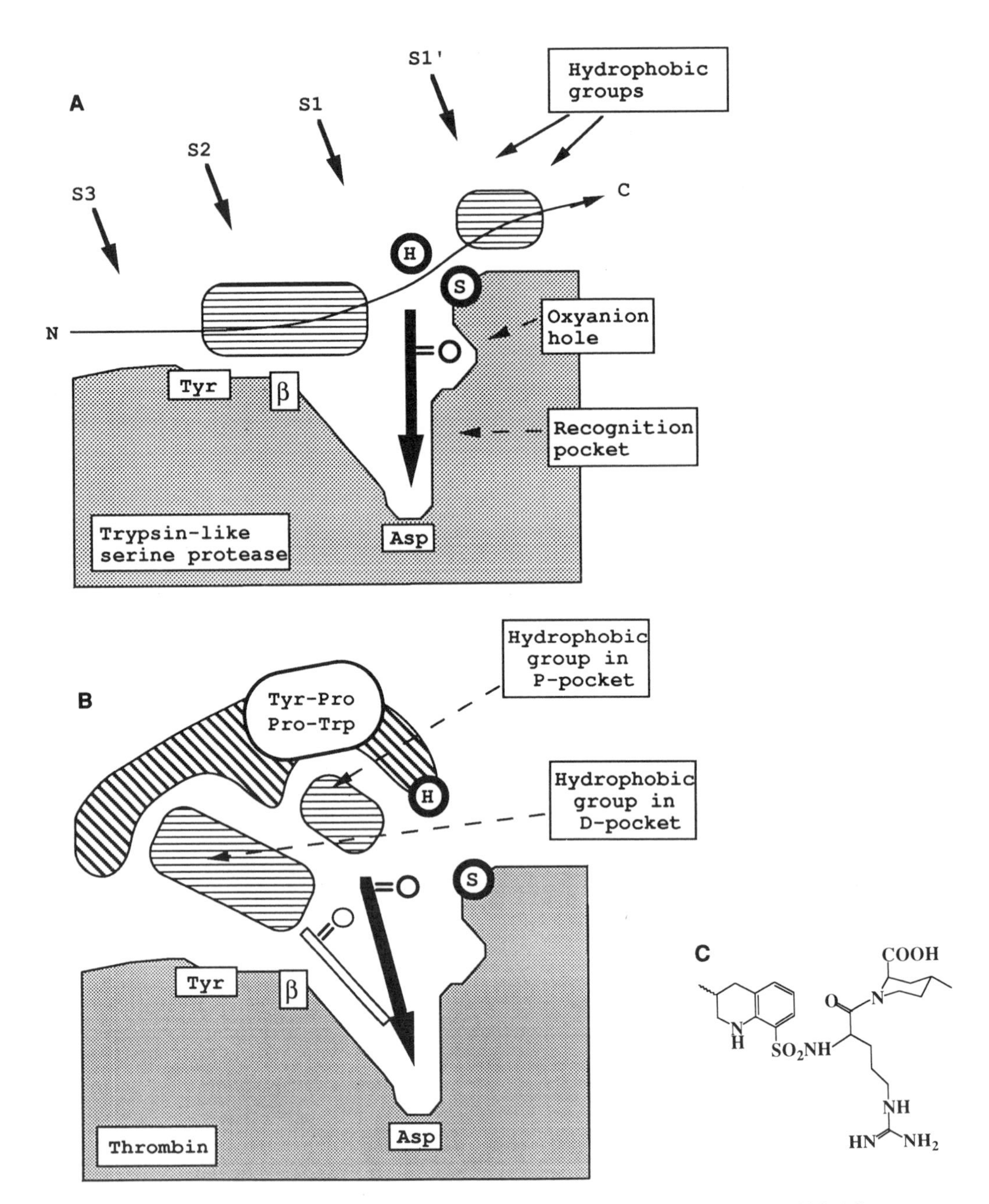

Figure 2A – Scheme of "substrate-like binding". The active site serine and histidine are shown as **S** and **H**.

B – Scheme of inhibitor binding. The filled arrow shows how the para-amidino-phenyl group of NAPAP lies in the recognition pocket, in contrast to the arginine moiety of MD-805 (unfilled).

C – Chemical structure of MD-805. For discussion of stereochemistry see text.

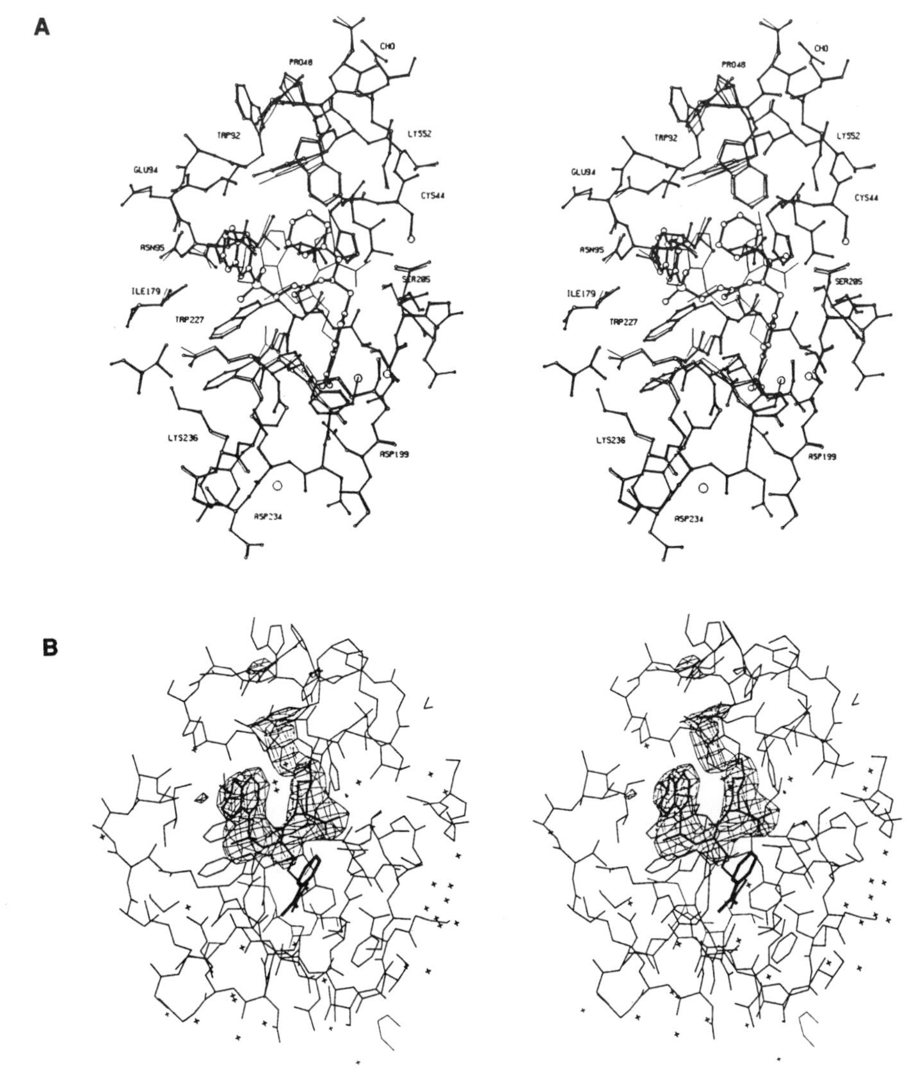

Figure 3A - The MD-805-thrombin and NAPAP-thrombin structures. The active site region of the refined structures is shown using a "ball and stick" representation for NAPAP (with the inhibitor itself bolder) and single line bonds for MD-805.

B - Observed difference electron density between MD-805 and benzamidine. The difference electron density between thrombin-MD-805 and thrombin-benzamidine computed using phases from the refined benzamidine complex and differences of observed amplitudes for all common reflections out to 3.16Å resolution. (The faint structures at lower front and bottom right come from the N- and C-termini of a neighbouring molecule in the crystal).

question of nomenclature in describing thrombin inhibitors: the S1, S2, etc. and S1′, S2′, etc. sites for an enzyme are normally well defined from binding studies on substrate analogues. In the thrombin case we have not until recently seen definitively how an analogue of the true substrate binds (if indeed we accept fibrinogen as the criterion). I, personally, think that we should be very careful in using the standard notation for thrombin, as only the S1 and S2 sites are unambiguous. We have developed our own notation from our inhibitor binding studies, which is shown in Figure 2, where we emphasize the contrast between the binding mode actually found for NAPAP, MD-805 and their analogues (inhibitor binding) and that previously modelled for such compounds on the (false) assumption that they are strictly substrate analogues (substrate-like binding). We use the notation "P-pocket" and "D-pocket" for the proximal and distal (as seen from the catalytic centre) binding pockets on thrombin for the hydrophobic inhibitor moieties.

The Binding to Thrombin of MD-805

We show in Figure 3A the binding of MD-805 to thrombin, in comparison to that of NAPAP. The coordinates are refined at 3Å resolution with some waters. For reference we show in Figure 3B the difference in electron density from which the initial model was derived - it can be seen that the interpretation is unambiguous although, as electron density for benzamidine has been subtracted away, the precise conformation of the arginine side chain of MD-805 cannot be seen. This was also not clear from a direct MD-805 synthesis, and we tried to make a symmetrical salt bridge to the aspartic acid at the bottom of the recognition pocket, before abandoning this in favour of the model shown.

We see from Figure 3 that MD-805 binds like NAPAP with the aryl-substituent in the D-pocket (Bode "aryl binding site") and the piperidine in the P-pocket. The pockets are in reality not independent, and all the evidence is that the two hydrophobic moieties interact strongly with each other when bound to thrombin and in solution as well, which is why a "substrate-like binding" does not occur. MD-805 and NAPAP are, however, very different from each other in the way they occupy the recognition pocket. The NAPAP para- amidinophenylalanine lies approximately in the P1 position, but the carbonyl group is well removed from the oxyanion hole - in fact, as we showed experimentally, it is only the D-stereoisomer which binds. MD-805, in contrast, has the natural L-arginine, but this, far from approaching the oxyanion hole, makes an anti-parallel ß interaction with glycine 228, just as does the glycine "spacer" of NAPAP. Thus, regarding hydrogen bonding to the protein, the MD-805 arginine makes the same interactions with thrombin as do the D-Phe of PPACK and Gly14 of the fibrinopeptide analogue. (If you will, those of a substrate amino acid in the P3 position). The structures of the NAPAP and MD-805 complexes thus make clear why the supposedly analogous para-amidinophenylalanine and arginine amino acids must have the opposite hand. MD-805 has also three other stereocentres. The quinoline is always present as a mixture of both stereoisomers, and we are not able to say which of these binds better to thrombin (or trypsin). The MD-805 piperidine has stereocentres at the 2 and 4 positions (carboxylate and methyl respectively). The originators of MD-805 synthesized all four stereoisomers of the piperidine and found to their surprise that whereas the (2R,4R) stereoisomer binds very well to thrombin (2R,4S) binds 10 times less well, (2S,4R) 100 times less well and (2S,4S) 10000 times less well. Further, the binding to trypsin, although about 250 times weaker, shows the same trend. These data were not understood in advance of the structure determination, but can now be explained by careful modelling (see our J.B.C. paper for details). We infer that the Tyr-Pro-Pro-Trp loop which gives thrombin its characteristic specificity by forming the 'roof' of the P-pocket is able to adjust its conformation to accommodate different hydrophobic groups in the P-pocket. This can be appreciated by comparing the structures of this loop when different inhibitors are bound. NAPAP and benzamidine have the same loop structure, but PPACK, with its small proline in the P-pocket pulls the tryptophan down and MD-805 with the larger piperidine pushes the loop up (Figure 3). The largest differences are about 1.5Å.

CONCLUSIONS

We have documented the binding of MD-805, NAPAP and similar compounds to thrombin. The challenge is to use our knowledge to develop new and interesting compounds with different biological profiles. I think we will see many of these in the near future.

I would like to thank the organizers of this meeting for their generous hospitality and to congratulate them on choosing just the right moment in time for a detailed discussion of thrombin inhibition. Our understanding in this field has reached the point where we can say with Churchill "This is not the end, nor even the beginning of the end, but it is the end of the beginning!".

MOLECULAR BASIS FOR THE INHIBITION OF THROMBIN BY HIRUDIN

Stuart R. Stone[1], Andreas Betz[1], Marina A. A. Parry[1],
Martin P. Jackman[2] and Jan Hofsteenge[2]

[1]Department of Haematology, University of Cambridge
MRC Centre, Hills Road, Cambridge CB2 ZQH, U.K.
[2]Friedrich Miescher-Institut
P.O. Box 2543, CH-4002 Basel, Switzerland

The Structure of Hirudin and of the Thrombin-Hirudin Complex

The amino acid sequence of the 65 residue major form of hirudin is shown in Figure 1. Hirudin was originally isolated from the salivary glands of the European medicinal leech Hirudo medicinalis by Markwardt[1]. The anticoagulant properties of leech saliva were, however, first described over 100 years ago by Haycraft[2]. The N-terminal region of the hirudin contains 6 cysteines that form three disulphide bridges: Cys6′-Cysl4′, Cys16′-Cys28′ and Cys22′-Cys39′. (Residues in hirudin are distinguished from those in thrombin by the use of primed numbers and the numbering of the sequence of thrombin is that of Bode et al.[3] which is based on chymotrypsin numbering). The C-terminal region of hirudin is particularly rich in acidic residues (Figure l) and the acidic nature of this region of hirudin is important for the formation of the thrombin-hirudin complex as will be discussed later. One of the charged residues found in this region of the natural hirudin is a sulphated tyrosine residue at position 63. This post-translational modification is not, however, found in recombinant molecules expressed in Escherichia coli or yeast. More than 20 different isoforms of hirudin have been isolated from H. medicinalis and sequenced[4-6] and in each of these isoforms the position of the disulphide bridges and the acidic nature of the C-terminal region is conserved. The residues that are conserved between the different isoforms are indicated in Figure 1 and many of these residues are involved in the binding of hirudin to thrombin[7-9].

Two-dimensional nuclear magnetic resonance (n.m.r.) studies indicate that in solution hirudin is composed of a compact N-terminal domain (residues 3-49) held together by the three disulphide bonds, and a disordered C-terminal tail (residues 50-65)[10,11,12]. The N-terminal domain is composed of a hydrophobic core of short ß-sheets and turns together with a "finger" subdomain of two anti-parallel ß-strands that does not display a fixed orientation with respect to the rest of the core.

The structure of thrombin-hirudin complex is presented in Figure 2. The structure of the N-terminal core of hirudin in the complex corresponds to those determined by two-

The Design of Synthetic Inhibitors of Thrombin
Edited by G. Claeson, *et al.*, Plenum Press, New York, 1993

dimensional n.m.r. However, in contrast to its disordered structure in solution, the C-terminal tail has a defined structure due to contacts with thrombin[7,8,9]. The structure illustrated in Figure 2 indicates that hirudin interacts with thrombin over an extended area. The N-terminus of hirudin is bound to the active site while the C-terminal region extends 35 Å across the surface of thrombin. The area of contact of hirudin with thrombin is about 1400 Å[9]. For comparison, the area of contact between bovine pancreatic trypsin inhibitor and trypsin is only 475 Å. The large number of favourable contacts between thrombin and hirudin over this extended area of interaction results in the formation of an extremely tight complex between thrombin and hirudin. (The values of the dissociation constant (K_d) of the complex are 10^{-14} M and 10^{-13} M for native and recombinant hirudin, respectively, which correspond to binding energies (ΔG_b) of 75 and 80 kJ mol^{-1} [13]. Hirudin inhibits thrombin by a novel mechanism. The N-terminal three residues of hirudin

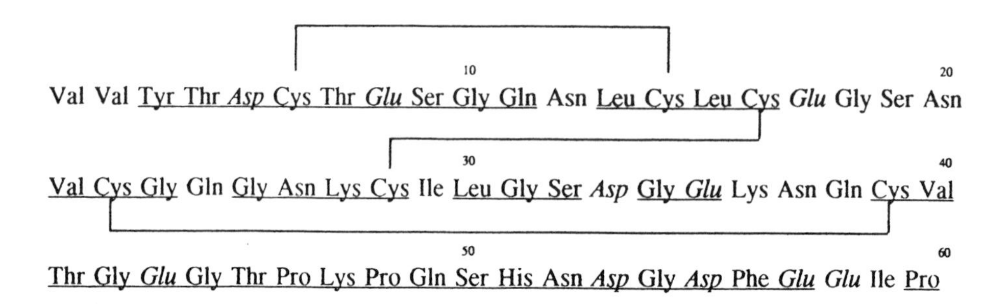

Figure 1. Sequence of hirudin variant 1. The sequence given in the three letter code is that determined by Bagdy et al.[46] and Dodt et al[47]. The disulfide bridges determined by Dodt et al.[48] are also given. Acidic residues are indicated by italics and residues invariant in other hirudin sequences are underlined[4].

are bound in the active-site cleft of thrombin, but the orientation of the polypeptide chain is opposite to that observed for other inhibitors. The polypeptide chain forms a parallel ß-strand with Ser214-Gly219 of thrombin, whereas in all other complexes of inhibitors with serine proteases an antiparallel interaction is observed[14]. The residues Val1′ and Tyr3′ occupy roughly the binding sites S_2, and S_3, (nomenclature of Schechter and Berger[15]), respectively, while Val2′ is bound at the surface of the primary specificity (S_1) pocket without penetrating it (Figure 3). The remainder of the N-terminal core region of hirudin closes off the active site of thrombin. The C-terminal tail of hirudin is bound to a surface groove that is rich in basic amino acids and has been termed the anion-binding exosite (Figure 4). The acidic residues in the C-terminal region of hirudin make several electrostatic interactions with the anion-binding exosite. In addition, numerous hydrophobic contacts occur between residues of the C-terminal tail and the anion-binding exosite.

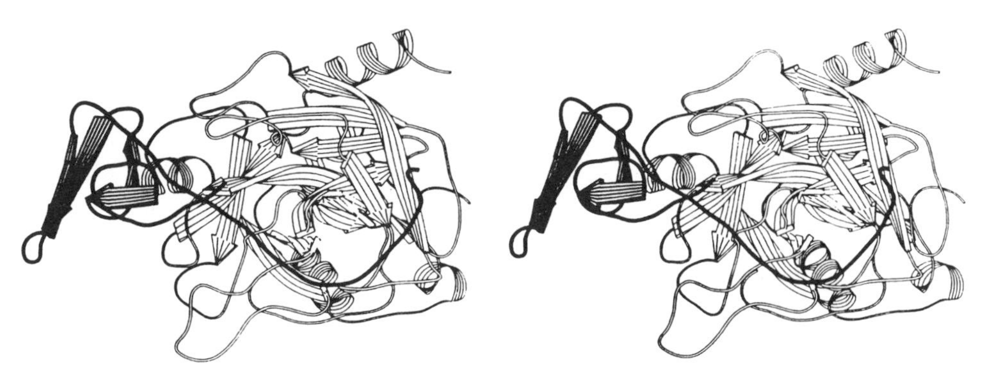

Figure 2. Ribbon drawing of the thrombin-hirudin complex. ß-sheets and α-helices are represented as arrows and coils, respectively. The hirudin structure is darker and is found in front of thrombin. It is composed of four short ß-sheets linked by loops and a long C-terminal tail that wraps around the thrombin molecule and binds to the anion-binding exosite. The N-terminus of hirudin binds in the active site of thrombin. The thrombin molecule consists of an A- and B-chain. The A-chain is composed of two helical portions and is found at the back of the B-chain while the B-chain consists mainly of ß-sheets. The structure represented is that of Grütter et al.[9] and the plot was made using the program of Priestle[49].

Mechanism for the Formation of the Thrombin-Hirudin Complex

The results of steady-state and rapid kinetic studies indicate that the formation of the thrombin-hirudin complex involves at least three steps as outlined in the following scheme:

$$E + I \underset{k_{-1}}{\overset{k_1}{\rightleftharpoons}} EI^1 \underset{k_{-2}}{\overset{k_2}{\rightleftharpoons}} EI^2 \underset{k_{-3}}{\overset{k_3}{\rightleftharpoons}} EI^3$$

Scheme 1

where E and I represent thrombin and hirudin, respectively. Studies using picomolar concentrations of hirudin indicated that the first step is rate-limiting under these conditions. Moreover, this step was found to involve an ionic interaction with a site distinct from the active site[16-19]. Thus, it was proposed that the rate-limiting step entailed the binding of the C-terminal region of hirudin to the anion-binding exosite[19]. By using stopped-flow spectrofluorometry to follow changes in the intrinsic fluorescence of thrombin upon the binding of hirudin, it was possible to demonstrate that the first step was indeed an ionic interaction involving the C-terminal region of hirudin. It was also possible to detect two subsequent intramolecular steps with rate constants of about $300s^{-1}$ and $50s^{-1}$. A step with a rate constant of $300s^{-1}$ was also observed for the binding of a C-terminal fragment of

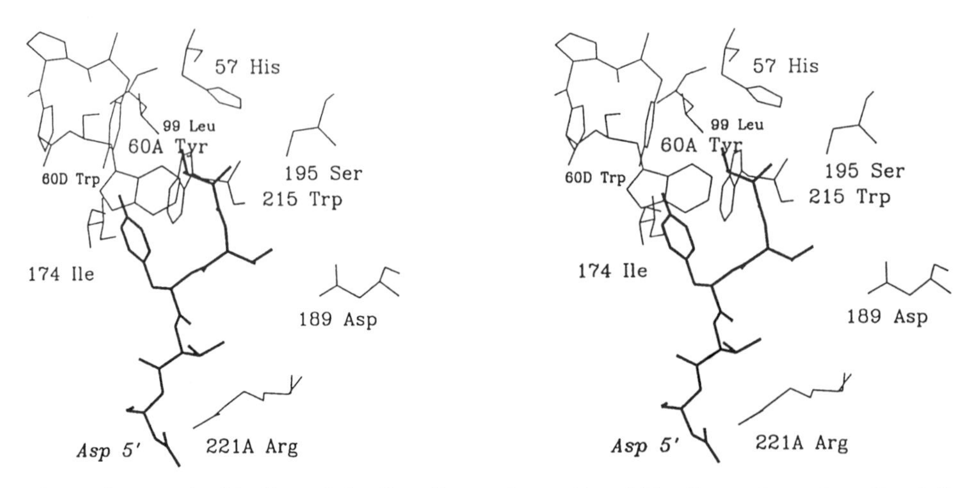

Figure 3. The binding of the first five amino acids of hirudin to the active-site cleft of thrombin. Hirudin is shown in thick lines. In general, only thrombin residues directly involved in interactions with hirudin are shown. The interactions of Val1′ and Val2′ were modelled from the structure of Rydel et al.[7,8].

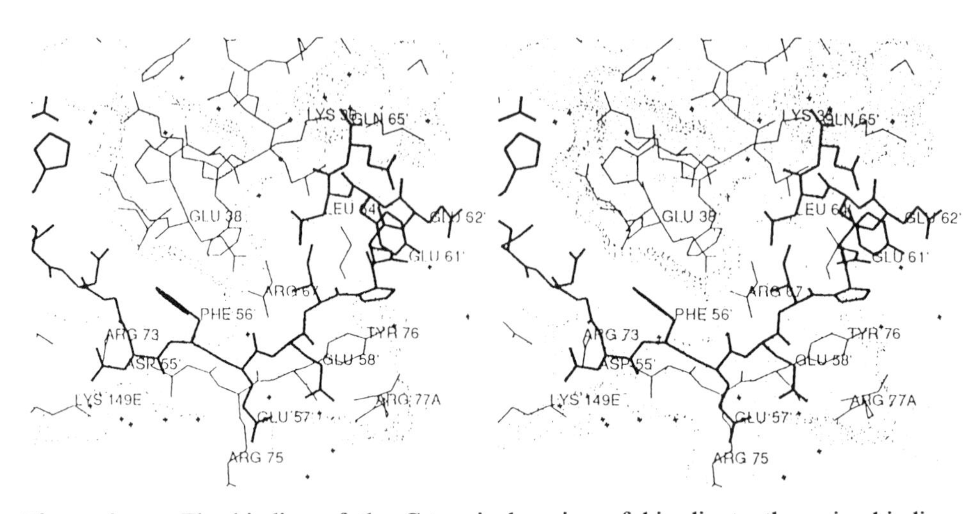

Figure 4. The binding of the C-terminal region of hirudin to the anion-binding exosite of thrombin. Residues of the anion-binding exosite of thrombin are displayed in thin connections together with the Connolly surface of this region of thrombin. Thick connections are used for hirudin. The structure is that of Rydel et al.[7,8].

hirudin consisting of residues 52 to 65, but the slower step was not observed[20]. It was, therefore, concluded, that the step with a rate constant of about $300s^{-1}$ corresponds to a conformational change induced by the binding of the C-terminal region of hirudin while the slower step probably represents the binding of the N-terminal region of hirudin to the active site. In summary, the formation of the thrombin-hirudin complex involves the interaction of the C-terminal region of hirudin with the anion-binding exosite which induces a conformational change in thrombin prior to the binding of the N-terminal region to the active site.

The extent to which the conformational change caused by the binding of the C-terminal region in the first step facilitates the binding of the N-terminal region of hirudin to the active site in a subsequent step has been investigated by using fragments of hirudin. Evidence that this conformational change affects the active site of thrombin comes from studies of the kinetics of cleavage of tripeptidyl substrates in the presence of C-terminal peptides. In the presence of these peptides, the kinetic parameters for cleavage of tripeptidyl chromogenic and fluorogenic substrates by thrombin are altered[21-24]. The binding of fragments of heparin cofactor II and the thrombin receptor to the anion-binding exosite have also been shown to affect the kinetics of cleavage of chromogenic substrates[24,25]. This change in the active site region favours the combination of an N-terminal core fragment of hirudin; the association rate constant for this fragment is 70% higher in the presence of the C-terminal peptide and its K_d value is 60% lower[20,21]. Thus, the binding of the C-terminal region causes a conformational change that affects the active site of thrombin and slightly facilitates the binding of the rest of the hirudin molecule. However, the amount of cooperativity observed for the binding of the two regions is small.

The conformational change induced by binding of the C-terminal region causes an enhancement in the fluorescence of one or more tryptophans in thrombin. Comparison of the crystal structures of D-Phe-Pro-ArgCH$_2$-thrombin and the thrombin-hirudin complexes suggests that this conformational change may correspond to a movement of the γ-loop of thrombin (Glu146-Gly150). In the structure of D-Phe-Pro-Arg CH$_2$ thrombin[3], Trpl48 in this loop occupies a solvent-exposed position at the entrance to the active-site cleft. In the hirudin-thrombin complex, the γ-loop is displaced to allow the binding of hirudin to the active-site cleft and Trp148 moves to a more buried location[7,8]. The resulting decrease in polarity of the Trpl48 environment on binding hirudin should lead to an enhancement in intrinsic fluorescence. It should be noted, however, that the position of Trp148 in free and complex thrombin in solution cannot be determined with great certainty from the crystal structures since the position of the γ-loop appears to be strongly influenced by crystal-packing interactions[8,26].

Electrostatic Interactions in the Thrombin-Hirudin Complex

Interaction of the complementary electrostatic fields created by the anion-binding exosite and the C-terminal region of hirudin appears to be the dominant force leading the formation of the initial complex between these two regions. Theoretical calculation of electrostatic potentials surrounding thrombin and hirudin indicates that they are asymmetric and complementary. The complementarity of these potentials would encourage an initial orientation of the molecules such that the anion-binding exosite and the C-terminal region of hirudin are juxtaposed[27]. The results of site-directed mutagenesis experiments support this hypothesis. The association rate constant for hirudin is markedly reduced by mutations that remove negative charges from the C-terminal region of hirudin and, thus, reduce the strength of the negative potential around this region of the molecule[19].

The contribution to binding energy made by the interaction of each of the negatively charged residues of hirudin between Asp55′ and Glu62′ as determined in site-

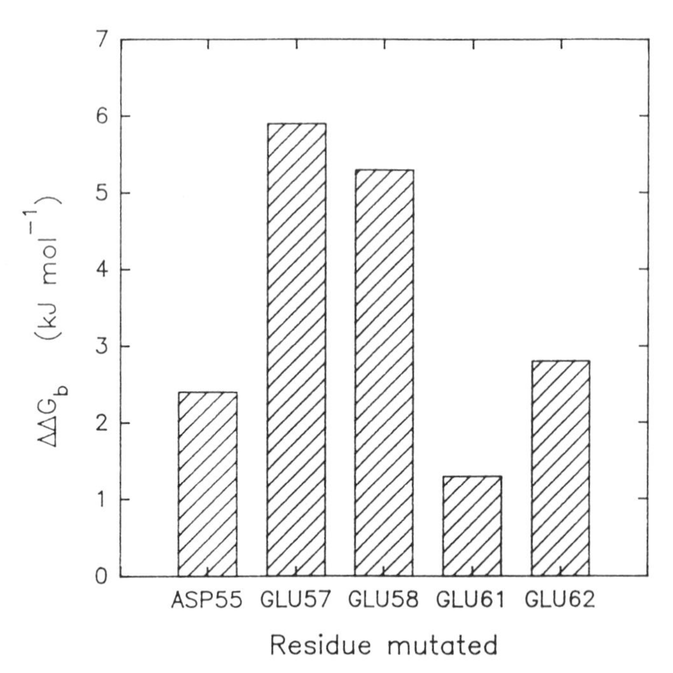

Figure 5. Effect of replacement of acidic residues in the C-terminal region of hirudin. The decrease in binding energy ($\Delta\Delta G_b$) caused by the replacement of the indicated residues by glutamine (residues 57, 58, 61 and 62) or asparagine (residue 55) was calculated from the data of Braun et al.[13] and Betz et al.[50].

directed mutagenesis experiments varies betweeen 1.3 and 5.9 kJ mol^{-1} (Figure 5). The results of these experiments seem at variance with the crystal structures of the thrombin-hirudin complex in which only Asp55′ and Glu57′ make direct electrostatic interactions (Figure 4); the side chains of Glu6l′ and Glu62′ point out into the solvent and, therefore, were not expected to make any contribution to binding energy[7,8]. However, calculations based on electrostatic potentials surrounding thrombin and hirudin indicate that the charges on Glu6l′ and Glu62′ make significant contributions to binding energy through their interactions with the positive potential created by the anion-binding exosite of thrombin[27]. Indeed, theoretical calculations of the effects of mutations on binding energy correspond well with the observed effects in all cases except for the mutation of Asp55′. This residue makes salt bridges with Arg73 and Lysl49E (Figure 4) and theoretical calculations indicate that these interactions should make a large contribution (14.2 kJ mol^{-1}) to binding energy whereas only a small contribution (2.4 kJ mol^{-1}) is observed experimentally (Figure 3). The salt bridge between Asp55′ and Lysl49E may, however, be a crystal artefact[26] and this would partially explain the discrepancy between the calculated and observed values[27].

Negatively charged residues within the N-terminal core domain also make salt bridges that contribute to the stability of the complex. In particular, Asp5′ and Glul7′ make salt bridges with Arg221A and Argl73, respectively[7-9]. Site-directed mutagenesis experiments indicate that the interactions with Asp5′ and Glu17′ contribute between 5 and 3 kJ mol^{-1} to binding energy[28,29]. Thus, although no one particular ionic interaction makes a very large contribution to binding energy, the association of thrombin and hirudin has a considerable electrostatic component due to the multiple electrostatic interactions that occur. Indeed, the contribution of ionic interactions to the binding energy as determined from the effect of ionic strength on complex formation[19] suggest that ionic interactions contribute about one-third of the binding energy at zero ionic strength. This value is ionic strength-dependent and decreases to about 20% at physiological ionic strength.

While the results of site-directed mutagenesis experiments indicate a significant role for the negatively charged residues in hirudin in the formation of its complex with thrombin, the role of the three positively charged lysines seems to be relatively minor. Prior to the determination of the crystal structure of the thrombin-hirudin complex, it was assumed that a basic residue in hirudin would be bound to the primary specificity pocket. Since good thrombin substrates often have a proline in the P_2 position[30,31], it was predicted that Lys47′ would be bound in the primary specificity pocket[32-34]. Site-directed mutagenesis experiments, however, failed to detect a crucial role for Lys47′ or any of the other basic residues[13,35,36]. All studies indicated a small decrease in inhibitory potency when Lys47′ was replaced by another amino acid. Rydel et al.[7,8] have suggested that hydrogen bonds formed by the N^{ε} of Lys47′ with the backbone carbonyl of Asp5′ and the Oγ of Thr4′ help to position the N-0γ terminal core of hirudin in the active-site cleft and, thus, the observed effects upon mutation of Lys47′ may be due to a suboptimal positioning of the N-terminal core with respect to the active site.

The only positively charged group in hirudin that appears to play an important role is the α-amino group of Val1′. This group forms hydrogen bonds with Ser195 and the main chain carbonyl of Ser214[7-9]. The strength of these hydrogen bonds is expected to be much greater if the α-amino group is charged[37]. In support of the proposal that a positive charge on the α-amino group increases the strength of these bonds, it was found that removal of the positive charge by acetylation reduces the binding energy by 23 kJ mol^{-1}. In comparison, acetimidation of the α-amino group, which adds a moiety of similar size to the acetyl moiety but maintains a positive charge, results in a decrease in binding energy of only 7 kJ mol^{-1} [38]. Thus, a positive charge at the N-terminus of hirudin appears to be important for optimal binding. Studies on the pH-dependence of the dissociation constant for hirudin also indicate that the formation of the complex is dependent on the α-amino group being protonated. It was found that a group with a pK_a of 8.4 must be protonated for optimal binding. This pKa value corresponds with that determined for the α-amino group of hirudin by titration with 2,4,6-trinitrobenzene-sulphonic acid[38]. Moreover, when the pH dependence of binding of a hirudin variant with an acetylated α-amino group was examined, the pK_a value of 8.4 was not observed[39].

Nonpolar Interactions in the Thrombin-Hirudin Complex

The crystal structures of the thrombin-hirudin complexes show that hirudin makes numerous nonpolar interactions with the active-site cleft and the anion-binding exosite[7-9]. In the active-site cleft, residues 1 and 3 of hirudin make numerous hydrophobic contacts; contacts between the three N-terminal amino acids of hirudin and thrombin's active-site cleft account for 22% of the intermolecular contacts less than 4 Å[7,8]. As shown in Figure 3, the side chain of the first amino acid makes numerous nonpolar interactions with thrombin residues including His57, Tyr60A, Trp60D and Leu99[7-9]. The second residue of hirudin is located at the edge of the primary specificity pocket but, unlike other inhibitors of serine proteinases, does not penetrate this pocket. Tyr3′ is buried in an hydrophobic pocket constituted by the residues Tyr60A, Ile174, Trp60D and Leu99 of thrombin (Figure 3). This binding site is occupied by the D-phenylalanine of the inhibitor D-Phe-Pro-ArgCH_2Cl and has been called the aryl-binding pocket[3,40].

The effects of site-specific substitutions of the first three amino acids of hirudin have been evaluated[28,41]. The consequences of different substitutions of Val1′ are shown in Figure 6; they indicate that nonpolar interactions play a major role in the binding of this residue. The binding energy (ΔG_b) was maintained when Val1′ was replaced by leucine but a decrease of 9.0 kJ mol^{-1} was seen with the serine substitution (Figure 6). While replacement of Val1′ by polar residues generally led a large decrease in binding energy, the mutant with arginine in the first position bound to thrombin almost as tightly as wildtype hirudin (Figure 6); molecular modelling cannot suggest a unique explanation for this observation.

In the second position of hirudin (Val2′), polar amino acids, with the exception of the negatively charged glutamate, are better accommodated than in the first (Figure 7). This can be ascribed to the more polar nature of the binding site for Val2′ (Figure 3). The mutant with arginine in the second position bound particularly well to thrombin; its dissociation constant that was 9-fold lower than that of wild-type recombinant hirudin.

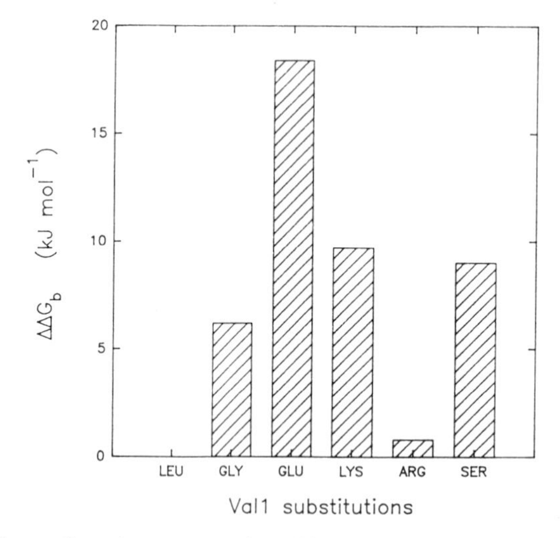

Figure 6. Effect of replacement of Val1 by various residues. The decrease in binding energy ($\Delta\Delta G_b$) caused by the replacement of Val1′ by the indicated residues was calculated from the data of Wallace et al.[38] and Betz et al.[28].

Molecular modelling provides a possible explanation for the increased potency of V2R. It is possible to model the side chain of Arg2′ such that it penetrates the primary specificity pocket of thrombin and makes a salt bridge with Asp189 of thrombin (Figure 8). In this model the guanidino moiety of the arginine occupies roughly the same position as that of the arginine in the structure of D-Phe-Pro-ArgCH$_2$-thrombin[3]. The orientation of the side chain in the modelled structure is, however, different to that observed for the arginine in the D-Phe-Pro-ArgCH$_2$-thrombin complex. Arg2′ would penetrate the primary specificity pocket from a different angle; the orientation of the side chain would be similar to that observed for the arginine in the thrombin-specific inhibitor MQPA ((2R,4S)-4-methyl-1-[N -(3-methyl-1,2,3,4-tetrahydro-8-quinoline-sulfonyl)-L-arginyl]-2-piperidine carboxylic acid)[40,42]. The hirudin V2K does not show an increased affinity for thrombin. The side chain of Lys2′ would not be able to penetrate as deeply into the primary specificity pocket as Arg2′ and the electrostatic interaction with Asp189 would not be as favourable.

The importance of the interaction of Tyr3′ with the aryl-binding pocket was suggested by the fact that a tyrosine is found in position 3 in all sequenced forms of hirudin[4]. The importance of the interactions made by Tyr3′ is confirmed by site-directed mutagenesis experiments. Elimination of the hydrophobic interactions made by the aromatic ring of Tyr3′ of hirudin results in a large loss of about 12 kJ mol^{-1} in binding energy[28,41].

Five of the 11 C-terminal residues of hirudin are involved in hydrophobic interactions with the anion-binding exosite[7,8]. In this respect, interactions with Phe56′ and with residues in the C-terminal 3_{10} helix are particularly noteworthy (Figure 4). Phe56′ is buried in a hydrophobic cleft formed by Phe34, Arg73 and Thr74[7]. In addition, the aromatic rings of Phe56′ and Phe34 of thrombin exhibit the edge-to-face configuration that is considered to be optimal for aromatic rings[43]. Pro60′ and Tyr63′ also make numerous hydrophobic contacts with thrombin. Pro60′ interacts with Tyr76 while Tyr63′ has nonpolar contacts with Leu65 and Ile82.

The role of hydrophobic interactions with Phe56′, Pro60′ and Tyr63′ has also been investigated by site-directed mutagenesis[44]. Phe56′ can be replaced by aromatic amino acids (tyrosine and tryptophan) without significant loss in inhibitory activity (Figure 9). Complete elimination of the aromatic ring by substitution of Phe56′ by alanine causes only a small decrease in binding energy (ΔG_b) of 2.0 kJ mol^{-1}. In contrast, replacement of this residue by larger hydrophobic amino acids causes larger decreases; for example, the mutant Phe56->Val displays a decrease in ΔG_b of 10.5 kJ mol^{-1} (Figure 9). These large decreases in binding energy are unexpected in view of the small decrease observed for the alanine substitution (Figure 9). No definitive possible explanation can be obtained by molecular modelling for the apparently incongruous results obtained with these substitutions. However, it has been possible to show experimentally that the valine substitution does not cause major structural changes leading to the disruption of ionic interactions in the hirudin-thrombin complex[44]. Crystal structures of mutant hirudins will be required to obtain a definitive explanation for the effects observed with the Phe56′ mutations.

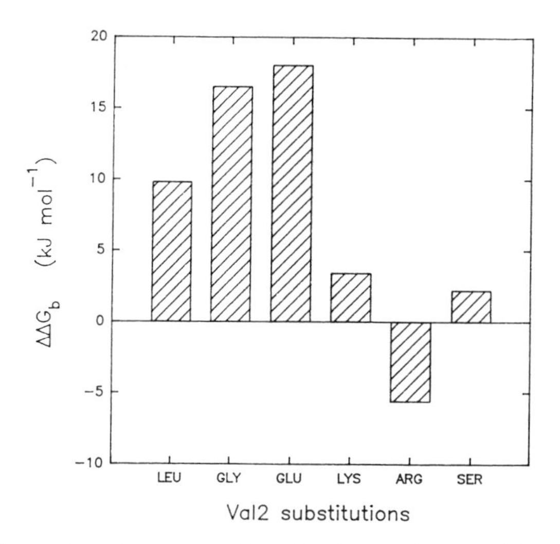

Figure 7. Effect of replacement of Val2′ by various residues. The decrease in binding energy ($\Delta\Delta G_b$) caused by the replacement of Val2′ by the indicated residues was calculated from the data of Wallace et al.[38] and Betz et al.[28].

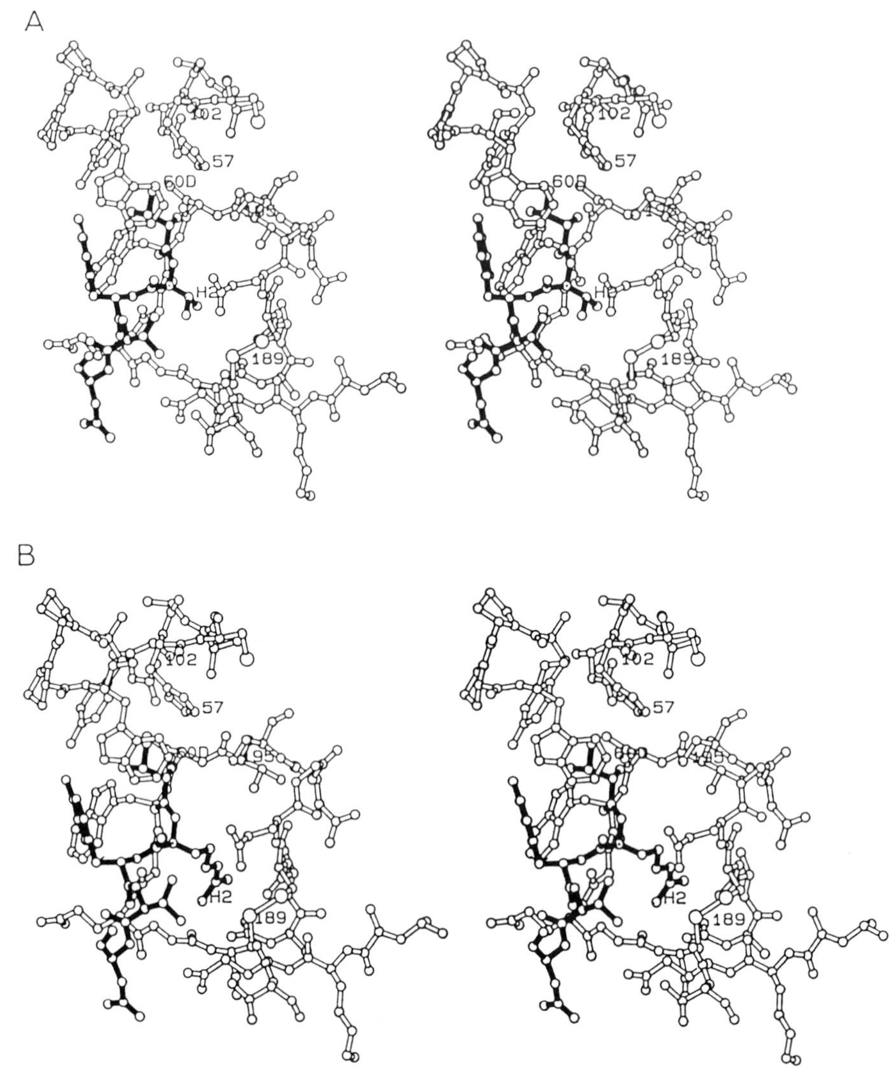

Figure 8. Modelled structure for the binding of the mutant with arginine in position 2 to the active site of thrombin. A. Binding of the N-terminal five residues of hirudin variant 1 to the active site. Filled connections are used for hirudin. The catalytic residues His57, Asp102 and Ser195 of thrombin are labelled together with Asp189 which is found at the bottom of the primary specificity (S_1) pocket. B. Modelled structure of the Arg2′ mutant. The arginyl side chain in position 2 can be modelled to make an electrostatic interaction with Asp189 in the primary specificity pocket.

Substitution of Pro60′ by alanine or glycine results in a decrease in binding energy (ΔG_b) of about 6 kJ mol^{-1}. Pro60′ has 7 contacts closer than 4 Å to Tyr76 of thrombin. The exchange of proline by alanine or glycine would lead to a loss of all these close contacts. In addition, a prolyl residue in position 60 would also stabilize the complex by reducing the entropy of the polypeptide backbone of the C-terminal region of unbound hirudin. A role for Pro60′ in initiating and positioning the C-terminal 3_{10} helix has also been suggested[8]. Thus, Pro60′ may have multiple functions and the decrease in binding energy observed upon mutation of Pro60′ is probably the consequence of an effect on more than one of these functions.

In contrast to Phe56′, Tyr63′ is not bound in a hydrophobic cleft, but it still participates in numerous hydrophobic interactions (Figure 4). Tyr63′ could be replaced by phenylalanine without any loss in binding energy (Figure 10). This result is consistent with the crystal structure in which the hydroxyl of Tyr63′ does not make any hydrogen bonds[8]. Tyrosine is necessary in position 63 of the native protein because its hydroxyl group is enzymatically sulphated in the leech and this post-translational modification causes the 10-fold lower Kd value observed with natural hirudin[9,35,45]. The complex of thrombin with a sulphated C-terminal fragment of hirudin indicates that the tyrosine-sulphate is involved in a network of charged hydrogen bonds[26].

Thus, on the basis of this structure, it is not unexpected that the introduction of another negatively charged group at position 63 (Glu) does not lead to an increase in binding energy (Figure 10); the exact position of the negatively charged group will be important for the formation of this hydrogen-bonding network. Substitution of a glutamate for Tyr63′ would result a loss of many of the hydrophobic contacts made by the aromatic ring of Tyr63′ and the loss of these contacts would explain the observed decrease in binding energy. The contribution of these contacts to binding energy can also be evaluated by examining the effects of the alanine substitution. The observed 2 kJ mol^{-1} decrease indicates that the hydrophobic contacts between the aromatic ring and Ile82 of thrombin make only a small contribution to binding energy (Figure 10). In comparison to the anomalous effects observed upon substitution Phe56′ by branched chain amino acids, the effects caused by replacement of Tyr63′ by such amino acids can be readily interpreted

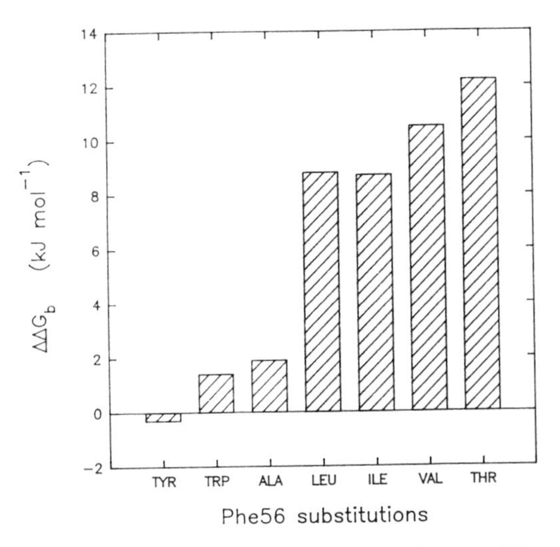

Figure 9. Effect of replacement of Phe56′ by various residues. The decrease in binding energy ($\Delta\Delta G_b$) caused by the replacement of Phe56′ by the indicated residues was calculated from the data of Betz et al.[44].

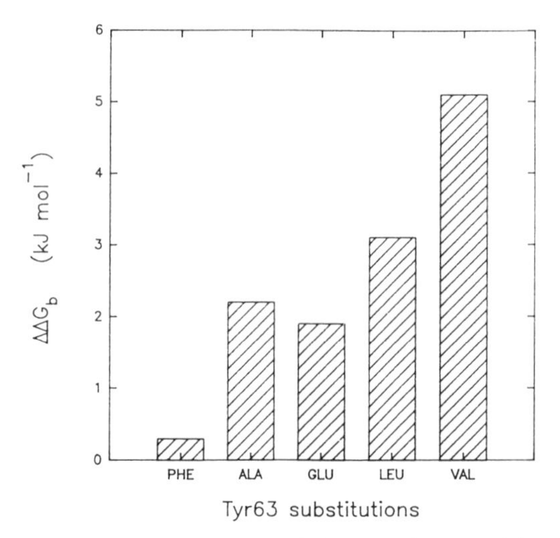

Figure 10. Effect of replacement of Tyr63′ by various residues. The decrease in binding energy ($\Delta\Delta G_b$) caused by the replacement of Tyr63′ by the indicated residues was calculated from the data of Betz et al.[44].

in terms of the crystal structure of the complex. Substitution of leucine or valine for Tyr63′ would lead to unfavourable contacts with either thrombin residues or with the hirudin backbone in the 3_{10} helix[44].

The results discussed above indicate that the formation of the thrombin-hirudin complex involves both electrostatic and hydrophobic interactions over extended areas of both molecules. The association of the molecules occurs in multiple steps. In the initial step, electrostatic interactions probably play a predominant role. However, hydrophobic contacts formed between complementary surfaces are also important in the stabilization of the final tight complex.

ACKNOWLEDGEMENTS

We thank Drs. W. Bode, A. Karshikov, M.G. Grütter, J.P. Priestle and U. Neumann for helpful discussions. Drs. W. Bode, J. Priestle, M.G. Grütter and A.M. Lesk also assisted with the production of figures of the thrombin-hirudin structure.

REFERENCES

1. F. Markwardt, Die Isolierung and chemische Characterisierung des Hirudins, Hoppe-Seylers Z. Physiol. Chem. 308:147 (1957).
2. J.B. Haycraft, Secretion obtained from the medicinal leech, Proc. R. Soc. London 36B:478 (1884).
3. W. Bode, I. Mayr, U. Baumann, R. Huber, S.R. Stone, and J. Hofsteenge, The refined 1.9 Å crystal structure of human α-thrombin: interaction with D-Phe-Pro-Arg chloromethylketone and significance of the Tyr-Pro-Pro-Trp insertion segment, EMBO J. 8:3467 (1989).
4. M. Scharf, M. Engels, and D. Tripier, Primary structures of new 'iso-hirudins', FEBS Lett. 255:105 (1989).

5. J. Dodt, N. Machleidt, U. Seemüller, R. Maschler, and H. Fritz, Isolation and characterization of hirudin isoinhibitors and sequence analysis of hirudin PA, Biol. Chem. Hoppe-Seyler 367:803 (1986).
6. R.P. Harvey, E. Degryse, L. Stefani, F. Schamber, J-P. Cazenave, M. Courtney, P. Tolstoshev, and J-P. Lecocq, Cloning and expression of a cDNA coding for the anticoagulant hirudin from the bloodsucking leech, Hirudo medicinalis, Proc. Natl. Acad. Sci. U.S.A. 83:1084 (1986).
7. T.J. Rydel, K.G. Ravichandran, A. Tulinsky, W. Bode, R. Huber, R, C. Roitsch, and J.W. Fenton II, The structure of a complex of recombinant hirudin and human α-thrombin, Science 249:277 (1990).
8. T.J. Rydel, A. Tulinsky, W. Bode, and R. Huber, Refined structure of the hirudin-thrombin complex, J. Mol. Biol. 221:583 (1991).
9. M.G. Grütter, J.P. Priestle, J. Rahuel, H. Grossenbacher, W. Bode, J. Hofsteenge, and S.R. Stone, Crystal structure of the thrombin-hirudin complex: a novel mode of serine protease inhibition, EMBO J. 9:2361 (1990).
10. G.M. Clore, D.K. Sukumaran, M. Nilges, J. Zarbock, and A.M. Gronenborn, The conformations of hirudin in solution: a study using nuclear magnetic resonance, distance geometry and restrained molecular dynamics, EMBO J. 6:529 (1987).
11. P.J.M. Folkers, G.M. Clore, P.C. Driscoll, J. Dodt, S. Köhler, and A.M. Gronenborn, Solution structure of recombinant hirudin and the Lys47-Glu mutant: a nuclear magnetic resonance and hybrid distance geometry-dynamical simulated annealing study, Biochemistry 28:2601 (1989).
12. H. Haruyama, and K. Wüthrich, The conformation of recombinant desulfato-hirudin in aqueous solution determined by nuclear magnetic resonance, Biochemistry 28:4301 (1989).
13. P.J. Braun, S. Dennis, J. Hofsteenge, and S.R. Stone, Use of site-directed mutagenesis to investigate the basis for the specificity of hirudin, Biochemistry 27:6517 (1988).
14. W. Bode, and R. Huber, Ligand binding: proteinase-protein inhibitor interactions, Curr. Opinion Struct. Biol. 1:45 (1991).
15. T. Schechter, and A. Berger, On the size of the active site in proteases, I. Papain, Biochem. Biophys. Res. Commun. 27:157 (1967).
16. S.R. Stone, and J. Hofsteenge, Kinetics of the inhibition of thrombin by hirudin, Biochemistry 25:4622 (1986).
17. S.R. Stone, and J. Hofsteenge, Recombinant hirudin: kinetic mechanism for the inhibition of human thrombin, Protein Eng. 4:295 (1991).
18. S.R. Stone, P.J. Braun, and J. Hofsteenge, Identification of regions of α-thrombin involved in its interaction with hirudin, Biochemistry 26:4617 (1987).
19. S.R. Stone, S. Dennis, and J. Hofsteenge, Quantitative evaluation of the contribution of ionic interactions to the formation of the thrombin-hirudin complex, Biochemistry 28:6857 (1989).
20. M.P. Jackman, M.A.A. Parry, J. Hofsteenge, and S.R. Stone, Intrinsic fluorescence changes and rapid kinetics of the reaction of thrombin with hirudin, J. Biol. Chem. 267:15375 (1992).
21. S. Dennis, A. Wallace, J. Hofsteenge, and S.R. Stone, Use of fragments of hirudin to investigate the thrombin-hirudin interaction, Eur. J. Biochem. 188:61 (1990).
22. M.C. Naski, J.W. Fenton II, J.M. Maraganore, S.T. Olson, and J.A. Shafer, The COOH-terminal domain of hirudin. An exosite-directed competitive inhibitor of the action of α-thrombin on fibrinogen, J. Biol. Chem. 265:13484 (1990).
23. T. Schmitz, M. Rothe, and J. Dodt, The mechanism of inhibition of α-thrombin by hirudin-derived fragments hirudin (1-47) and hirudin (45-65), Eur. J. Biochem. 195:251 (1991).

24. L. Liu, T.H. Vu, C.T. Esmon, and S.R. Coughlin, The region of the thrombin receptor resembling hirudin binds to thrombin and alters enzyme specificity, J. Biol. Chem. 266:16977 (1991).
25. G.L. Hortin, and B.L. Trimpe, Allosteric changes in thrombin's activity produced by peptides corresponding to segments of natural inhibitors and substrates, J. Biol. Chem. 266:6866 (1991).
26. E. Skrzypczak-Jankun, V.E. Caperos, K.G. Ravichandran, and A. Tulinsky, Structure of the hirugen and hirulog 1 complexes of α-thrombin, J. Mol. Biol. 221:1379 (1991).
27. A. Karshikov, W. Bode, A. Tulinsky, and S.R. Stone, Electrostatic interactions in the association of proteins: an analysis of the thrombin-hirudin interaction, Protein Sci. 1:727 (1992).
28. A. Betz, J. Hofsteenge, and S.R. Stone, Interactions of the N-terminal region of hirudin with the active-site cleft of thrombin, Biochemistry 31:4557 (1992).
29. S.R. Stone, and A. Betz, unpublished results.
30. J-Y. Chang, Thrombin specificity. Requirement for apolar amino acid adjacent to the thrombin cleavage site of polypeptide substrates, Eur. J. Biochem. 151:217 (1985).
31. G. Claeson, L. Aurell, G. Karlsson, and P. Friberger, Substrate structure and activity relationship, in New Methods for Analysis of Coagulation using Chromogenic Substrates (I. Witt, ed.), de Gruyter, Berlin (1977).
32. P. Walsmann, and F. Markwardt, Biochemische and pharmakologische Aspekte des Thrombin-inhibitors Hirudin, Pharmazie 36:653 (1981).
33. J.W. Fenton II, Thrombin interactions with hirudin, Semin. Thromb. Hemostasis. 15:265 (1989).
34. P.H. Johnson, P. Sze, R. Winant, P.W. Payne, and J.B. Lazar, Biochemistry and genetic engineering of hirudin, Semin. Thromb. Hemostasis 15:302 (1989).
35. J. Dodt, S. Köhler, and A. Baici, Interaction of site specific hirudin variants with α-thrombin, FEBS Lett. 229:87 (1988).
36. E. Degryse, M. Acker, G. Defreyn, A. Bernat, J.P. Maffrand, C. Roitsch, and M. Courtney, Point mutations modifying the thrombin inhibition kinetics and antithrombin activity *in vivo* of recombinant hirudin, Protein Eng. 2:459 (1989).
37. A.R. Fersht, The hydrogen bond in molecular recognition, Trends Biochem. Sci. 12:301 (1987).
38. A. Wallace, S. Dennis, J. Hofsteenge, and S.R. Stone, Contribution of the N-terminal region of hirudin to its interaction with thrombin, Biochemistry 28:10079 (1990).
39. A. Betz, J. Hofsteenge, and S.R. Stone, pH dependence of the interaction of hirudin with thrombin, Biochemistry 31:1168 (1992).
40. W. Bode, D. Turk, and J. Stürzebecher, Geometry of binding of benzamidine- and arginine-based inhibitors Nα-(2-naphthyl-sulphonyl-glycyl) DL-p-amidino-phenylalanyl-piperidine (NAPAP) and (2R,4R)-4-methyl1-[Nα-(3-methyl-1,2,3,4-tetra- hydro-8-quinolinesulphonyl)-L-arginyl]-2-piperidine carboxylic acid (MQPA) to human thrombin, Eur. J. Biochem. 193:175 (1990).
41. J.B. Lazar, R.C. Winnant, and P.H. Johnson, Hirudin: Amino-terminal residues play a major role in the interaction with thrombin, J. Biol. Chem. 266:685 (1991).
42. D.W. Banner, and P. Hadvary, Crystallographic analysis at 3.0 Å-resolution of the binding to human thrombin of four active site-directed inhibitors, J. Biol. Chem. 266:20085 (1991).

43. S.K. Burley, and G.A. Petsko, Aromatic-aromatic interaction: a mechanism of protein structure stabilization, Science 229:23 (1985).
44. A. Betz, J. Hofsteenge, and S.R. Stone, Role of hydrophobic residues in the C-terminal region of hirudin in the formation of the thrombin-hirudin complex, Biochemistry 30:9848 (1991).
45. J. Dodt, S. Köhler, T. Schmitz, and B. Wilhelm, Distinct binding sites of Ala^{48}-$hirudin^{1-47}$ and Ala^{48}-$hirudin^{48-65}$ on α-thrombin, J. Biol. Chem. 265:713 (1990).
46. D. Bagdy, E. Barabas, L. Graf, T.E. Peterson, and S. Magnusson, Hirudin, Methods Enzymol. 45:669 (1976).
47. J. Dodt, H. Müller, U. Seemüller, and J-Y. Chang, The complete amino acid sequence of hirudin, a thrombin specific inhibitor, FEBS Lett. 165:180 (1984).
48. J. Dodt, U. Seemüller, R. Maschler, and H. Fritz, The complete covalent structure of hirudin. Localisation of the disulfide bonds, Biol. Chem. Hoppe-Seyler 366:379 (1984).
49. J.P. Priestle, RIBBON: a stereo cartoon drawing program for proteins, J. Appl. Cryst. 21:572 (1988).
50. A. Betz, J. Hofsteenge, and S.R. Stone, Role of ionic interactions in the formation of the thrombin-hirudin complex, Biochem. J. 275:801 (1991).

BIOPHYSICAL STUDIES OF INTERACTIONS OF HIRUDIN ANALOGS WITH BOVINE AND HUMAN THROMBIN BY ESR AND FLUORESCENCE LABELLING STUDIES

Lawrence J. Berliner and Judith K. Woodford[1]

Department of Chemistry, The Ohio State University
120 West 18th Avenue, Columbus, Ohio 43210-1173, U.S.A.

[1]Current address: Department of Pharmacology and
Medicinal Chemistry, University of Cincinnati
Cincinnati, Ohio 45267-0004, U.S.A.

INTRODUCTION

The interactions between thrombin and synthetic or natural (polypeptide) inhibitors has been well characterized at the atomic level by xray crystallography as evidenced by the spectacular progress outlined elsewhere in this book. In some cases, however, elements of structural disorder in the x-ray picture are clouded by uncertainties as to the origins of this disorder: multiple conformers of the protein crystal packing effects or real conformational changes resulting from the enzyme:inhibitor interaction.

The NMR work of Scheraga and co-workers[1] with bovine thrombin and various hirudin and C-terminal inhibitor analogs showed clearly that certain regions of the inhibitor (and/or the enzyme) were too mobile to resolve by the transferred nuclear Overhauser effect method (TRNOE), which identifies inhibitor protons that bind to the protein. Furthermore, the rapid kinetics measurements of Stone and Hofsteenge[2] has strongly suggested multiple binding steps in thrombin:hirudin complex formation. The goal of the studies outlined in this chapter was the application of ESR spin labelling and fluorescence spectroscopy as probes of local, dynamic structural changes on thrombin in solution.

ESR Spin Labelling

The general approach to this technique, which has been outlined earlier[3,4], utilizes the active site probes depicted in Figure 1A. In the case of a nitroxide spin label, analysis

The Design of Synthetic Inhibitors of Thrombin
Edited by G. Claeson, *et al.*, Plenum Press, New York, 1993

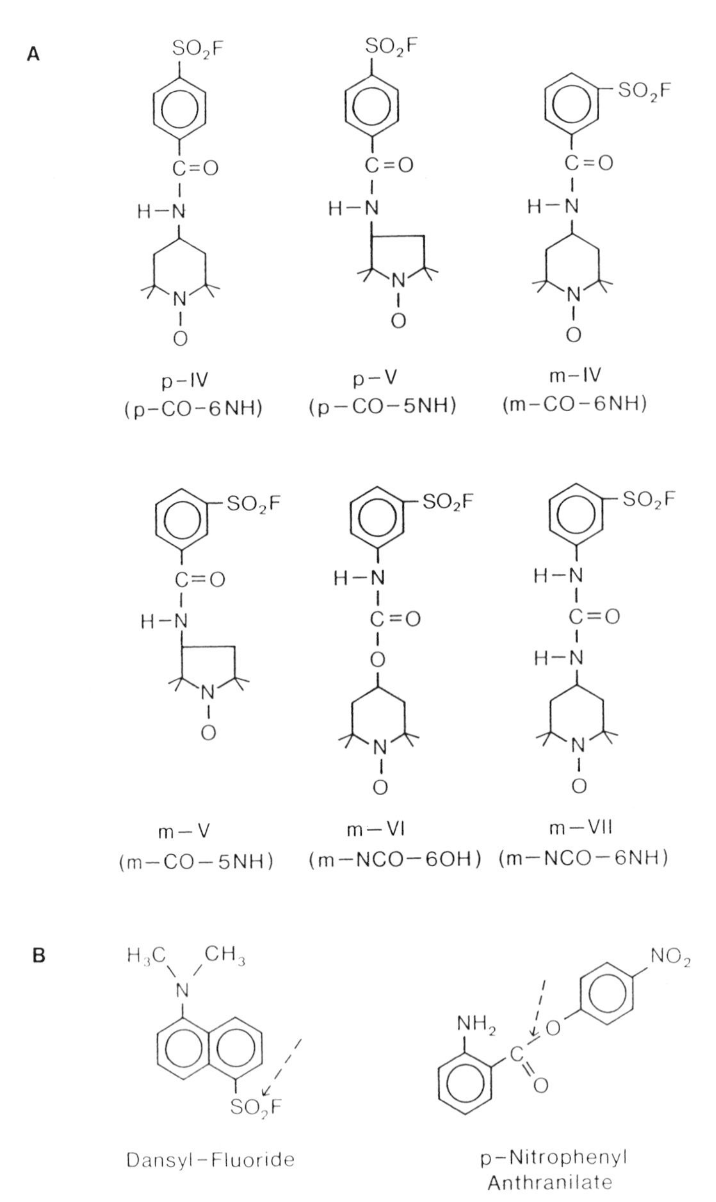

Figure 1 Structures of (A) sulfonyl fluoride spin labels and (B) fluorescent probes used in this study. The spin label nomenclature is described in Berliner[4]. The arrows (↓) denote the bond attacked by Ser195. p-IV (*p*-CO-6NH), 4-(2,2,6,6-tetramethyl piperidine-1-oxyl)-p-(fluorosulfonyl)-benzamide; p-V (*p*-CO-5NH), 4-(2,2,5,5-tetramethyl-pyrrolidine-1-oxyl)-p-(fluorosulfonyl) benzamide; m-IV (*m*-CO-6NH, 4-(2,2,6,6-tetramethyl-piperidine-1-oxyl)-m-(fluorosulfonyl)benzamide; m-V (*m*-CO-5NH), 3-(2,2,5,5-tetramethyl-pyrrolidine-1-oxyl)-m-(fluorosulfonyl)-benzamide; m-VI (*m*-NCO-6OH), N-(*m*-fluorosulfonyl-benzene)-4-0-(2,2,6,6-tetramethyl piperidine-1-oxyl) carbamate; m-VII (*m*-NCO-6NH), N-(*m*-[fluorosulfonyl]-phenyl)-4-N-(2,2,6,6-tetramethyl-piperidine-1-oxyl) urea. From Nienaber and Berliner[21].

of the ESR spectrum related the mobility and hence the local interactions between the nitroxide ring and the protein. These interactions reflect primarily hydrophobic van der Waals contacts and some contributions from steric restrictions at the binding locus. While specific assignments of those protein residues involved in this interaction are (currently) impossible to assign, changes in this region which alter protein-nitroxide interactions are detectable with high sensitivity. The ESR spin labeling method utilizes extrinsic cyclic nitroxide probes. These are synthetically "designed" to contain groups that are either specifically reactive or have high affinity for the enzyme system of interest. Most of the work done in the thrombin field have involved fluorosulfonyl-phenyl nitroxides, which contain the reactive, serine195 specific sulfonyl fluoride group. The result is a covalent sulfonate linkage to the active serine which geometrically resembles a tetrahedral intermediate in the typical amide catalyzed mechanism by serine proteases. These probes, which are of dimensions of 10 to 15 Å in length, are very sensitive to both ligands binding in the substrate binding pocket as well as conformational changes elicited from distant points in the molecule to the active site region. For example, the binding of apolar, allosteric effectors for thrombin were first observed by Berliner and Shen[5]. In a more physiologically relevant example Musci et al[6] observed conformational changes at the active site upon complexation of thrombomodulin to human thrombin.

Fluorescence Labelling

In the case of the fluorescent probes shown (Figure 1B), both will reflect subtle changes in active site polarity, solvent accessibility, tryptophan energy transfer and several other factors which effect the quantum yield and emission maximum. While some dynamical information is obtainable from lifetime and polarization measurements, the steady state fluorescence spectrum is already highly sensitive to small conformational changes.

Fluorescence spectroscopy offers a multitude of physical parameters to monitor upon the interaction of a protein with an effector (i.e. hirudin analog) of interest. From simple steady-state fluorescence experiments, one can monitor the quantum yield (relative fluorescence intensity), the excitation or emission maximum position (λ_{ex}, λ_{em}) the polarization anisotropy or lifetime, where appropriate. Techniques such as simple quantum yield measurements can detect the change in solvent accessibility of a particular fluorophore (or exposure of a fluorophore to a nearby electron rich residue (i.e., Cys, carboxylates, basic residues, etc.) with unusually high sensitivity. The conformational changes can be both global or local, however, the fluorescing group is reporting in the end a change to its local environment. In more sophisticated approaches one can incorporate both donor and acceptor fluorophores for accurate distance measurements between two well placed fluorophores in a biological system. One of these fluorophores can be an intrinsic tryptophan residue or an externally introduced label.

The beauty of intrinsic fluorescence spectroscopy is the lack of any need to introduce extrinsic fluorophores. That is, aromatic residues, principally tryptophan, contribute to the fluorescence emission observed with a excitation in the 280-295 nm range. Again, the fluorescence parameters are multifold as mentioned above. The only uncertainties arise when the protein contains multiple fluorescing aromatic residues, i.e. four or five tryptophans where all contribute collectively to the fluorescence emission.

Here, the results are almost always of a general, global nature unless a single tryptophan can be "isolated" and characterized by virtue of its unique fluorescence lifetime or excitation band.

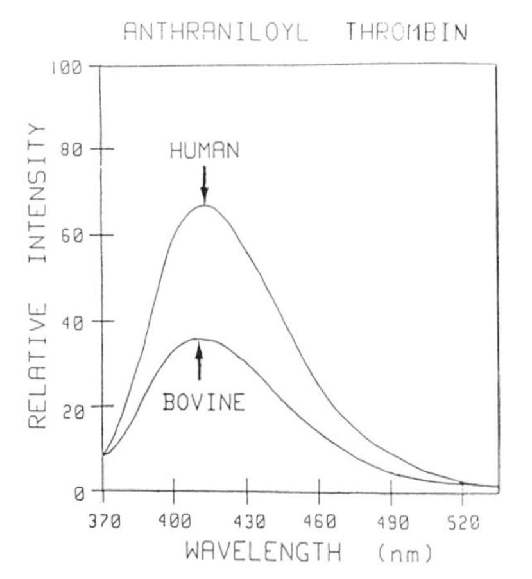

Figure 2A Fluorescence emission spectra of 2 μM anthraniloyl human and bovine α-thrombins, respectively. Conditions were pH 6.5, 0.05 M Tris-HCl, 0.75 M NaCl, 25°C, λ_{ex}=350nm. Both dansyl and anthraniloyl human α-thrombin gave larger quantum yields relative to the bovine enzyme. We did not quantitate this difference exactly; however, as it depends on a precise estimation of the percent protein labelled during the labelling reaction (ca 80-90% dansyl and 20-30% anthranilate). From Nienaber and Berliner[21].

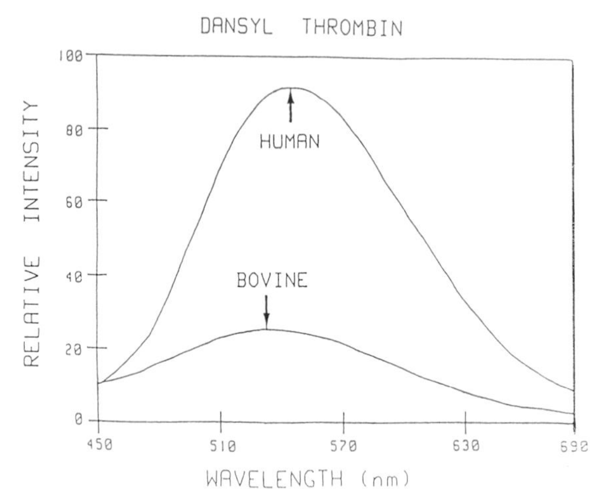

B Fluorescence emission spectra of 2 μM dansyl human and bovine α-thrombins. Conditions were pH 6.5, 0.05 M Tris-HCl, 0.75 M NaCl, 25°C, λ_{ex}=360nm (human), and 357 nm (bovine). From Nienaber and Berliner[21].

Methodology: Detection and Characterization of Conformational Changes upon Hirudin Binding

The crystal structure of hirudin-thrombin complexes[7,8] as well as complexes between hirudin C-terminal fragments (hirugens) with both human and bovine thrombin define at best one restricted, low energy conformation of this binary complex[9]. In some cases, crystallographers have reported no electron density or a blur of multi-conformers, particularly in the regions of the two flexible insertion loops. Thus the need for highly sensitive solution methods is obvious. The techniques used in the work presented in this chapter center mainly on fluorescence spectroscopy (intrinsic and extrinsic labeling studies, and electron spin resonance (ESR spin labelling)). The two methods are uniquely suited for detecting small, yet significant differences in structure and conformation between various thrombin forms, various thrombin species, thrombin-ligand complexes. As an example, Figures 2A and 2B show fluorescence emission spectra for bovine and human thrombin, respectively, covalently labeled at the active serine with p-nitrophenyl anthranilate and dansyl fluoride. In both cases the fluorophores are virtual substrate intermediates. With the high level of sequence homology, and now x-ray crystal structures reported for both proteins[10,11] we know that the differences at the three dimensional static level are slight at most. Yet both the quantum yields and emission maxima of the active site labeled thrombin species in solution (Figures 2A, 2B) are quite different! Actually, several other pieces of indirect evidence have reported unusual differences in their behaviors, such as reactivity with some exosite benzamidine affinity label analogs[12], the fibrinogen clotting behavior of proteolytically cleaved bovine thrombin to its ß-form (10% residual activity), vs. human ß-thrombin which has essentially zero activity[13,14]. Another feature was their differential binding behavior with bovine pancreatic trypsin inhibitor. Bovine thrombin complexed with a K_i value only two to three orders of magnitude larger than that for trypsin while human thrombin was 10^6 to 10^7 orders larger[15], (Rowand and Berliner, unpublished results). These differences are further demonstrated in the ESR spectra shown in Figure 3 for m-V and p-V labelled human and bovine thrombins, respectively. Here the nitroxide is reporting its interactions with the structural environment about 10Å distant from serine 195 (covalent phenylsulfonyl linkage). Again, in these cases we note that the bovine ESR spectra show different nitroxide mobility, i.e. different interactions between the nitroxide and the enzyme, as exemplified by the hyperfine extrema splittings $2T_{||}$ (denoted by arrows). Some of these differences were speculated to be attributed to a difference in flexibility in the ß-insertion loop (residues 59-61), containing Trp 60D and ß-insertion loop 145-150. In particular, in the latter loop a major substitution occurs at residue 149E, from Lys in human to the oppositely charged Glu in bovine thrombin, respectively[10].

Ligand Binding

The nitroxide spin labels shown in Figure 1A have been shown to be sensitive to two general regions of the (human) thrombin active site. That is, while all of the fluorosulfonylphenyl nitroxides are identical in structure excepting the nitroxide moiety and its connecting linkage, we may model the entire set (Figure 1A) covalently bound to Ser 195, near the basic binding pocket, with the various nitroxide substituents occupying in the end, only two general regions of the extended active site, one sensitive to the binding of apolar ligands, such as indole; the other apparently interacting with either the insertion loop (residues 145-150) and/or the ß-insertion loop (residues 59-60). It is from this "vantage point" that the binding effects of hirudin analogs were monitored. In the case of active site fluorophores (eg., dansyl-, anthraniloyl-) changes in quantum yield or emission wavelength are sensitive to subtle conformational effects upon binding benzamidine, at the

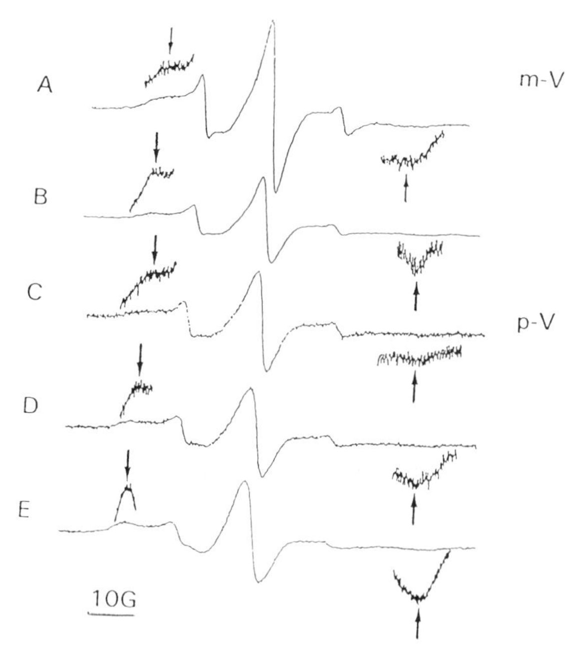

Figure 3 X-band ESR spectra of active site spin labelled human and bovine α-thrombins (A) m-V labelled human α-thrombin; (B) m-V bovine α-thrombin, (C) *p*-V human α-thrombin in the presence of saturating sucrose; (D) p-V bovine α-thrombin in the presence of saturating sucrose; (E) *p*-V bovine α-thrombin in the presence of 10 mM indole, and saturating sucrose. Conditions were pH 6.5, 0.05 M Tris-HCl, 0.75 M NaCl, 20 °C. Protein concentrations were typically 0.07-0.08 mM. Arrows (↓) indicate points used in determining the maximum hyperfine splitting ($2T_{||}$). From Nienaber and Berliner[21].

P1 specificity pocket, one notes shifts in the mobility of active site conformational probes. Figure 4 shows a plot of the decrease in emission intensity for dansyl α-thrombin upon titration with benzamidine. The apparent K_d for benzamidine was 66±12 mM, about 70 times the K_i for clotting or amidase activity, which accounts for a steric shift of the fluorophore over the specificity pocket to allow entry of the aromatic benzamidine inhibitor[16].

Interactions of Active Site Labelled Thrombins and Recombinant Hirudins and Hirudin Analogs

In order to address Stone and Hofsteenge's[2] kinetic results with direct physical evidence, we examined the binding of intact recombinant hirudin HV2-Lys47, (Transgene, Strasbourg, France) and two C-terminal fragments, which we have named 12-mer and 21-mer (Marion Merrell-Dow, Cincinnati, OH). Their respective sequences are shown in Table 1.

Upon titration of human anthraniloyl-α-thrombin with 12-mer, 21-mer and intact hirudin, we observed the fluorescence emission changes depicted partially in Figure 5. For

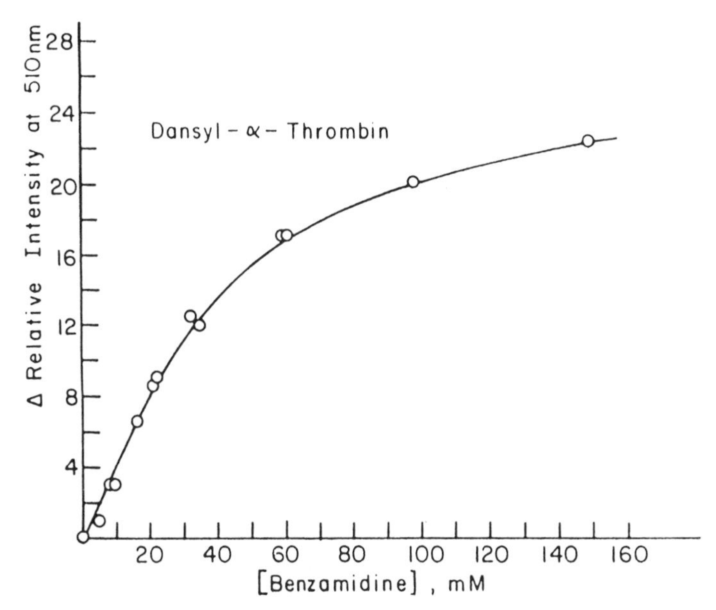

Figure 4 Decrease in emission intensity for a benzamidine titration of 2.4 μM dansyl human α-thrombin. Conditions were pH 6.5, 0.05 M sodium phosphate, 0.75 M NaCl, 25±1 °C. The excitation wavelength was 345 nm. From Berliner and Shen[5].

Table 1 Sequences of HV2-Lys47, and 21-mer and 12-mer C-terminal fragments.

HV2-Lys47

Ile-Thr-Tyr-Thr-Asp-Cys-Thr-Glu-Ser-Gly-Gln-Asn-Leu-Cys-Leu-Cys-Glu-Gly-Ser-Asn-Val-Cys-
1 22
Gly-Lys-Gly-Asn-Lys-Cys-Ile-Leu-Gly-Ser-Asn-Gly-Lys-Gly-Asn-Gln-Cys-Val-Thr-Gly-Glu-Gly-
23 44
Thr-Pro-Lys-Pro-Glu-Ser-His-Asn-Asn-Gly-Asp-Phe-Glu-Glu-Ile-Pro-Glu-Glu-Tyr-Leu-Gln
45 65

21-mer

Thr-Pro-Lys-Pro-Gln-Ser-His-Asn-Asp-Gly-Asp-Phe-Glu-Glu-Ile-Pro-Glu-Glu-Tyr-Leu-Gln
45 65

12-mer

Gly-Asp-Phe-Glu-Glu-Ile-Pro-Glu-Glu-Tyr-Leu-Gln
54 65

both C-terminal fragments, a small but significant decrease in quantum yield was observed (10%) while the thrombin:HV2 Lys 47 complex showed a net enhancement of anthraniloyl emission intensity (J.R. Rowand and L. J. Berliner, to be published). In all cases the emission maximum was blue shifted by about 3 nm for the saturated complex. Thus, although neither the 12-mer or 21-mer can overlap the active site, as deduced from crystal structure analysis[7,8,9] their binding to the hirudin C-terminal anion exosite on thrombin elicits a conformational change at the active site, which is identical for both the 12- and 21-mer:human thrombin complex. Thus, we might surmise that the conformational change itself, is similar for both C-terminal analogs. On the other hand, HV2-Lys47 essentially "smothers" the catalytic site, resulting here (and in the case of dansyl thrombin as well) in an enhancement which possibly derives from blocking the access of H_2O molecules. A quite unusual result was found with HV2-Lys47:human α-thrombin, which opens some enticing speculation. That is, when following a titration the ligand, a suggestively sigimoidal curve is observed with a half-transition point corresponding to a stoichiometry HV2-Lys47: anthraniloyl-α- thrombin $\leq$ 0.7:1.0. The position of this transition was essentially unchanged if one or mixed unlabeled with labeled thrombin, diluted the solutions 1:1. Instead the result "appears" to be consistent with a tight complex representing, on average, two thrombin molecules per HV2-Lys47. Normally one expects these titrations to fit a perfectly straight line up to 1:1 stoichiometry followed by a zero slope straight line at higher ratios of hirudin to thrombin, such as that observed for HV2-Lys47:dansyl human α-thrombin (Rowand and Berliner, to be published). The only plausible explanation is that two thrombin molecules bind to one HV2-Lys47 monomer at present derives from titrations of high concentrations of C-terminal fragments (12-mer, 21-mer) with dansyl-human $\propto$-thrombin which gave evidence for a second albeit weaker binding site for these fragments. While this phenomena showed up only with the anthraniloyl derivative, we note similar observations with native thrombin: hirudin complexes by CD spectroscopy reported by Kono et al[17].

When a parallel study with both active site fluorophores was carried out with bovine α-thrombin, similar results were found, although the emission spectrum of dansyl bovine α-thrombin was not enhanced vs. the human thrombin derivative; in fact, although the λ_{em} was substantially blue shifted (7-13 nm) upon binding 21-mer or 12-mer, the emission intensity was unchanged or slightly quenched, respectively. On the other hand, the complex HV2-Lys47: dansyl bovine-α-thrombin complex was enhanced as also found with the anthraniloyl derivatives of both bovine and human thrombin (J. K. Rowand and L. J. Berliner, to be published).

The results with active with spin labeled derivatives were more conclusive regarding conformationally induced changes since the ESR spectrum reflects essentially only nitroxide motion. Figure 6 depicts ESR spectra for *m*-VII labeled bovine thrombin respectively. The outer hyperfine extrema (denoted by arrows) reflect the overall tumbling rate of the nitroxide, where an increase in this splitting, $2T_{||}$, correlates with slower motion. First, if one compares human (not shown) vs. bovine m-VII thrombin alone, the larger $2T_{||}$ for bovine (51.6 Gauss vs. human, 49.1 Gauss) exemplifies their subtle structural differences discussed early. Upon saturating with 12- or 21-mer only slight changes, if any, were observed in the $2T_{||}$, although all were within experimental accuracy. However, complexation with HV2-Lys47 resulted in a major immobilization of the piperidinyl nitroxide moiety, with only a slight amount residual motion. Presumably, the large hirudin molecule has sterically hindered nitroxide motion but could also contribute to structural changes which favor stronger hydrophobic interactions with the enzyme surface. In the case of the label *p*-V (data not shown) complexation of the spin labelled human form with 12-mer or 21-mer resulted in label immobilization; the $2T_{||}$ changed from 49.5 to 60 Gauss (data not shown). The corresponding HV2-Lys47: human thrombin complex was even more immobilized, $2T_{||} = 63.3$ Gauss (data not shown).

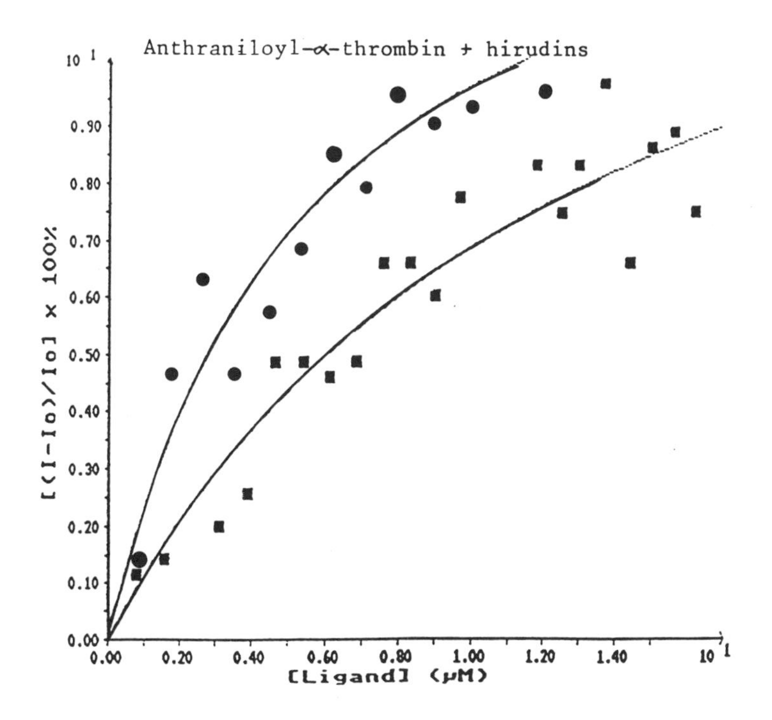

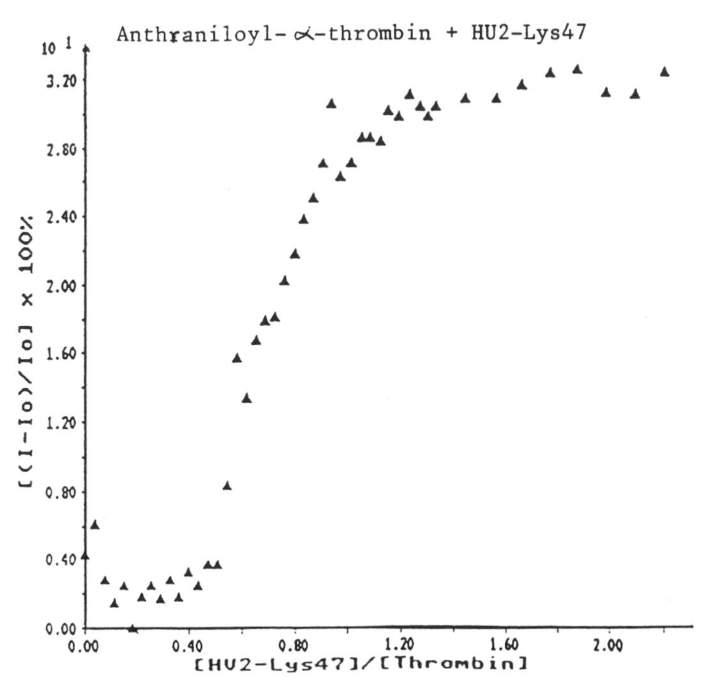

Figure 5 Extrinsic emission fluorescence intensity titration curves of anthraniloyl-α-thrombin with 12-mer (●), 21-mer (■), and HV-2-Lys47 ( ). Thrombin concentrations were 2.0 μM. λ_{ex}=340 nm. Conditions were 0.75 M NaCl, 0.05 M Tris, pH 6.5, 25±1°C.

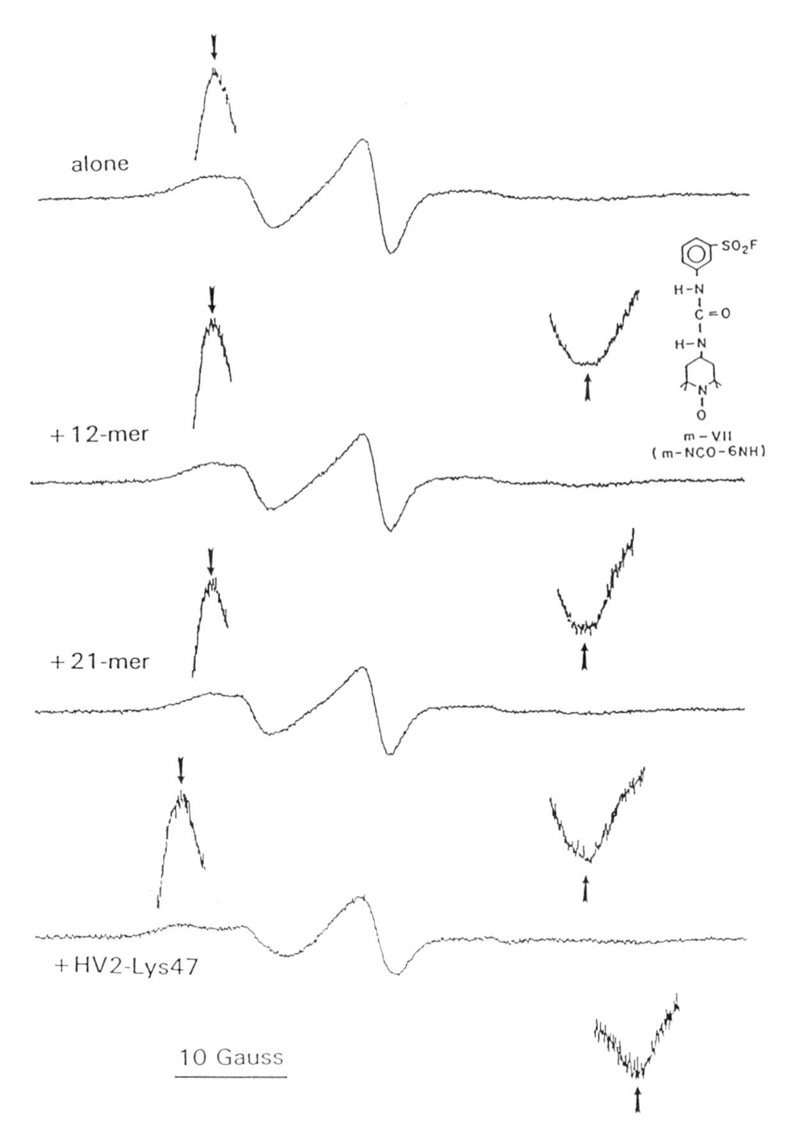

Figure 6 X-band ESR spectra of bovine *m*-VII-α-thrombin in the absence and presence of 12-mer($\approx 100\ \mu$M), 21-mer($\approx 100\ \mu$M), and HV2-Lys47 ($\approx 70\ \mu$M). Thrombin concentration was 50-60 μM. The arrows indicate the hyperfine extrema, the positions of the outermost peaks. Conditions were 0.75M NaCl, 0.05 M Tris, pH 6.5, $20 \pm 2^{\circ}$C.

On the other hand, *p*-V labelled bovine thrombin only showed changes upon complexation with HV2-Lys47 as also found for *m*-VII thrombin (Figure 7).

What are the structural explanations for the difference in sensitivity to the binding of C-terminal hirudin fragments? The *p*-V label was shown earlier to reside in a binding locus on human thrombin that was sensitive to the binding of apolar ligands[18]. On the other hand, the *m*-VII label bound to another locus which was significantly altered between human α- and γ-thrombin. Secondly, this region occupied by the *p*-V label is structurally linked to the binding of 12-mer and 21-mer to human thrombin, but has been lost in the case of the bovine derivative. Upon comparing any major structural differences between the two species that is also implicated in hirudin C-terminal binding Lys149E, which forms a salt bridge with Asp55 of HV2-Lys47, is critical since it is substituted by a Glu in bovine thrombin. That is, one expects no interaction (i.e., repulsion) between residue 149E of bovine thrombin and Asp55 of hirudin (or its 12-mer and 21-mer C-terminal analogs). Consequently, the *p*-V label is interacting directly with loop 145-150 and/or is perturbed as a result of binding interactions between hirudin analogs and human thrombin. Preliminary molecular graphics of our spin labels with thrombin models shows that the nitroxide can potentially interact with either insertion loop (145-150,59-61) which distinguishes thrombin from the simpler pancreatic serine proteases.

Distinguishing Between Hirudin Isoinhibitors

To date three hirudin isoinhibitors of almost identical primary structure have been reported from medicinal leeches, their relative amounts varying with geographic origin of the leech. Yet, any physical characterization of these species has been limited principally to HV2 Lys 47. We were curious as to whether the dynamic solution probes described earlier might be sensitive to these small differences in sequence upon complexing various labelled thrombins[22]. Indeed some marked differences were found, allowing one to distinguish these isoinhibitors on the basis of the fluorescence or ESR properties of labeled thrombins as probes. The scheme below lists the primary sequences of the two recombinant hirudin forms, HV1 and HV2-Lys47 (Table 2). These differ by eight residues which are indicated in bold. Figure 7 depicts ESR spectra for HV1 and HV2-Lys47, respectively complexed with *m*-VII- and *p*-III-ϵ-labelled-thrombins, where the α-form is cleaved at Ala 149A. The differences in immobilization are quite striking between HV2-Lys47 and HV1 when comparing the hyperfine extrema $2T_{\|}$ (indicated by arrows). The larger splitting of HV2-Lys47 indicates a greater nitroxide immobilization vs. the HV1- complex. Addition of the two hirudin isoinhibitors to *m*-VII- and *p*-III-labelled thrombins also showed similar discrimination[22].

The data suggest that structural differences exist between the HV1 and HV2-Lys47-thrombin complexes. These are monitored in the active site region but may also be conformationally elicited from a more distant binding site. The larger $2T_{\|}$ observed for HV2-Lys47:spin labelled thrombin complexes in the majority of the ESR results suggested that the binding interactions with HV2-Lys47 result in a greater nitroxide immobilization than with HV1. Since the nitroxide moieties of these labels, eg. *p*-III and *m*-VII, are known to occupy spatially distinct regions in the extended active site of thrombin[18], the differences may be localized to the phenylsulfonyl binding locus rather than the region occupied by the nitroxide group. Another possibility is that HV2-Lys47 more completely 'blankets' the active site restricting the nitroxide rotation more severely[22].

The results would suggest that loop 145-150, containing both the ϵ- and ζ-thrombin cleavages, may be interacting differently with the two hirudin isoinhibitors[19,20]. Due to insertions both loops 145-150 and 70-80 contain 11 residues, while loop 59-61 contains 12. From examination of the N-terminus of hirudin, we note that in the crystallographic structure of HV2-Lys47:human α-thrombin complex, Ile 1 was involved

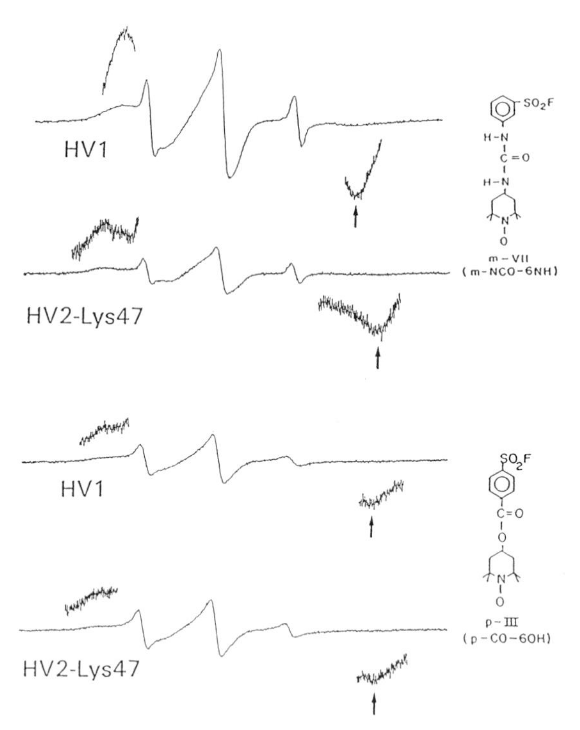

Figure 7 X-band ESR spectra of spin labelled human ϵ-thrombin (50-60 μM) complexed with HV1 (≈70 μM) and HV2-Lys47 (≈70 μM), respectively. The upper spectra are m-VII-ϵ-thrombin; the lower spectra are p-III-ϵ-thrombin. All spectra were measured at pH 6.5, 0.05 M Tris, 0.75 M NaCl, 20±2°C. Protein concentration was typically 50-90 μM. From Rowand and Berliner[22].

Table 2 Primary sequences of HV1 and HV2-Lys47.

HV1	Val-Val-Tyr-Thr-Asp-Cys-Thr-Glu-Ser-Gly
HV2-Lys47	Ile-Thr-Tyr-Thr-Asp-Cys-Thr-Glu-Ser-Gly
	1 10
HV1	Gln-Asn-Leu-Cys-Leu-Cys-Glu-Gly-Ser-Asn
HV2-Lys47	Gln-Asn-Leu-Cys-Leu-Cys-Glu-Gly-Ser-Asn
	11 20
HV1	Val-Cys-Gly-Gln-Gly-Asn-Lys-Cys-Ile-Leu
HV2-Lys47	Val-Cys-Gly-Lys-Gly-Asn-Lys-Cys-Ile-Leu
	21 30
HV1	Gly-Ser-Asp-Gly-Glu-Lys-Asn-Gln-Cys-Val
HV2-Lys47	Gly-Ser-Asn-Gly-Lys-Gly-Asn-Gln-Cys-Val
	31 40
HV1	Thr-Gly-Glu-Gly-Thr-Pro-Lys-Pro-Gln-Ser
HV2-Lys47	Thr-Gly-Glu-Gly-Thr-Pro-Asn-Pro-Glu-Ser
	41 50
HV1	His-Asn-Asp-Gly-Asp-Phe-Glu-Glu-Ile-Pro
HV2-Lys47	His-Asn-Asn-Gly-Asp-Phe-Glu-Glu-Ile-Pro
	51 60
HV1	Glu-Glu-Tyr-Leu-Gln
HV2	Glu-Glu-Tyr-Leu-Gln
	61 65

in non-polar interactions at the catalytic site[7]. (In HV1 the substitution of a Val for this residue should be inconsequential). Residue 2 binds at the edge of the specificity pocket and does not appear to contribute much to the overall binding. The C-terminus showed that residues 31-36 were disordered[7]. Residue 53 was also found to have poorly defined electron density in the crystal.

The two residues which appear to be potentially critical are: Lys24 which is involved in a hydrogen bond to thrombin through a water molecule (a substitution of Gln24 should alter this interaction); and Glu49 which in HV2-Lys47 forms an ion pair with His51 of thrombin. In HV1 the corresponding Gln49 disallows an ion pair. While residue 49 lies near the catalytic site, residue 24 resides nearer to thrombin loops 59-61 and 70-80.

REFERENCES

1. H.A. Scheraga, K. Gibson, and F. Ni, NMR studies of thrombin-fibrinopeptide and thrombin-hirudin complexes, in: "Thrombin: Structure and Function," Berliner, L. J. (ed.), Plenum Publishing, New York, NY. pp.63 (1992).

2. S.R. Stone, J. and Hofsteenge, Kinetics of the inhibition of thrombin by hirudin, Biochemistry 25:4622 (1986).
3. L.J. Berliner, Using the spin label method in enzymology, in: "Spectroscopy in Biochemistry, Vol. 2," J. E. Bell, ed. CRC Press, Boca Raton, Florida pp. 1 (1980).
4. L.J. Berliner, ESR and fluorescence studies of thrombin active site conformation, in: Thrombin: Structure and Function, L.J. Berliner, ed. Plenum Publishing, New York, NY pp.87 (1992).
5. L.J. Berliner, and Y.Y.L. Shen, Active site fluorescent labelled dansyl and anthraniloyl human thrombins, Thromb. Res. 12:15 (1977).
6. G. Musci, L.J. Berliner, and C.T. Esmon C.T, Evidence for multiple conformational changes in the active center of thrombin induced by complex formation thrombomodulin: an analysis employing nitroxide spin labels, Biochemistry 27:769 (1988).
7. T.J. Rydel, K.G. Ravichandran, A. Tulinsky, W. Bode, R. Huber, C. Roitsch, and J.H. Fenton Jr., The structure of a complex of recombinant hirudin and human α-thrombin, Science 249:277 (1990).
8. T.J. Rydel, and A. Tulinsky, Refined structure of the hirudin-thrombin complex, J. Mol. Biol. 221:583 (1991).
9. E. Skrzypczak-Jankun, V.E. Carperos, K.G. Ravichandran, and A. Tulinsky, Structure of the hirugen and hirulog 1 complexes of α-thrombin, J. Mol. Biol. 221:1379 (1991).
10. W. Bode, R. Huber, T.J. Rydel, and A. Tulinsky, X-ray structure of human thrombin: structure of hirudin:thrombin complex, in: "Thrombin: Structure and Function," Berliner, L. J. (ed.), Plenum Publishing, New York, NY. pp.3 (1992).
11. W. Bode, I. Mayr, U. Baumann, R. Huber, S.R. Stone, and J. Hofsteenge, The refined 1.9 Å crystal structure of human α-thrombin: Interaction with D-Phe-Pro-Arg chloromethylketone and significance of the Tyr-Pro-Pro-Trp insertion segment, EMBO. J. 8:3467 (1989).
12. S.A. Sonder, and J.W. Fenton, Jr. Differential inactivation of human and bovine α-thrombins by exosite affinity-labelling reagents, Thromb. Res. 32:623 (1983).
13. R.L. Lundblad, M.E. Nesheim, D.L. Straight, S. Sailor, J. Bowie, J.W. Jenzano, J.D. Robert, and K.G. Mann, Bovine α- and ß-thrombin is not a consequence of reduced affinity for fibrinogen, J. Biol. Chem. 259:6991 (1984).
14. R. Lottenberg, J.A. Hall, J.W. Fenton, Jr. and C.M. Jackson, The action of thrombin on peptide p-nitroanilide substrates: hydrolysis of tos-gly-pro-arg-pNA and D-phe-pip-arg-pNA by human α- and γ- and bovine α- and ß-thrombins, Thromb. Res. 28:313 (1982).
15. Y. Sugawara, J.J. Birktoft, and L.J. Berliner, Human α- and y-thrombin inhibition by trypsin inhibitors supports predictions from molecular graphics experiments, Seminars in Thrombosis and Hemostasis 12:209 (1985).
16. L.J. Berliner, and Y.Y. Shen, Physical evidence for an apolar binding site near the catalytic center of human α-thrombin, Biochemistry 16:4622 (1977b).
17. S. Kono, J.W. Fenton, Jr. and G.B. Villanueva, Analysis of the secondary structure of hirudin and the conformational change upon interaction with thrombin, Arch. Biochem. Biophys. 267:158 (1988).
18. L.J. Berliner, R.S. Bauer, T-L. Chang, J.W. Fenton II, and Y.Y.L. Shen, Active site topography of human coagulant (α) and noncoagulant (γ) thrombins, Biochemistry 20:1831 (1981).
19. M.S. Brower, D.A. Walz, K.E. Garry, and J.W. Fenton, J.W. II, Human neutrophil elastase alters human α-thrombin function: Limited proteolysis near the γ-cleavage site results in decreased fibrinogen clotting and platelet-stimulatory activity, Blood 69:813 (1987).

20. D.V. Brezniak, M.J. Brower, J.I. Witting, D.A. Walz, and J.W. Fenton II, Human α- to ζ-thrombin cleavage occurs with neutrophil cathepsin G or chymotrypsin while retaining fibrinogen clotting activity, Biochemistry 29:3536 (1989).
21. V.L. Nienaber, and L.J. Berliner, Conformational differences between human bovine thrombins as detected by electron spin resonance and fluorescence spectroscopy, Thromb. Haemostas. 65:40 (1991).
22. J.K. Rowand, and L.J. Berliner, Structural differences in active site-labelled thrombin complexes with hirudin isoinhibitors, J. Protein Chem. 11:483 (1992).

pH-DEPENDENT BINDING CONSTANTS FOR THE INHIBITION OF THROMBIN BY TRANSITION STATE ANALOGS

Manfred Philipp[a,b], Ling-Hao Niua[b], Tushini DeSoyza[b], Göran Claeson[c], and Rainer Metternich[d]

[a]Chemistry Department, Lehman College/CUNY, Bronx, N.Y. 10468
[b]Biochemistry Ph.D. Program, Graduate & University Center of CUNY
[c]Thrombosis Research Institute, Manresa Road, London SW3 6LR
[d]Preclinical Research, Sandoz Pharma, CH-4002 Basel

INTRODUCTION

Transition state analogs provide a unique view into the course of enzyme catalysis by providing a model of the enzyme's state during the reaction. Studying the pH dependencies of binding by such inhibitors promises a better understanding of the course of catalysis, especially when these dependencies are compared to those of substrates and other inhibitors.

One family of transition state analog inhibitors of serine proteases are the boronic acids. As with boric acid, boronic acids are Lewis acids that ionize by addition of a hydroxyl group from water[1].

$$R\text{-}B(OH)_2 + H_2O \rightleftharpoons R\text{-}B(OH)_3^- + H^+$$

This mode of ionization facilitates their reversible esterification to alcohols[2]. Diols and other bidentate and tridentate compounds form complexes with boronic acids[3-6] that are, in some cases, stable and crystallizable[7]. The formation of such reversible complexes has led to boronic acids' utility in the separation and analysis of various diol-containing compounds[8-10]. Binding and catalysis by boronic acids to various ligands has been of interest in this laboratory and elsewhere[11-14].

The first reports of boronic acid binding to proteases included the pH profile of inhibition, which showed a pK near 7[15]. Further study demonstrated that the more complete pH profile is bell-shaped, with pK_1 near seven and pK_2 near the pK of the boronic acid used[16]. The interpretation of pK_2 was dictated by the observation that it corresponds to the various pK values of different substituted benzeneboronic acids. The pK near seven for all of these inhibitions is close to the pK seen in second-order acylation rate constants (k_2/K_s) for slow substrates. Thus, the acidic, trigonal boronic acids are the active inhibitors which bind to the alkaline form of the enzyme active site. Boronic acid inhibitions have the form of an acid-base reaction.

The first crystal structure analyses confirmed the suggestion that boronic acids, once bound, are esterified to the active site serine hydroxyl as a complex containing a tetrahedral, anionic boron atom[17]. The boron atom has been also observed to exist in a coordinate covalent complex with the active site imidazole[18-20]. NMR evidence has been gathered to support both types of complex[21-23].

Interpretation of pK_2 values for boronic acid inhibitions relates to the anionic nature of serine protease active sites above pH 7. That the active site is anionic has been shown for a number of serine proteases using a variety of techniques. For subtilisin and chymotrypsin, for instance, while neutral substrates bind in a pH-independent manner, specific peptide carboxylate anions show a single-ionization pK near seven with binding below this pK[24,25]. This pK is clearly that of the enzyme active site, not the substrate. For most serine proteases, peptide anion binding constants below pH 7 are similar to those of analogous neutral peptide amides[24-26]. Thus, the inference is that peptide anions are repelled by an anionic active site at pH values above seven, but bind normally to a neutral active site at pH values below seven. Molecular modeling calculations[27] also support the idea of an anionic active center in the trypsins. Thus, pK_2 for boronic acid-serine protease binding profiles must result from charge repulsion between the boronate anion in solution and the anionic active site.

The observed pK_1 may have two different origins. One is the inability of the serine OH in the protonated active site to act as a nucleophile towards the trigonal boronic acid, in the same way that the serine OH in the protonated active site cannot react with the trigonal carboxamide substrate. Such kinetic control in binding has been observed in peptide aldehydes[28]. Another origin for pK_1 is the inability of the protonated imidazolium cation to complex to boronic acids in cases where a coordinate covalent B-N bond is formed.

Other anionic inhibitors have also shown pH-dependent binding constants not seen in substrate binding profiles. The best-known examples are the arsonic acids, which bind only at acid pHs[29].

The pH-dependent behaviour of thrombins towards substrates has been studied extensively using nitroanilide[30,31] and nitrophenyl esters[32], and has been shown to be very dependent on NaCl concentration.

MATERIALS AND METHODS

Materials

Bovine plasma thrombin was purchased from Sigma (1000 NIH units per vial) as a lyophilized powder. After reconstitution with 1 ml water, the solution had a pH of 6.5 in 0.15 M NaCl and 0.05 M sodium citrate. Bovine serum albumin (BSA) was purchased from Sigma. The stock solution was made in water (3 mg/ml).

The fluorogenic thrombin substrate, N-t-Boc-Valyl-Prolyl-Arginine-7-AMC-HCl[33], was purchased from Sigma. A 3.9 mM stock solution was made in N,N dimethyl-formamide. The chromogenic thrombin substrate, D-Phe-Pipecolyl-Arg-pNA (S-2238)[34], without mannitol solubilizer, was a gift of KabiVitrum. A stock solution of 3.36 mM was made in methanol.

1-(Z-D-Phenylalanyl-Prolineamido)-1-dihydroxyborono-4-methoxybutane(madeas the pinanediol ester of the boronic acid and abbreviated as Z-D-Phe-Pro boroMpg OPin) was synthesized at KabiVitrum and at Sandoz. Stock solutions of 7.88 mM were made in methanol.

Z-D-Phenylalanyl-Prolyl-Arginine was synthesized at KabiVitrum. A 2 mg/ml stock solution was made in 0.1 M pH 7.0 phosphate buffer. N^{α}-(2-Naphthalene- sulfonylglycyl) -4-amidino-D,L-Phenylalaninepiperidide (abbreviated NAPAP)[36] was purchased from Sigma. The stock solution of 1 mM was made in water. Leupeptin (Ac-Leu-LeuArgH)[37] was purchased from Sigma. The stock solution of 19.14 mM was made in methanol. Phenylarsonic acid was purchased from Pfaltz & Bauer. A stock solution of 0.743 M was made in methanol.

I = 0.1 M buffers used in the enzyme assays were prepared according to the Biochemists' Handbook[38]. These were acetate for pH 4.6 to 5.4, phosphate for pH 6.0 to 7.8, diethylbarbiturate for pH 8.6, and bicarbonate for pH 9.0 to 10.0. Aside from the small amount arising from the enzyme stock solution, the buffers were free of chloride.

Methods

Thrombin concentrations were determined by titration with p-nitrophenyl p'-guanidinobenzoate[39]. Kinetic assays were performed in 0.1 M ionic strength buffers containing 10 μg/ml bovine serum albumin (BSA) at 26°C. Reactions were started by adding enzyme to mixtures of substrate and inhibitor.

The type of inhibition for each inhibitor was determined by Lineweaver-Burk plots using initial rate data. Substrate concentrations ranged from around 0.5 Km to 3 K_m.K_i values for thrombin inhibitors were measured at each pH by comparing the pseudo-first-order hydrolysis rates of the fluorogenic thrombin substrate for inhibited and uninhibited reactions. K_m values of the thrombin substrates at each pH were obtained from Lineweaver-Burk plots.

When the fluorogenic thrombin substrate was used, reaction progress curves were observed on the SLM SPF-500C spectrofluorometer equipped with a temperature control system. The emission wavelength was 440 nm, the excitation at wavelength was 350 nm. Initial rates for the hydrolysis of the p-nitroanilide substrate were determined on the Vmax kinetic microplate reader at 405 nm.

NAPAP was subjected to spectrophotometric titration[40]. Analysis of the titration curve was done using a program to be published elsewhere.

Generation of theoretical pH-profiles was done using a Lotus-123 spreadsheet which will also be described elsewhere. Parameters for NAPAP's theoretical pH profile (Figure 4) were determined using nonlinear regression analysis in SigmaPlot.

RESULTS AND DISCUSSION

pH-Dependent Substrate Kinetics

k_{cat}/K_mm for N-t-Boc-Valyl-Prolyl-Arginine-7-AMC. Under the conditions of this bovine thrombin behaves as a normal serine protease in its pH profile of very close to that seen for trypsin and chymotrypsin and is clearly a single proton ionization. The observed pK_2 of 9.7 is also similar to that seen for other mammalian serine proteases. pK_2 is not, however, an essential feature of serine proteases since some, like subtilisin, lack it[26].

Analysis of these data and those listed by Lottenberg et al. for the k_{cat}/K_m of nitroanilides shows similar pK values. A reanalysis of their data shows no evidence for the transition near pH 8.5 that they inferred. Since pH profiles of k_{cat}/K_m for slow substrates reflect free enzyme and substrate ionizations[41], it is expected that these profiles for different substrates should be identical. Since pH profiles of K_m include ionizations of the enzyme-substrate complex, there is no theoretical requirement that they be identical for different substrates.

K_m for N-t-Boc-Valyl-Prolyl-Arginine-7-AMC. The pH profile for log ($1/K_m$) in Figure 1 shows no discrete transition between pH 5 and pH 10 and a low (slope = 0.17) sensitivity to pH. Therefore, the pK of 7.3 in k_{cat}/K_m arises in a catalytic step. A nearly flat pH profile for substrate binding constants is not without precedence for thrombin; Ks values for nitrophenyl esters are pH-independent above pH 5[32]. However, it is apparent that the pH profile of K_m seen here is at variance with others in the literature[30-32]. Since a different substrate is used in this case, we confirmed that a substrate used in earlier studies, D-Phe-Pipecolyl-Arg pNA, shows the variations of K_m with pH seen there. Thus, the shape of the K_m-pH profile depends on the substrate chosen.

pH-Dependent Inhibitor Binding

pH-dependent studies were performed on a variety of different reversible inhibitors of thrombin, some that are clearly transition state analogs, some that may bind tightly for other reasons. These include the tight-binding peptide boronic acids, a peptide aldehyde, a peptide carboxamide, an arsonic acid, and a control peptide carboxylate anion. These were selected to give a broad spectrum of inhibitor types. Some of these have been studied before at a single pH. The pK values seen in these profiles are most appropriately compared with those of k_{cat}/K_m for slow substrates, since they show the pK's of the free enzyme.

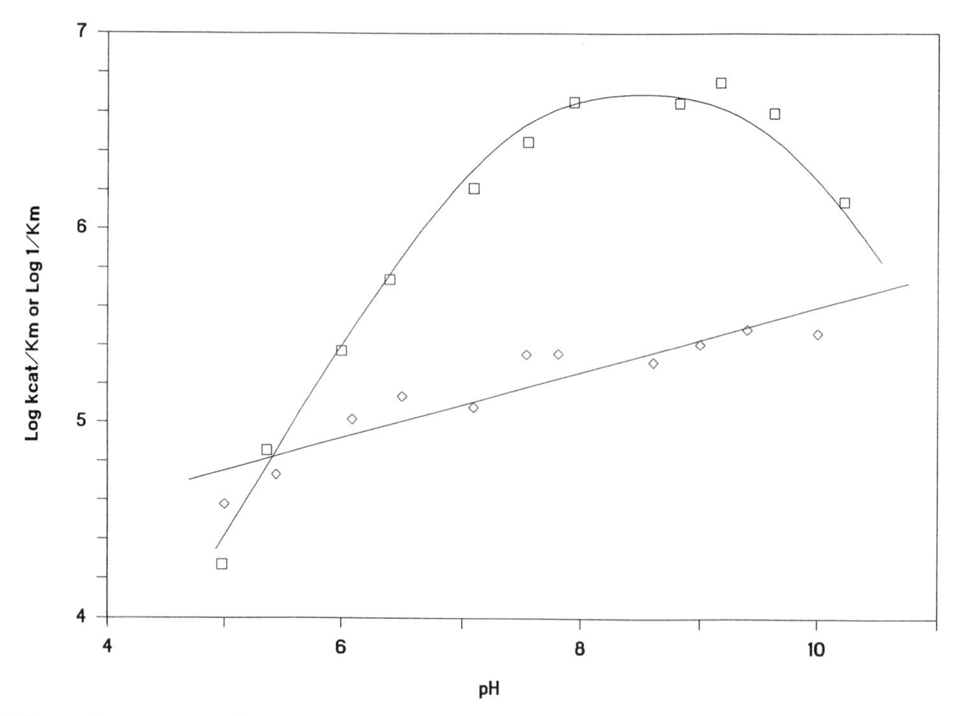

Figure 1 pH profile for the bovine thrombin-catalyzed hydrolysis of N-t-Boc-Valyl-prolyl-arginine-7-AMC. The theoreticl curve for k_{cat}/K_m (□) is drawn using $pK_1=7.3$, $pK_2=9.7$ and $k_{cat}/K_m(Lim)=5.6x10^6 M^{-1}sec^{-1}$. The plot of log $1/K_m$) (¢) vs. pH is compared to a slope of +0.17. K_m is in M, k_{cat}/K_m is in $M^{-1}sec^{-1}$. Reaction conditions are given in the text.

A Peptide Boronic Acid. A boronic acid with unionized side chains, Z-D-Phe-ProboroMpg-OPin, (where boroMpgOPin represents the pinanediol ester[42] of methoxypropylboroglycine) was studied at various pH values. Boroglycine has the -B(OH)2 group replacing the C-terminal carboxylate. The synthesis of this compound will be presented elsewhere.

Z-D-Phe-Pro-boroMpg OPin is a competitive thrombin inhibitor that binds rapidly and reversibly. It does not exhibit two binding modes, seen for some boronic acids by Kettner et al.[35]. With thrombin, this compound shows a bell-shaped pH profile (Figure 2) much like that seen earlier for simpler boronic acids. Together with the competitive nature of the inhibition, this suggests that the inhibitor binds to the active site in a manner similar to that seen for other boronic acid-serine protease pairs. The value of pK_1, 7.2 is essentially the same as that for pK_1 in kcat/Km. The apparent value of pK_2, 8.6, is much lower than pK_2 for substrate hydrolysis, showing that this value is not due to the enzyme but must originate in boronic acid ionization. More data are needed in order to determine the number of protons involved in pK_2.

Peptide Carboxylates In order to compare the various inhibitors used here with a similar unsubstituted peptide, a pH profile was prepared for a peptide carboxylate anion. Figure 3 shows the pH profile for Z-D-Phe-Pro-Arg in binding to thrombin. This pH profile is qualitatively similar to that seen for other serine proteases but significantly different in that

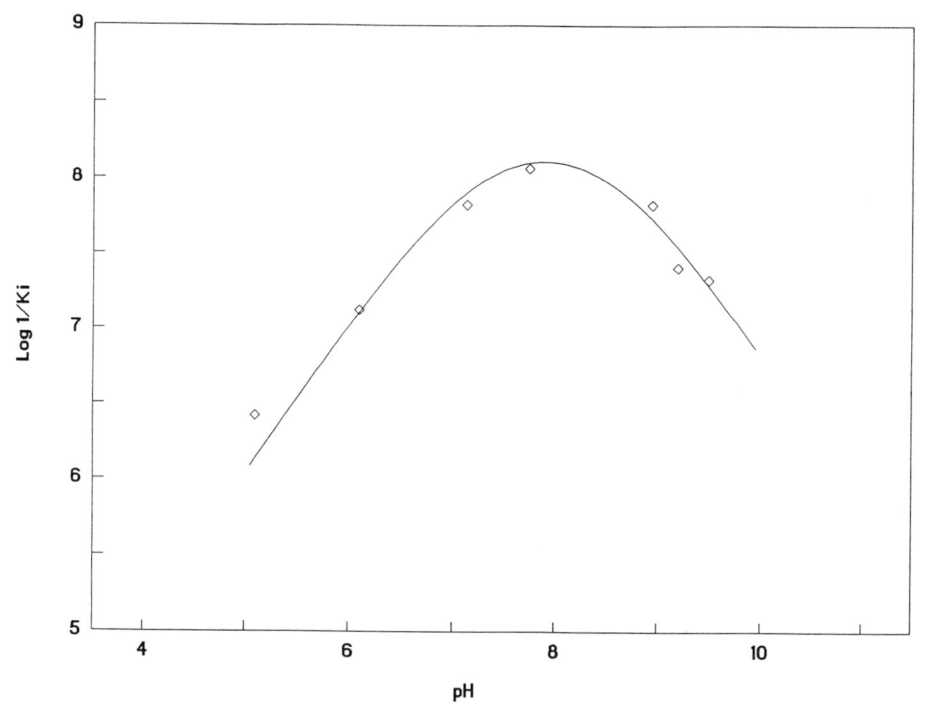

Figure 2 pH profile for inhibition of bovine thrombin by Z-D-Phe-Pro-boroMpgOPin. The theoretical curve for Z-D-Phe-Pro-boroMpgOPin is drawn using pK_1=7.2, pK_2=8.6, and K_1 (Lim)=5.5 nM. In the figure, K_1 is in M. Reaction conditions are given in the text.

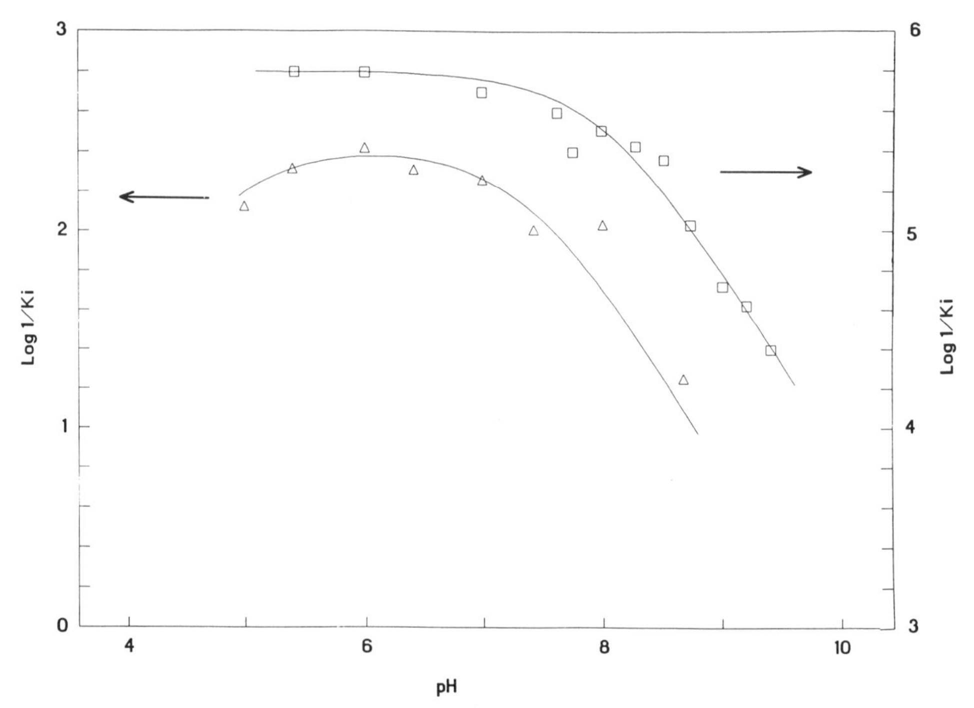

Figure 3 pH profile for inhibition of bovine thrombin by Z-D-Phe-Pro-Arg (□) and by benzenearsonic acid (Δ). The theoretical curve for Z-D-Phe-Pro-Arg (using the right axis) is drawn using pK=8.0, and K_1(Lim)=1.6μM. The theoretical curve for benzenearsonic acid (using the left axis) is drawn using pK_1=4.8, pK_2=7.4, and K_1(Lim)=3.8mM. In the figure, K_1 is in M. Reaction conditions are given in the text.

the pK of 8.0 is higher than the pK_1 seen in substrate hydrolyses. This suggests that the peptide, while binding competitively, binds in a configuration that places the arginine carboxylate anion away from the active site serine. Wrong-way binding is one possibility.

A Peptide Carboxamide The example of this class of inhibitors studied was Nα(2-Naphthalenesulfonyl-glycyl) -4-amidino-D,L-Phenylalaninepiperidide (NAPAP), first studied by Stürzebecher et al.[36]. This compound might have been expected to bind either in nearly pH-independent manner around pH 7 as does the 7-AMC substrate used here, or in a pH dependent manner like the nitroanilide substrates. Its pH dependent K_i (Figure 4) perhaps reflects the closer resemblance between the piperidide and the nitroanilide groups.

The presence of the pK near 7.3 demonstrates that the active-site histidine side chain must be unionized for the inhibitor to bind. In the crystal structure for the NAPAP-thrombin complex solved by Bode et al.[43], the active site's histidine side chain is bound to the inhibitor's piperidide group. Protonation of the histidine clearly destroys this association. This explanation has also been applied to the differential binding behavior of substrates[41]. The alkaline part of the profile is, at this stage, still under active study. It appears to show both the enzyme's pK_2 and the pK resulting from the ionization of the inhibitor's sulfonamide NH group. The data suggest that a third ionizable group may be

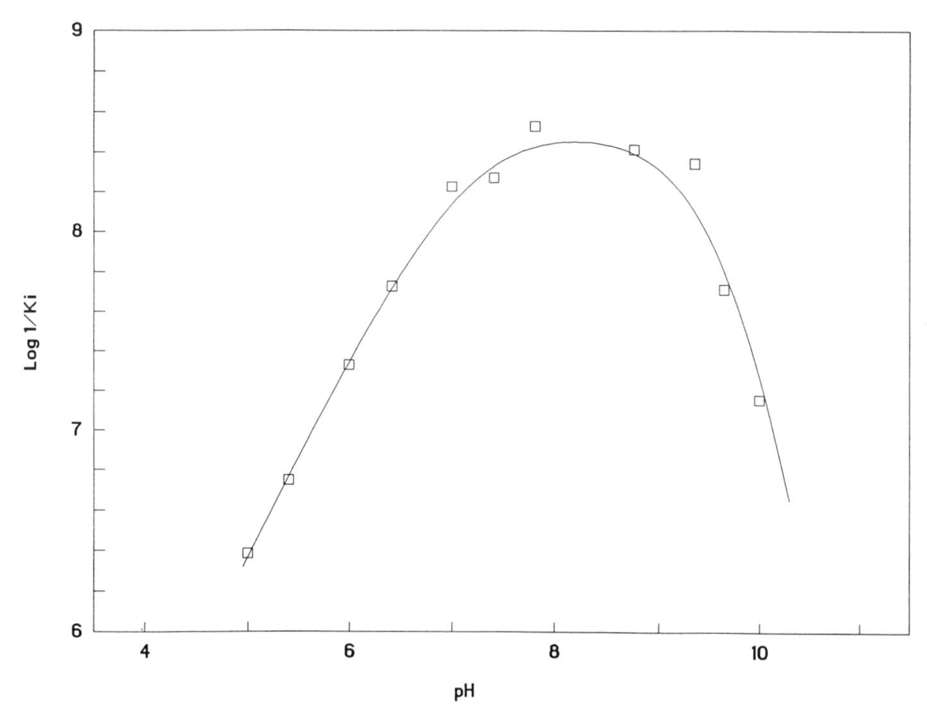

Figure 4 pH profile for inhibition of bovine thrombin by Nα-(2-naphthyl-sulphonyl-glycyl)-D,L-*p*-amidinophenylalanyl-piperidine (NAPAP). The theoretical curve is drawn using $pK_1=7.2$, a triple ionization $pK_2=9.87$, and $K_i(\text{Lim})=2.7\text{nM}$. In the figure, K_i is in M. Reaction conditions are given in the text.

involved. The thrombin pH profiles for these compounds in this pH region will be discussed in a future publication. The hibitor's sulfonamide pK, measured by spectrophotometric titration at 247 nm, is 9.4, close to the value of 9.9 previously determined for L-1-chloro-3-tosylamido-4-phenyl-2 butanone[40] (TPCK).

Leupeptin Ac-Leu-Leu-ArgH (Leupeptin)[37] was studied in connection with thrombin. This compound is a less potent thrombin inhibitor than the peptide aldehyde described by Bajusz and coworkers in this volume and in[44]; leupeptin inhibits in the micromolar range. The pH profile of inhibition is seen in Figure 5. The single-proton pK_1 of 7.3 is clearly diagnostic for the active site, and the pK_2 of 9.6 is close to that seen in substrate hydrolysis. The data do not extend far enough to determine the proton number for pK_2.

In contrast, other peptide aldehyde-serine protease systems show no pronounced pH dependence near pH seven[28]. The pK of 7.3 seen here shows that leupeptin's binding to thrombin's active site is controlled by the active site imidazole, with binding depending on a neutral imidazole.

This pK may indicate that the hemiacetal, which should be anionic immediately after attack by the serine oxygen atom, remains so and forms an ion-pair with the active site imidazolium ion[28]. The other possibility is that leupeptin's aldehyde carbon atom forms a complex with the nitrogen atom of the active site imidazole. Such a complex could not readily form with the protonated imidazolium ion. Formaldehyde inhibits serine proteases via such a complex[45].

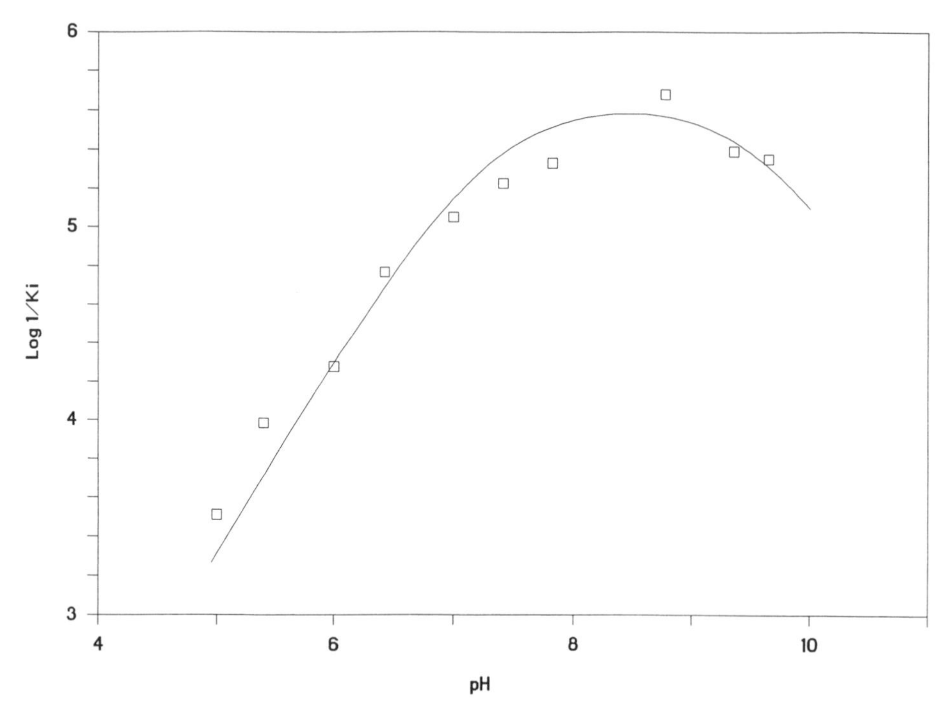

Figure 5 pH profile for inhibition of bovine thrombin by leupeptin (Ac-Leu-Leu-ArgH). The theoretical curve is drawn using pK_1=7.3, pK_2=9.6, and K_I(Lim)=2.3 230M. In the figure, K_I is in M. Reaction conditions are given in the text.

Benzenearsonic Acid Arsonic acids are potential transition state analog inhibitors studied by Glazer[29]. These compounds inhibit slowly, and are believed to form a tetrahedral monoester to the active site serine OH group. Like the boronic acids, arsonic acids reversibly esterify to alcohols. In contrast to boronic acids, arsonic acids are tetrahedral Bonsted acids. Arsonic acids are anionic throughout the physiological pH range. They are similar to the phosphonates studied by Claeson and coworkers and discussed elsewhere in this volume. Figure 3 shows that benzenearsonic acid binds to thrombin with a pK of 7.4, similar to that seen for substrate hydrolysis (Figure 1). Like Z-D-Phe-Pro-Arg, the arsonic acid exhibits better binding on the acid side of the pK. This is similar to results obtained with arsonic acid inhibitions of trypsin and chymotrypsin, and shows that the anionic inhibitor cannot bind to the enzyme until the anionic active site is neutralized by protonation of the active site imidazole.

The curve parameter's pK_1 of 4.8 is obtained in a pH region where the enzyme is not very active and is not considered to be significant.

CONCLUSIONS

The thrombin pK of 7.3 seen in substrate k_{cat}/K_m values is reflected in the pH K_i profiles for the thrombin inhibitors studied here. The presence of this pK indicates the importance of the active site imidazole in these enzyme-inhibitor complexes.

While the neutral inhibitors used here show better binding on the alkaline side of this pK, the anionic compounds tested bind more poorly in alkaline pHs. The binding of benzenearsonate to the active site is clearly blocked by the active site's anionic charge at pH values above seven.

These results suggest that future work on thrombin inhibitors should not emphasize anionic inhibitors that are repelled from the active site at physiological pH values.

ACKNOWLEDGEMENTS

This work was supported the Thrombosis Research Trust, the PSC-CUNY Research Foundation, and the National Institutes of Health (1 R15 DK37885-01.)

REFERENCES

1. J.P. Lorand, and J.O. Edwards, Polyol complexes and structure of the benzeneboronate ion, J. Org. Chem. 24:769 (1959).
2. H. Steinberg, and D.L. Hunter, Preparation and rate of hydrolysis of boric acid esters, Ind. Eng. Chem. 16:1677 (1957).
3. P.A. Sienkiewicz, and D.C. Roberts, pH-Dependence of boronic acid-diol affinity in aqueous solution, J. Inorg. Nucl. Chem. 42:1559 (1980).
4. R. Pizer, and L. Babcock, Mechanism of the complexation of boron acids with catechol and substituted catechols, Inorg. Chem. 16:1677 (1977).
5. L. Babcock, and R. Pizer, Dynamics of boron acid complexation reactions, Inorg. Chem. 19:56 (1980).
6. For a review of early chemical work on boron acids and their complexes, see P.H. Hermans, Über die Konstitution der Borsäuren und einiger Ihrer Derivate, Anorg. Allgem. Chem. 142:83 (1924).
7. R.L. Letsinger, and I. Skoog, Organoboron compounds. IV. Aminoethyl diarylborinates, J. Amer. Chem. Soc. 77:2491 (1955).
8. H.L. Weith, J.L Wiebers, and P.T Gilham, Synthesis of cellulose derivatives containing the dihydroxyl group and a study of their capacity to form specific complexes with sugars and nucleic acid components, Biochem. 9:4396 (1970).
9. H. Seliger, E. Rössner, G. Aumann, V. Genrich, M. Holupirek,T. Knäble, and M.Philipp. Sequence-specific cooligo-condensation of mononucleotides using protecting groups with affinity properties, Makromol. Chem. 176:2915 (1975).
10. H. Seliger, B. Haas, M. Holupirek, T. Knäble, G. Tödling, and M. Philipp, Nonstepwise methods in the preparation of building blocks for polynucleotide synthesis, Nucl. Acids Res. 7:191 (1980).
11. R.L. Letsinger, and D.B. McLean, Cooperative functional group effects in reactions of boronarylbenzimidazoles, J. Amer. Chem. Soc. 85:2230 (1963).
12. G. Wulff, W. Vesper, R. Grobe-Einsler, and A. Sarhan,Enzyme-Analog Polymers, The Synthesis of polymers containing chiral cavities and their use for the resolution of racemates, Makromol. Chem., 178:2799(1977).
13. G. Rao, and M. Philipp, Boronic acid-Catalyzed hydrolyses of salicylaldehyde imines, J. Org. Chem. 56:1505 (1991).
14. G. Rao, and M. Philipp, Boronic acids catalyze the hydrolysis of mandelonitrile, in The Bioorganic Chemistry of Enzymatic Catalysts, V. D'Souza and J. Feder, eds., CRC Press, 1991.

15. V.K. Antonov, T.V. Ivaniva, I.V. Berezin, and K. Martinek, n-Alkylboronic acids as bifunctional reversible inhibitors of α-chymotrypsin, FEBS Lett. 7:23 (1970).
16. M. Philipp, and M.L. Bender, Inhibition of serine proteases by arylboronic acids, Proc. Nat. Acad. Sci. 68:478(1971).
17. D.A. Matthews, R.A. Alden, J.J. Birktoft, S.T. Freer, and J. Kraut, X-ray crystallography study of boronic acid adducts with subtilisin BPN' (Novo), J. Biol. Chem.250:7120 (1975).
18. R. Bone, D. Frank, C.A.Kettner, and D.A. Agard, Structural analysis of specificity: α-Lytic protease complexes with analogues of reaction intermediates, Biochem. 28:7600 (1989).
19. R. Bone, A.B. Shenvi, C.A.Kettner, and D.A. Agard, Serine protease mechanism: Structure of an inhibitory complex of α-lytic protease and a tightly bound peptide boronic acid, Biochem. 26:7609 (1987).
20. L.H. Takahashi, R. Radhakrishnan, R.E. Rosenfeld, Jr., and E.F. Meyer, Jr., Crystallographic analysis of the inhibition of porcine pancreatic elastase by a peptidylboronic acid, Biochem. 28:7610 (1989).
21. G. Robillard and R.G. Schulman, Polarization of histidine-57 by substrate analogs and competitive inhibitors, J.Mol. Biol. 86:541 (1974).
22. W. Bachovchin, W.Y.L. Wong, S. Farr-Jones, A.B. Shenvi, and C.A. Kettner, Nitrogen-15 NMR spectroscopy of the catalytic-triad histidine of a serine protease in peptideboronic acid-inhibitor complexes, Biochem. 27:7689 (1988).
23. J.E. Baldwin, T.D.W. Claridge, A.E. Derome, C.J. Schofield,and B.D. Smith, ^{11}B-NMR studies on an aryl boronic acid bound to chymotrypsin and subtilisin, Bioorg. Med. Chem. Lett, 1:9 (1991).
24. C.H. Johnson, and J.R. Knowles, The binding of inhibitors to α-chymotrypsin at alkaline pH, Biochem. J. 103:428 (1967).
25. H.R. Bosshard, and A. Berger, The topographical differencesin the active site region of α-chymotrypsin, subtilisin Novo, and subtilisin Carlsberg, Biochem. 13:266 (1974).
26. M. Philipp, and M. L. Bender, Subtilisin and thiol-subtilisin, Molec. and Cell. Biochem., 51:5 (1983).
27. K. Soman, A.S. Yang, B. Honig, and R. Fletterick,Electrical potentials in trypsin isozymes, Biochem.28:9918 (1989).
28. W.P. Kennedy, and R.M. Schultz, Mechanism of association of a specific aldehyde transition-state analog to the active site of α-chymotrypsin, Biochem. 18:349 (1979)
29. A.N. Glazer, Inhibition of serine esterases by phenyl-arsonic acids, J. Biol. Chem. 243:3693 (1968).
30. E. Di Cera, R. De Cristofaro, D.J. Albright, and J.W.Fenton II, Linkage between proton binding and amidase activity in human α-thrombin: effect of ions and temperature, Biochem., 30:7913 (1991).
31. R. Lottenberg, J. Hall, M. Blinder E.P. Blinder, and C.M. Jackson, The action of thrombin on p-nitroanilide substrates, Biochim. Biophys. Acta 742:539 (1983).
32. P. Ascenzi, E. Menegatti, M. Guarneri, F. Bartolotti, and E. Antonini, Catalytic properties of serine proteases. 2. Comparison between human urinary kallikrein and human urokinase, bovine ß-trypsin, bovine thrombin, and bovine α-chymotrypsin, Biochem., 21:2483 (1982).
33. R. Lottenberg, U. Christensen, C.M. Jackson, and P.Coleman, Assay of coagulation proteases using peptide chromogenic and fluorogenic substrates, Meth. in Enz. 80:341 (1981).

34. G. Claeson, and L. Aurell, Small synthetic peptides with affinity for proteases in coagulation and fibrinolysis, Ann. NY Acad. Sci. 370:798 (1981).
35. C. Kettner, L. Mersinger, and R. Knabb, The selective inhibition of thrombin by peptides of boroarginine, J Biol. Chem. 265:18289 (1990).
36. J. Stürzebecher, F. Markwardt, B. Voigt, G. Wagner, and P. Walsmann, Cyclic Amides of N-α-arylsulfonyl-aminoacylated 4-amidinophenylalanine--tight binding inhibitors of thrombin, Thomb. Res. 29:35 (1983).
37. A.H. Umezawa, Enzyme Inhibitors of Microbial Origin,University Park Press, Maryland, 1972.
38. C. Long, ed. Biochemists' Handbook, Van Nostrand, Princeton, New Jersey (1961), pp. 30-42.
39. T. Chase, Jr. and E. Shaw, Titration of trypsin, plasmin, and thrombin with p-nitrophenyl-p'-guanidinobenzoate HCl, Meth. in Enz. XIX:20 (1970).
40. F.J. Kézdy, A. Thompson, and M.L. Bender, Studies on the reaction of chymotrypsin and L-1-chloro-3-tosylamido-4-phenyl-2-butanone, J. Amer. Chem. Soc. 89:1004 (1967).
41. H. Hirohara, M. Philipp, and M.L. Bender, Binding rates, O-S substitution effects, and the pH dependence of chymotrypsin reactions, Biochem. 16:1573 (1977).
42. D.S. Matteson, K.M. Sadhu, and G.E. Lienhard,(R)-1-Acetamido-2-phenylethaneboronic acid. A specific transistion-state analogue for chymotrypsin, J Amer. Chem. Soc. 103:5241 (198l).
43. W. Bode, D. Turk, and J. Stürzebecher, Geometry of binding of the benzamidine and arginine-based inhibitors Nα-(2-naphthyl-sulphonyl-glycyl)-D,L-p-amidinophenylalanyl-piperidine (NAPAP) and (2R,4R)-4-methyl-1-[Nα(3-methyl-1,2,3,4-tetrahydro-8-quinolinesulphonyl)-L-arginyl]-2-piperidine carboxylic acid (MPQA) to human α-thrombin, Eur. J Biochem. 193:175 (1990).
44. S. Bajusz, Å. Barabás, P. Tolnay, E. Szell, and D. Bagdy, Inhibition of thrombin and trypsin by tripeptide aldehydes, Int. J. Pept. Prot. Res. 12:217 (1978).
45. C.J. Martin, and M.A. Marini, Spectral detection of the reaction of formaldehyde with the histidine residues of α-chymotrypsin, J. Biol. Chem. 242:5736 (1967).

THE COMPARISON OF AN INTERIM TERTIARY PREDICTED MODEL OF BOVINE THROMBIN AND THE X-RAY STRUCTURE OF HUMAN THROMBIN

E. Platt

Proteus Molecular Design Ltd.
Proteus House
48 Stockport Road
Marple, Cheshire, SK6 6AB,UK

The comparison of an interim computer model of the tertiary structure of the B-chain of bovine α-thrombin predicted by homology with α-chymotrypsin[1] and the refined 1.9 Å structure of human alpha-thrombin[2,3] is reported here. The refinement of the surface loops is incomplete, as (in the interest of demonstrating objectivity) modelling was stopped when completion of the X-ray crystallographic structure[2] was announced. Modelling of the loops is now in progress.

The alignment between bovine thrombin and α-chymotrypsin was obtained by use of the structural alignment routine, STALM[4] and a sequence alignment routine COREBLOCK[4]. In addition to the 2 sequences and the known crystal structure of α-chymotrypsin(1), the sequences and x-ray crystal structures of elastase and trypsin[5,6] were used. The aligned blocks obtained are given in Table 1.

After the alignment stage, the basic method involves 5 stages, only the first 4 of which were completed before the x-ray structure was published[2]. The model used in the comparison is the one obtained after stage 4.

Stage 1 The use of RMS (root mean square) fitting of the aligned blocks, followed by the RMS + Energy calculations[7] (see Table 1).

Stage 2 Energy minimisation with an end-to-end Cα-Cα distance constraint for the loop sections.

Stage 3 Building up of the complete model for the various building blocks using RMS, then RMS + Energy calculations, followed finally by Energy with distance contraints calculations.

Stage 4 Then 3 picoseconds of Momentum Directed Minimisation at 4° K were performed[8].

The Design of Synthetic Inhibitors of Thrombin
Edited by G. Claeson, *et al.*, Plenum Press, New York, 1993

Stage 5 Extensive simulation of the thrombin model with water and counterions under periodic boundary conditions. This stage was not completed before the comparison of the model and x-ray structure was performed[3].

Table 1 The conserved aligned blocks of thrombin and α-chymotrypsin by residue number and the RMSR deviations for these aligned blocks, where M is the model, X is the X-ray structure and C is the α-chymotrypsin X-ray structure.

Thrombin	alpha -chymotrypsin	RMSR deviations MIX	MIC
1-20	12-31	1.09	1.22
24-44	34-54	0.92	0.56
58-67	59-68	1.22	1.14
75-91	75-91	1.00	0.97
97-121	96-120	1.00	0.88
132-145	128-141	0.98	0.93
157-174	146-163	1.28	1.25
181-190	170-179	1.19	0.60
198-212	182-196	1.07	1.16
219-230	201-212	1.50	0.95
236-254	218-236	0.90	1.27

RESULTS

The rms deviation of the interim model by the rotation method (RMSR) of the c-alpha backbone carbon atoms is 3.58 Å (See Figure 1). The RMSR deviation of the c-alpha backbone carbon atoms of the aligned blocks or core between the model and the x-ray was 1.35 Å.

CONCLUSIONS AND DISCUSSION

The results of of the interim model indicate that the core region of thrombin, as determined by the aligned blocks given in Table 1, is predicted to an accuracy of that of the X-ray structure[9], whereas as expected the unfinished loops show large deviations from the X-ray structure indicating that more accurate methods of predicting loop structures are required. Indeed, the supposition, that extensive simulation including water and counterions, should be the next step in loop refinement is supported by this finding. These conclusions are supported by studies that have been carried out elsewhere[9].

This study shows that core features can be routinely, automatically and reliably predicted. The refinement process with inclusion of a water treatment is currently being undertaken and will be reported in the near future along with a more detailed account of the methods used, hopefully before detailed experimental coordinates become publically available.

REFERENCES

1. H. Tsukada, and D.M. Blow, J. Mol. Biol. 184:703 (1985).
2. W. Bode, I. Mayr, U. Baumann, R. Huber, S.R. Stone, and J. Hofstenge, EMBO J. 8:3467 (1989).
3. Comparison preformed by D.Turk, Max-Planck Institut for Biochemie, 8033 Martinsried Bei Munchen, F.R.G.
4. Prometheus Suite of Programs. Copyright Proteus Molecular Design Ltd., Proteus House , 48 Stockport Rd., Marple, Stockport, Cheshire, SK6 6AB. STALM written by A Marsden & RV Fishleigh and COREBLOCK written by RV Fishleigh.
5. D.L. Hughes, L.C. Sieker, J. Bieth, and L. Dimicoli, J. Mol. Biol. 162:645 645.
6. J.L. Chambers, and R.M. Stroud, Acta Crystallogr. 35B:1861 (1979).
7. B. Robson, and E. Platt, J. Mol. Biol. 188:259 (1986).
8. B. Robson, and E. Platt, J. Computer-Aided Mol. Design 1:17 (1987).
9. E. Platt, and B. Robson, Proceedings (Section B - Biological Sciences) of the Royal Society of Edinburgh), 99B:123 (1992).

DESIGN OF NOVEL TYPES OF THROMBIN INHIBITORS BASED ON MODIFIED D-PHE-PRO-ARG SEQUENCES

G. Claeson, S. Elgendy, L. Cheng, N. Chino, C.A. Goodwin, M.F. Scully and J. Deadman

Thrombosis Research Institute
Manresa Road
London SW3 6LR, UK

INTRODUCTION

In 1956, Bettelheim[1] showed that fibrinopeptide A (FPA) competitively inhibits the reaction between thrombin and fibrinogen, that is the amino acid sequence, FPA, preceding the bond split by thrombin has affinity for the enzyme.

In attempts to synthesize shorter imitations of FPA we found a peptide with only three amino acids, Phe-Val-Arg, which in the form of its methyl ester, was shown by Blombäck et al[2,3] to have thrombin inhibition properties similar to that of FPA. The same sequence was also used for the first chromogenic peptide substrate, Bz-Phe-Val-Arg-pNA[4,5]. This sequence, Phe-Val-Arg, has later been improved; the protected N-terminal Phe was with advantage replaced by unprotected D-Phe and when Val was replaced by Pro or its six-membered analogue, Pip, the sequence functioned better, especially in combination with N-terminal D-Phe[6,7,8].

Table 1 Some examples of thrombin inhibitors based on the sequence D-Phe-Pro-Arg.

Inhibitor	Reference
D-Phe-Pro-Arg-H	Bajusz[6]
D-Phe-Pro-Arg-CH_2Cl	Kettner and Shaw[13]
D-Phe-Pro-Arg($COCH_2$)-Gly-piperidide	Szelke[14], Scully
D-Phe-Pro-Arg($COCH_2$)-Hir$^{49-65}$	DiMaio[15]
D-Phe-Pro-Arg-boroArg	Kettner[16]
D-Phe-Pro-Arg-boroIrg*	Kettner[16]

*Irg is the isothiouronium analogue of Arg.

The Design of Synthetic Inhibitors of Thrombin
Edited by G. Claeson, *et al.*, Plenum Press, New York, 1993

The improved sequence D-Phe-Pro-Arg/D-Phe-Pip-Arg has been widely used in substrates for thrombin with various chromophoric and fluorophoric groups attached to the C-terminal Arg[8,9,10] e.g. the thrombin substrate S-2238, D-Phe-Pip-Arg-pNA, has kinetic constants, Km 1.6 μM, kcat 95 s^{-1} [11] which can be favourably compared to those of the natural substrate fibrinogen, Km 7.2 μM, kcat 84 s^{-1} [12]. D-Phe-Pro-Arg has also been used as a base for several different types of thrombin inhibitors, some examples of which are shown in Table 1. This paper and the following papers from our group will describe our work on further modifications of both ends of the D-Phe-Pro-Arg sequence in order to obtain substrates and inhibitors of thrombin with improved properties.

Modification of the N-Terminal end of D-Phe-Pro-Arg

In the course of our attempts to explain some puzzling test results with the chromogenic susbtrates shown in Table 2 (Claeson, Aurell, Karlsson and Friberger, unpublished results) we got an idea how the N-terminus of the peptide sequence possibly could be improved. The replacement of Bz-L-Phe by H-L-Phe in these substrates resulted in decreased reaction rate by thrombin. However, the corresponding removal of the protecting group of the D-form, Bz-D-Phe changed to H-D-Phe, gave the opposite effect, an increased reaction rate by thrombin. Similar results have been obtained with the corresponding L- and D-forms of Phe-Val-Arg-pNA[8] and also with the thrombin inhibitor Phe-Pro-Arg-H (Claeson, Lundin & Mattson, unpublished results).

Table 2 Relative reaction rate of thrombin with substrates containing N-terminal L and D amino acids with and without N-protection.

Substrate		Relative reaction rate of thrombin
Bz-L-Phe-Pro-Arg-pNA		++
H-L-Phe	"	+
Bz-D-Phe	"	+++
H-D-Phe	"	++++

One possible explanation of these contradictory results with D and L containing substrates could be given in the following way (see Figure 1). This figure illustrates, as a cartoon, the binding of the varying N-terminal ends of the substrates to the apolar binding area of thrombin. The C-terminal ends (Pro-Arg-pNA) are identical for the four different substrates and assumably bind in approximately the same way. Therefore, the different thrombin affinities of the peptides are determined only by the different binding of the N-terminal end of the molecule. If we further assume that the Bz-L-Phe substrate binds with both aromatic groups to the apolar binding area, the removal of the Bz group leaves the peptide with only one aromatic binding group, and furthermore, the formed free amino group does not like the apolar environment and a much worse substrate should result. In the case of the H-D-Phe substrate, the free amino group points into the water phase, away from the apolar surface, which stabilizes the binding, resulting in an excellent substrate. In this case, N-benzoylation of the D-Phe introduces a lipophilic aromatic group in the water environment, and some of the good effect of the free amino group is lost. This hypothetical model was put up long before the X-ray structure and molecular modelling of thrombin and its inhibitors were reported. It is a very simple model, but it explains the behaviour of the variations of these chromogenic peptide substrates, and it further predicts that the best peptide sequence, H-D-Phe-Pro-Arg, could be made even better if the N-terminal end contained an additional, lipophilic/aromatic group which could bind to the, in this case, free area where the benzoyl group of the Bz-L-Phe susbtrate binds. Consequently, we decided to replace Phe with synthetic amino acids containing two aromatic rings.

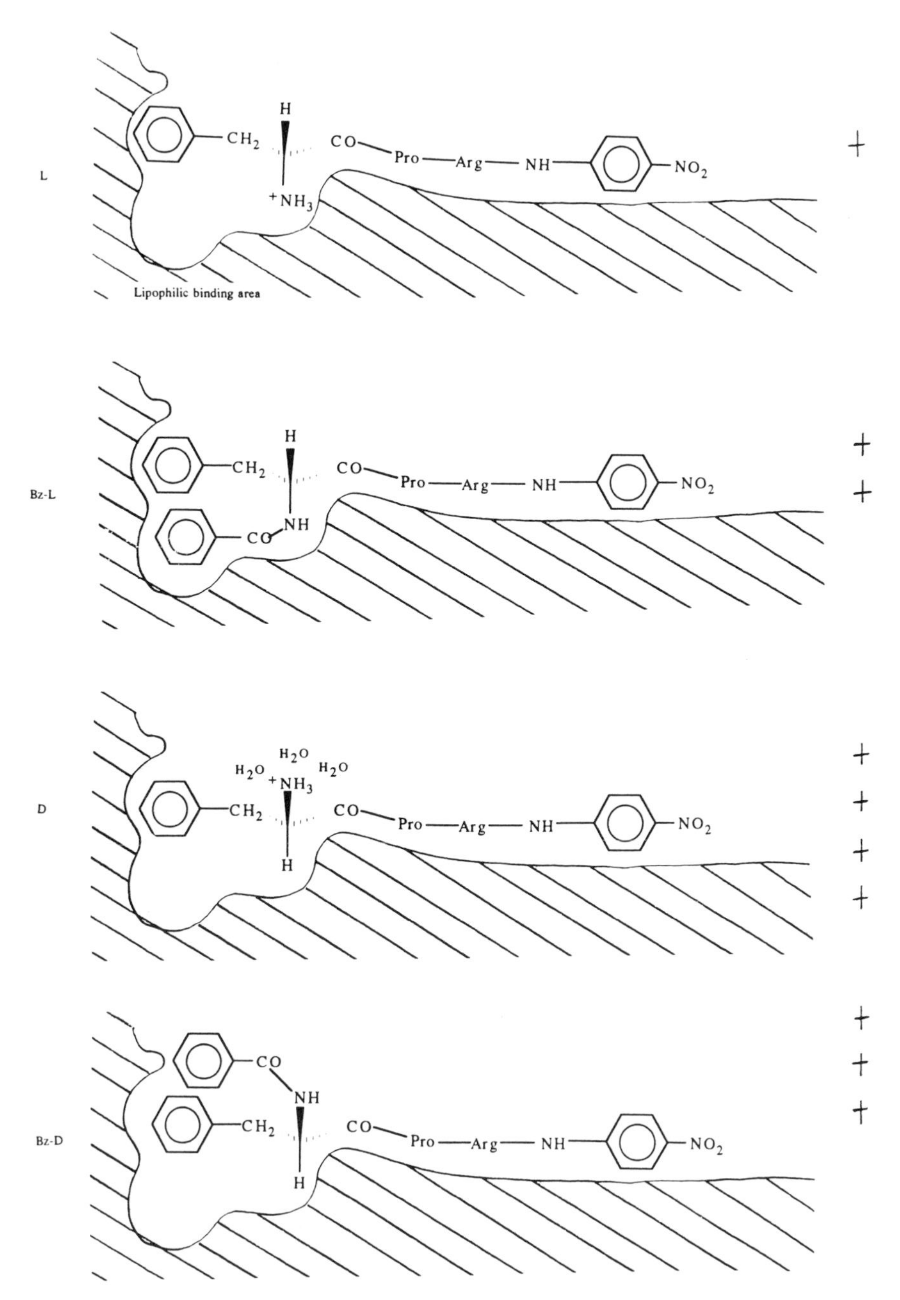

Figure 1 Hypothetical binding between thrombin and the N-terminal end of the substrates shown in Table 2.

Table 3 The effect of exchange of Phe for Dpa in a tripeptide aldehyde.

Aldehyde	Ki(μM)	TT(μM)	BP(%)**
Boc-D-Phe-Pro-Arg-H	0.1	5	40
Boc-DL-Dpa-Pro-Arg-H	0.03	3	100

* Concentration (μM) needed to double the thrombin time (TT).
** % of normal blood pressure (BP) after administration of 4 mg/kg iv to cats.

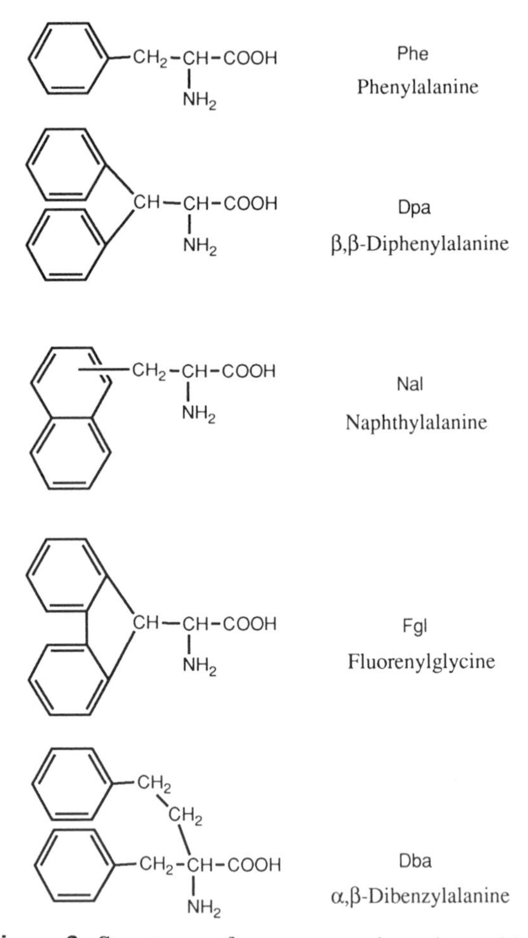

Figure 2 Structures of some aromatic amino acids.

Several new as well as earlier known such amino acids were synthesized and introduced as N-terminal amino acid in peptide aldehydes of the type Aa-Pro-Arg-H (Claeson, Lundin & Mattson, to be published). Most of the synthesized amino acids did not improve the inhibitor compared to our standard, the Bajusz aldehyde D-Phe-Pro-Arg-H. However, we found a few of the amino acids (Figure 2) to be interesting and especially the novel ß,ß-diphenylalanine (Dpa).

When D-Phe was replaced by DL-Dpa, the tripeptide aldehyde, see Table 3, showed an improved inhibition constant, a better inhibition of the plasma coagulation and, surprisingly, the blood pressure lowering side-effect was reduced compared to our standard inhibitor[17]. Most likely, the D-form of Dpa would show even more favourable values.

Modification of the C-Terminal end of D-Phe-Pro-Arg

During our work with peptides containing a C-terminal amino boronic acid, we made an attempt to exchange a chlorine atom for a guanidino group in order to obtain the boronic acid analogue of Arg (II). The main reaction product was, however, a compound (I) where a methoxy group had replaced the chlorine atom (Figure 3).

This new compound I was tested and found to have a higher Ki than the corresponding boroArg inhibitor II, but it was nevertheless an excellent inhibitor of thrombin with Ki in the low nanomolar range[18] (Table 4).

Another interesting property of the inhibitor I is its very high selectivity for thrombin[18]. Inhibition constants of plasmin and urokinase is, for example, several powers of ten higher than that of thrombin. The boroArg inhibitor II does not show the same selectivity. The inhibitor I also shows neglectibly low drop of blood pressure, which for inhibitors containing an Arg or Arg-like S1 moiety, can be a serious side-effect[17,19].

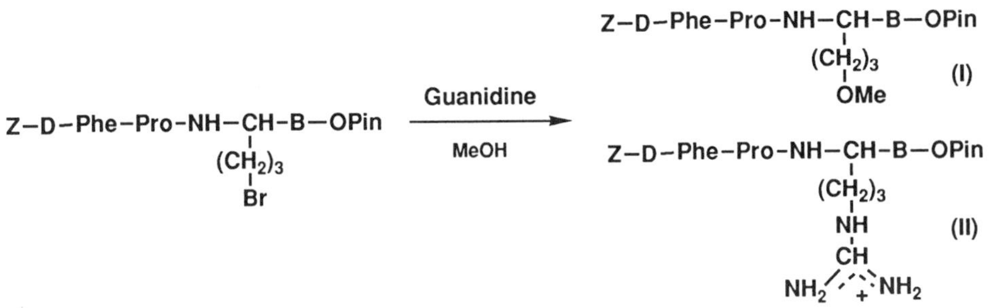

Z–D–Phe–Pro–NH–CH(–$(CH_2)_3$–Br)–B–OPin —(Guanidine, MeOH)→

Z–D–Phe–Pro–NH–CH(–$(CH_2)_3$–OMe)–B–OPin (I)

Z–D–Phe–Pro–NH–CH(–$(CH_2)_3$–NH–CH(NH$_2$)$^+$NH$_2$)–B–OPin (II)

Figure 3

Table 4 Inhibition constants and effect on plasma coagulation of thrombin inhibitors I and II.

	Ki	TT	(μM) APTT	Heptest
I	0.007	2.8	3.4	0.6
II	0.001	0.7	0.5	0.1

DISCUSSION

The good Ki value of compound I was unexpected as natural substrates of thrombin are also known to require an Arg at the S1 position making it possible for its positively charged guanidino group to interact with the negatively charged carboxyl group of Asp-189 located at the base of the specificity pocket. In the case of compound I, the methoxypropyl side chain has no charge, and electrostatic attraction cannot contribute to the binding of the inhibitor.

The above described properties of this inhibitor I, high affinity and selectivity, suggest that its binding to thrombin is mainly determined by the good fit of the N-terminal lipophilic D-Phe-Pro to the apolar binding area of thrombin[20] as well as by the boronic acid binding to the catalytic site Ser/His[21,22]. The absence of the Arg side chain and as a consequence the loss of electrostatic attraction between the guanidino group and the carboxylic group (Asp-189 of thrombin) must be a more significant factor for other proteases for which the D-Phe-Pro- part was not designed and to which this part does not bind well, thus resulting in bad inhibition constants for these kinds of proteases which in turn means that the inhibitor I has good selectivity for thrombin.

The results from the chemical manipulations of the D-Phe-Pro-Arg structure described in this paper seem very promising and prompted us to extend this type of studies. We have therefore further varied the side chain of the C-terminal amino acid with the intention to find out how the size and steric requirements of neutral side chains affect the properties of the inhibitor. We have also continued studies of the effect of new lipophilic amino acids at the N-terminal end of the inhibitor. We also hope that the new compounds can be suitable for NMR and X-ray studies of their thrombin-inhibitor complex and can shed light on the binding between enzyme and inhibitor.

The following papers from our group will describe the synthesis of new inhibitors containing chemically modified D-Phe-Pro-Arg sequences incorporated in the different types of inhibitors mentioned in Table 1. Peptide inhibitors with a C-terminal amino phosphonic acid, a novel type of thrombin inhibitor, will also be described.

REFERENCES

1. F.R. Bettelheim, Clotting of fibrinogen II. Fractionation of peptide material liberated, Biochem. Biophys. Acta 19:121 (1956).
2. B. Blombäck, M. Blombäck, P. Olsson, L. Svendsen, B. Thuresson af Ekenstam, and G. Claeson, Process for preparing peptides presenting biological activities, US 3,826,793 (1974).
3. B. Blombäck, M. Blombäck, P. Olsson, L. Svendsen, and G. Åberg, Synthetic peptides with anticoagulant and vasodilating action, Scand. J. Clin. Lab. Invest. 24, Suppl.107:59 (1969).
4. B. Blombäck, M. Blombäck, G. Claeson, and L. Svendsen, Substrates for diagnostic use, with high susceptibility to thrombin and other proteolytic enzymes of the type peptidohydrolases, US 3,884,896 (1975).
5. L. Svendsen, B. Blombäck, M. Blombäck, and P. Olsson, Synthetic chromogenic substrates for determination of trypsin, thrombin and thrombin-like enzymes, Thromb. Res. 1:267 (1972).
6. S. Bajusz, E. Barabas, P. Tolnay, E. Szell, and D. Bagdy, Inhibition of thrombin and trypsin by tripeptide aldehydes, Int. J. Pept. Prot. Res. 12:217 (1978).
7. G. Claeson, L. Aurell, . Karlsson, and P. Friberger, Substrate structure and activity relationship, in New Methods for Analysis of Coagulation using chromogenic Substrates, I. Witt ed. Walter de Gruyter, Berlin (1977).

8. G. Claeson, and L. Aurell, Small synthetic peptides with affinity for proteases in coagulation and fibrinolysis: An overview, Ann. New York Acad. Sci. 370:798 (1981).
9. J. Fareed, H L. Messmore, J.M. Walenga, and E.W. Bernes Jr, Synthetic peptide substrates in hemostatic testing, CRC Crit. Rev. Clin. Lab. Sci. 19:71 (1983).
10. H.C. Hemker, in Handbook of Synthetic Substrates for the Coagulation and Fibrinolytic Sy, Kluwer, Boston (1983).
11. R. Lottenberg, U. Christensen, C.M. Jackson, and P.L. Coleman, Assay of coagulation proteases using peptide chromogenic and fluorogenic substrates, Meth. Enzym. 80:341 (1981).
12. D.L. Higgens, S.D. Lewis, and J.A. Shafer, Steady state kinetic parameters for the thrombin-catalyzed conversion of human fibrinogen to fibrin, J. Biol. Chem. 258:9276 (1983).
13. C. Kettner, and E. Shaw, D-Phe-Pro-Arg-CH_2Cl - a selective affinity label for thrombin, Thromb. Res. 14:969 (1979).
14. M. Szelke, and D.M. Jones, Enzyme inhibition, US 4,638,047-A (1987).
15. J. DiMaio, F. Ni, B. Gibbs, and Y. Konishi, A new class of potent thrombin inhibitors that incorporates a scissile pseudopeptide bond, FEBS, 282:47 (1991).
16. C. Kettner, L. Mersinger, and R. Knabb, The selective inhibition of thrombin by peptides of boroarginine, J. Biol. Chem. 265:18289 (1990).
17. Ch. Mattson, E. Eriksson, and S. Nilsson, Anticoagulant and antithrombotic effect of some protease inhibitors, Folia Haemat. Leipzig, 109:43 (1982).
18. G. Claeson, M. Philipp, R. Metternich, E. Agner, T. Desoyza, M.F. Scully, and V.V. Kakkar, New peptide boronic acid inhibitors of thrombin, Thromb. Haemostas. 65:1289 (1991).
19. B. Kaiser, J. Hauptman, and F. Markwardt, Studies of the pharmacodynamics of synthetic thrombin inhibitors of the basically substituted N-α-arylsulfonylated phenylalanine amide type, Die Pharmacie, 42:119 (1987).
20. W. Bode, I. Mayr, U. Baumann, R. Huber, S.R. Stone, and J. Hofsteenge, The refined 1.9 A crystal structure of human α-thrombin: interaction with D-Phe-Pro-Arg chloromethyl ketone and significance of the Tyr-Pro-Pro-Trp insertion segment, EMBO J. 8:3467 (1989).
21. R. Bone, A.B. Shenvi, C.A. Kettner, and D.A. Agard, Serine protease mechanism: structure of an inhibitory complex of α-lytic protease and a tightly bound peptideboronic acid, Biochem. 26:7609 (1987).
22. W.W. Bachovchin, B. Wyl-wong, S. Farr-Jones, A.B. Shenvi, and C. Kettner, ^{15}N NMR spectrometry of the active-site histidyl residue of serine proteases in complexes formed with peptide boronic acid inhibitors, Biochemistry 27:7689 (1988).

CHEMISTRY AND BIOLOGY OF THE PEPTIDE ANTICOAGULANT D-MePhe-Pro-Arg-H (GYKI-14766)

S. Bajusz

Institute for Drug Research
Budapest, Hungary

INTRODUCTION

The story of D-MePhe-Pro-Arg-H goes back to the mid-70s when its parent compound D-Phe-Pro-Arg-H (GYKI-14166) was discovered[1,2].

In a search for specific inhibitors of thrombin, we studied the peptide aldehydes derived from the thrombin cleavage sites of clotting factors. Since thrombin cleaves after Val-Arg and Pro-Arg portions (Table 1), Gly-Val-Arg-H and Gly-Pro-Arg-H were prepared together with their Phe analogues because Phe-Val-Arg-OMe was known at that time as a peptide ester substrate of thrombin[3]. It appeared soon that the thrombin-fibrinogen reaction could be inhibited by Gly or Phe-Pro-Arg-H much better than by their analogues having the Val-Arg sequence of the substrate. Thus it seemed advantageous if the residues in positions P3-P2 of the inhibitor, e.g. Phe-Pro, differed from those found in the primary cleavage site of the substrate, i.e. Gly-Val. This finding prompted us to introduce D residues into P3, i.e. preparation of D-Xaa-Pro-Arg-H structures. A D-Xaa, when selected properly, might bind to an otherwise non-operational site of thrombin. Such an extra bond formed between an L residue of the enzyme and the D-Xaa of the inhibitor can hardly be disrupted by the native, all-L substrate, fibrinogen. Of this series, D-Phe-Pro-Arg-H proved to be not only a powerful but also a specific inhibitor of thrombin.

It is worth noting that the N-acyl derivatives were much less specific for thrombin. For instance, Boc-D-Phe-Pro-Arg-H (GYKI-14451)[4] showed about as high antithrombin activity as the free peptide but were more inhibitory in the fibrin-plasmin reaction[5].

D-Phe-Pro-Arg-H also exerted its anticoagulant and antithrombotic activity *in vivo* upon both parenteral and oral applications[6-13].

As for the high affinity for thrombin of the tripeptide sequence D-Phe-Pro-Arg, a recent X-ray study may be recalled[14], according to which the Bzl ring of D-Phe is accommodated in a part of a hydrophobic cavity of thrombin which is of just the suitable size and architecture to enclose it.

The Design of Synthetic Inhibitors of Thrombin
Edited by G. Claeson, *et al.*, Plenum Press, New York, 1993

Table 1 Development of D-Phe-Pro-Arg-H.

Thrombin Cleavage Sites of Clotting Factors

	P_3 P_3 P_1 P_1' P_2' P_3'
Fibrinogen α-chain:	-Gly-Val-Arg-↓-Gly-Pro-Arg-
Factor II/XIII:	-Ile/Val-Pro-Arg-↓-Ser/Gly-

Inhibitory action of tripeptide aldehydes, P_3-P_2-Arg-H, on clotting of fibrinogen by thrombin

Peptide P_3-P_2-	Relative potency	Peptide P_3-P_2	Relative potency
Gly-Val-Arg-H	1	D-Ala-Pro-Arg-H	469
Gly-Pro-Arg-H	9	D-Val-Pro-Arg-H	1273
Phe-Val-Arg-H	5	D-Ale-Pro-Arg-H	2144
Phe-Pro-Arg-H	67	D-Phe-Pro-Arg-H	7370

Since the discovery of D-Phe-Pro-Arg-H, many analogues have been prepared, e.g. the corresponding chloromethyl ketone[15] and the nitrile[16], the Agm[17] and some Arg [$COCH_2$]Gly pseudopeptide amide analogues[18] as well as the analogous boroArg peptides[19]. Introduction of the D-Phe-Pro-Arg into hirudin fragments[20-22] was also significant.

In contrast to biology, the chemistry of D-Phe-Pro-Arg-H was far less favourable because of its limited stability. I summarize here the chemical feature of free dipeptidyl arginine aldehydes, the way that led to stable analogs, and the biological properties of the compound, D-MePhe-Pro-Arg-H, selected for further development.

CHEMISTRY

The acetate salt of D-Phe-Pro-Arg-H[1,2] was found to be rather unstable in neutral aqueous solution, rapidly losing its initial high antithrombin activity (t½ 15 days at 5°C and ≈3 days 40°C), and only the sulfate salt (GYKI-14166)[23] possessed significant stability in cool solution (5°C) for a prolonged period (2-6 months). It was revealed (Figure 1) that inactivation of the free tripeptide aldehyde (I) was associated with the formation of a less polar compound which proved to be a heterocyclic compound (3)[24-25]. As evidenced by this structure, both protons of the terminal amino group were lost during an intramolecular condensation. This fact suggests that N substitution would result in stable analogues. Indeed, the N-Boc derivative was stable, however, far less specific. If thrombin specificity of the free peptide is due to its basic amino terminus, one may expect that N-alkylation will result in stable and specific analogues because the terminal amino group of such analogues is both substituted and basic.

The various N-alkyl tripeptide aldehydes prepared[24,25] were stable, no transformation into heterocyclic compounds could be observed. As for their antithrombin properties, D-MePhe-Pro-Arg-H was most similar to the parent compound.

The last steps of the synthesis of D-Phe/D-MePhe-Pro-Arg-H (Figure 2). Let us concentrate on the reactions involving the C-terminal aldehyde function. Due to its high reactivity, peptide aldehydes mainly exist in the form of amino cyclols when the guanidino is blocked. After deblocking, the aldehyde hydrate (A) forms first (70-90%) then it gives three further equilibrium structures: two amino cylols (B and B′) and aldehyde (C). At equilibrium, forms A, B, B′ and C are present in a ratio of about 40:41:16:3. It has to be noted that this is the safest way for the synthesis of peptidyl arginine aldehydes. When using Fmoc, and so basic conditions for deprotection, epimerization will occur because the Arg-H moiety is prone to racemization. In case of tBoc protection, final deblocking will be made with acids. Such treatment again leads to decomposition.

As illustrated in Figure 3, treatment with TFA or HCl in EtOAc of a peptidyl arginine aldehyde, either free or protected, leads to the formation of a dehydrated compound (D). If some water is also present, dipeptide amide (E) is split off. In anhydrous HCOOH, D is only formed in a few per cent, and also the free tripeptide aldehyde exists in form of the two amino cyclols, B and B′. In HPLC at pH <3 (Figure 4) the components of D-MePhe- Pro-Arg-H seem to be eluted in order of B′, A and B. The aldehyde form C cannot be detected. The minor peak (about 2%) may arise from D-MePhe-Pro-D-Arg-H being present as impurity.

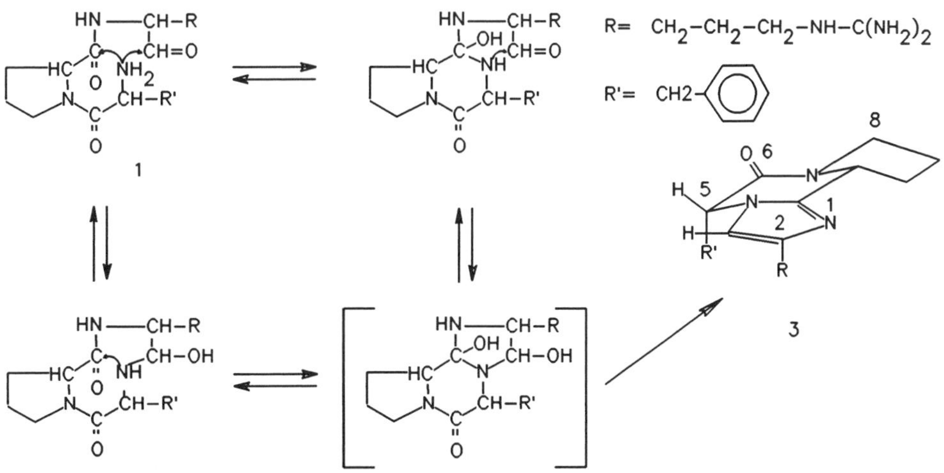

R= $CH_2-CH_2-CH_2-NH-C(NH_2)_2$

R'= CH2-(phenyl)

Figure 1 Transformation of D-Phe-Pro-Arg-H in aqueous solution.

Forms A, B and B′ of D-MePhe-Pro-Arg-H could be separated by HPLC using a 0.005M sulfuric acid/MeCN mixture. When freeze-dried, B was stable for several days while A and BI remained unchanged for only 24 hours. In solution at pH >6, equilibrium composition was formed from each structure in less than 10 minutes.

To assess initial activities (Table 2), fractions were assayed within 1 min after dissolution. Samples were dissolved either in phosphate buffer (column 3) or in distilled water (column 4) and diluted with the same solvent. A potency order of A > mixture [A+B+B′+C] > B > B′ was obtained in both cases. Comparing data of columns 3 and 4 reveals that the mixture as well as A and B shows increased potency in phosphate buffer. It may be explained by the fact that phosphate, like acetate, catalyzes all the reactions of the aldehyde function. Interaction of phosphate with the aldehyde could be detected by NMR. However, we have no reason at present for the exceptional behaviour of B′. According to the data of column 3 and the equilibrium composition (column 2), structures A, B and BI contribute to the potency of D-MePhe-Pro-Arg-H in about 90%, which would indicate a potency of 49 for C. The high potency of A gives evidence of the finding that freshly prepared and immediately isolated D-MePhe-Pro-Arg-H preparations, which contain A in 70-90%, show 1.3-1.5 times higher antithrombin activity than those

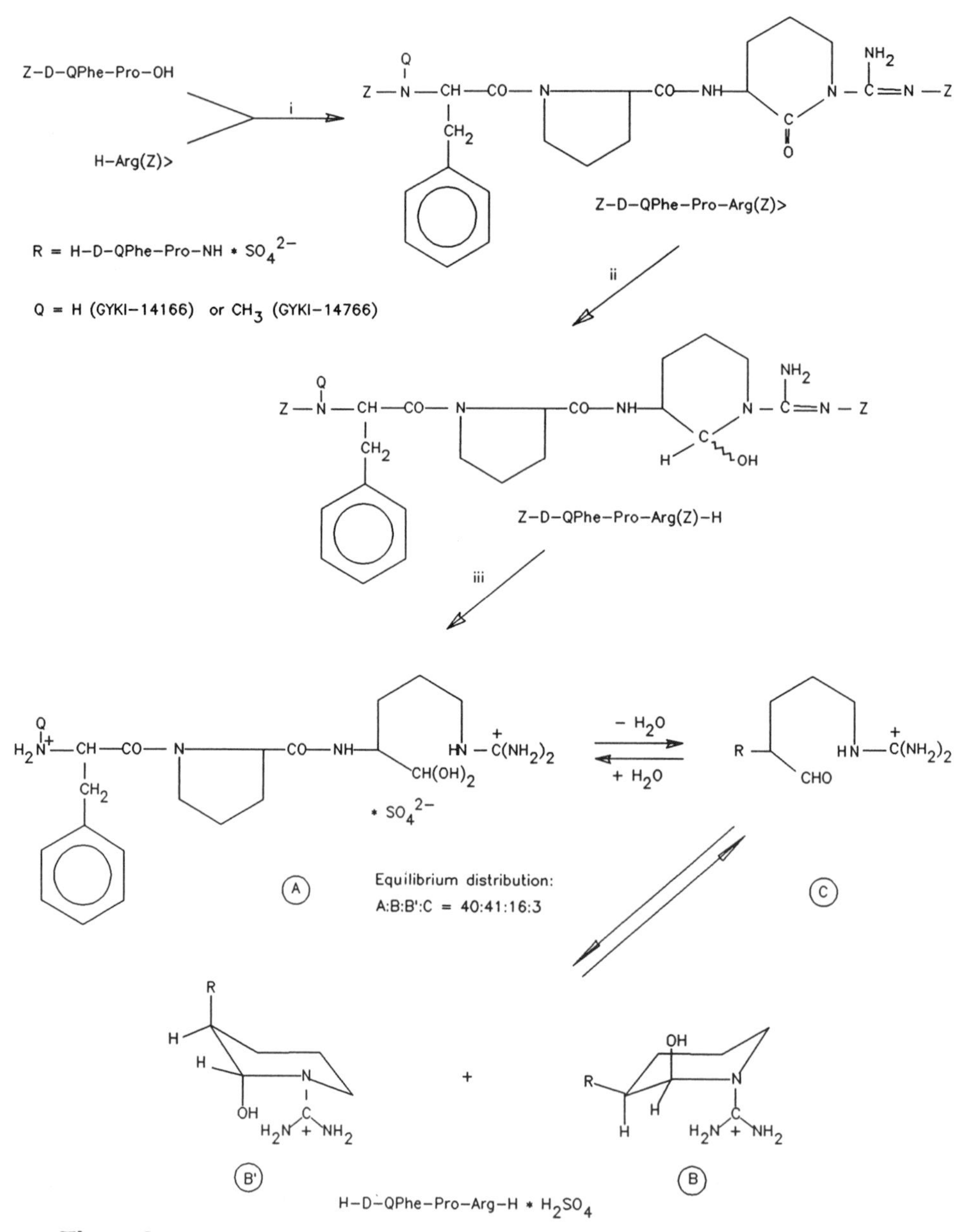

Figure 2 Last steps of the synthesis of D-Phe- and D-MePhe-Pro-Arg-H. i, activation of -COOH by i-BuOCOCl/NMM; ii, reduction by $LiAlH_4$ in THF, iii, deprotection with H_2/Pd/C in EtOH/H_2O/H_2SO_4

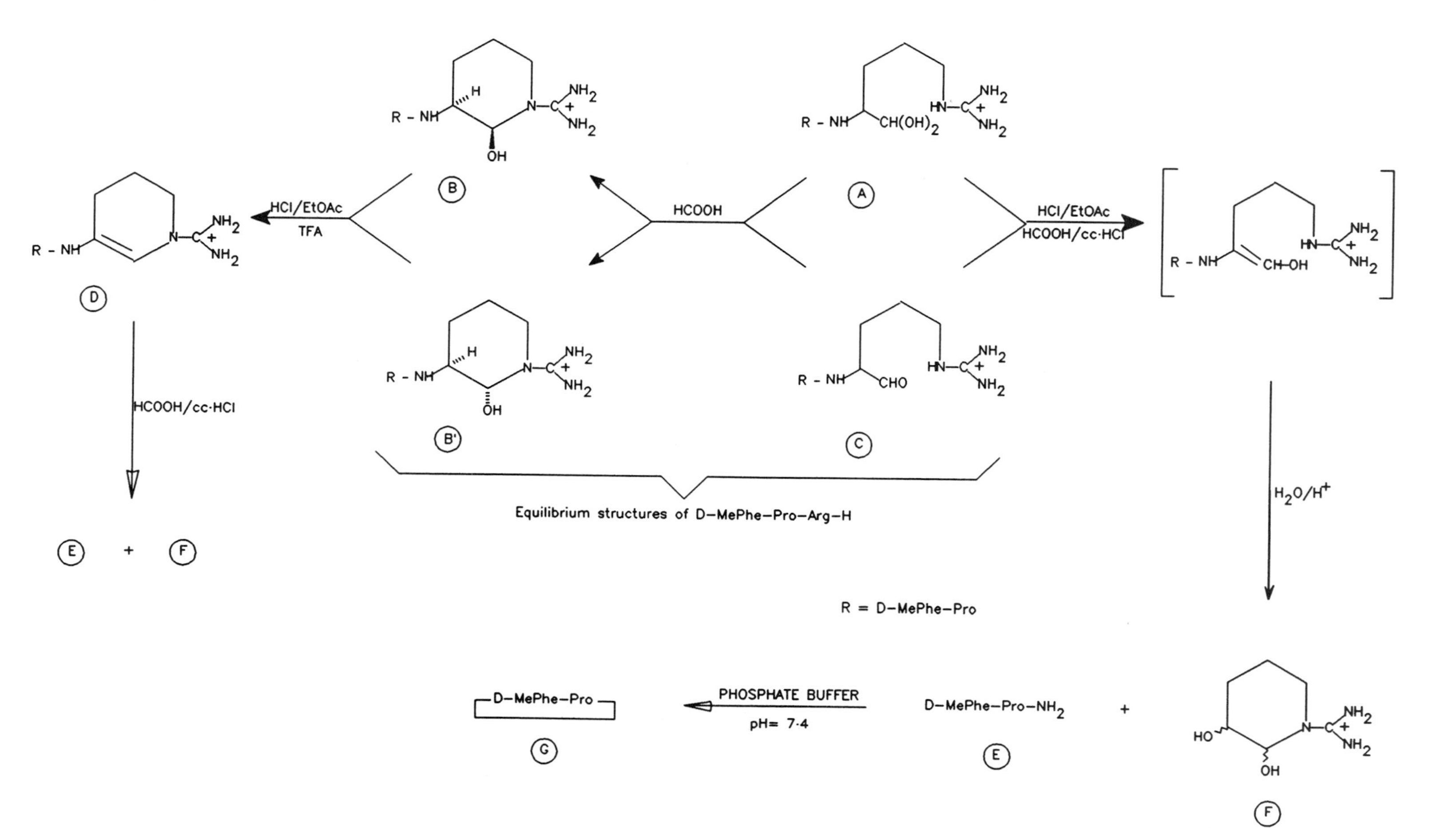

Figure 3 Transformation of D-MePhe-Pro-Arg-H under acidic conditions.

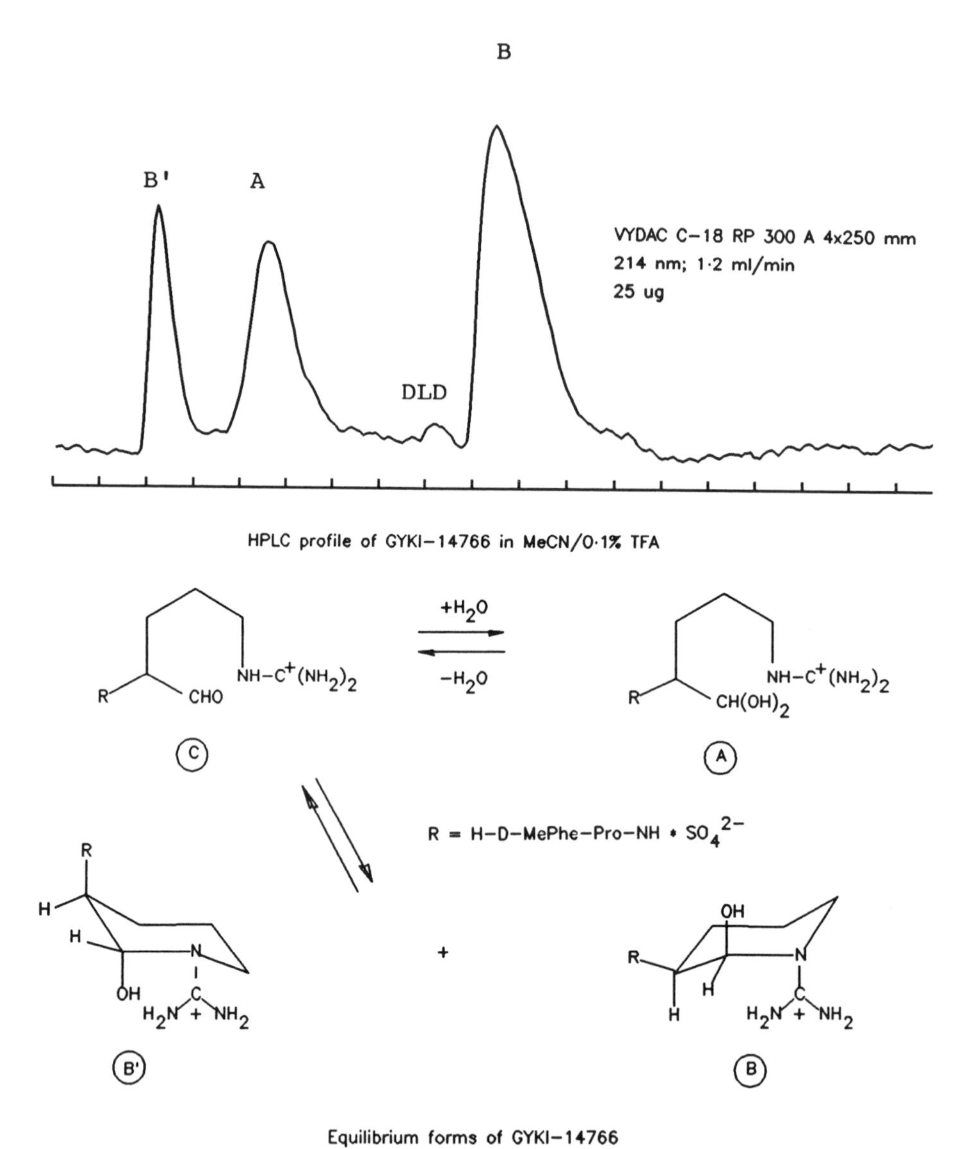

Figure 4 HPLC pf D-MePhe-Pro-Arg-H at PH < 3 and assignation of peaks to equilibrium structures: B′-A-B.

having the structures in equilibrium distribution. Nevertheless, due to the fast equilibrium at physiological pH, the existence of D-MePhe-Pro-Arg-H in equilibrium forms does not seem significant from a practical point of view.

BIOLOGY

The anticoagulant activities and inhibition of thrombin-mediated platelet aggregation by D-MePhe-Pro-Arg-H and its parent compound, as compared to heparin, in various *in vitro* assays are summarized in Table 3[24-27].

In systems containing platelet poor or normal plasma (APTT and TT), more or less similar data were obtained for the peptides and heparin. However, activities of heparin in whole blood (WBCT) and in platelet rich plasma (PC) were much lower than those of the peptides. The fact that the clotting of antithrombin III deficient patient plasma could only be delayed by the peptides is obvious and points out their advantage over heparin.

Table 2 Antithrombin activity of D-MePhe-Pro-Arg-H [A+B+BI+C] and its equilibrium structures A, B and B.

Structure	Phosphate buffer[a] % Present in the mixture	Phosphate buffer[a] Potency $1/I_{50}$[c]	Water[a] Potency $1/I_{50}$[c]
[A+B+B'+C]	100	13.33	5.00
A, hydrate	41.6	19.06	7.87
B, amino acid	42.2	9.70	3.92
B', amino cyclol	13.7	0.60	3.13
C, aldehyde	2,5	49[d]	

[a] Solvent used for solution/dilution of samples.
[b] Determined by NMR. [c]I_{50}, amount of sample (μg/ml) required for doubling the thrombin time of human plasma.
[d] Potency of C was calculated.

Comparison of the peptide, a recombinant hirudin and heparin in two *in vitro* assays[28] (Figure 5) indicated that the peptide and hirudin are about equipotent.

The *in vivo* anticoagulant activities of the peptide as measured by the WBCT, APTT and TT assays after i.v. infusion and sc, im or oral administration to rabbits in various doses are shown in Figure 6[25,26,29]. The required therapeutic range, as defined for monitoring heparin therapy, is indicated by the parallel dotted lines for each assay. From these data it emerges that the iv, sc and im activities of the peptide are rather similar. It is particularly apparent when comparing the relative clotting times measured during and after the three-hour iv infusion of the peptide at a dose of 1 mg/kg/h, i.e. 3 mg/3 h (open circles) with those produced by the same 3 mg/kg injected in a single dose subcutaneously

Table 3 Anticoagulant and antiplatelet effects of D-Phe-Pro-Arg-H (1) and D-MePhe-Pro-Arg-H (2) as compared to heparin in various *in vitro* assays.

	I_{50}[b]		
Assay[a]	1	2	Heparin[c]
APTT	0.24	0.44	0.38
TT, #2 normal plasma	0.032	0.04	0.10
TT, #2 antithrombin III deficient patient plasma	0.032	0.04	no activity
WBCT	0.085	0.09	0.39
PA inhibition	0.010	0.013	0.50

[a] Fresh human blood (WBCT) and citrated plasma (platelet poor in APTT, normal in TT, platelet rich in PA) were used.
[b]Amount of agents (μg/reaction mixture) required for doubling the clotting time (WBCT, APTT, TT) and for inducing 50% inhibition of PA, respectively.
[c] Commercial preparation, 142 USP U/mg.

(x). Secondly, according to the APTT, the medium dose of 10 mg/kg produced acceptable anticoagulation for 3 hours, an effect similar to that produced by an im dose of 2 mg/kg, and somewhat higher than that resulting from the 3 mg/kg dose given sc. Thirdly, the TT assay measured the anticoagulant activities of the peptide well over the therapeutic range. This phenomenon may indicate that "the therapeutic range" of anticoagulation with peptides is other than that of non-peptide agents, e.g. heparins.

A comparison of the pharmacodynamics of D-MePhe-Pro-Arg-H and a recombinant hirudin in monkeys was made by Dr. Fareed[28].

Figure 7 presents the results obtained with 1 mg/kg peptide and 2500 U/kg (about 0.3 mg/kg) hirudin given iv and sc, respectively. The peptide, unlike hirudin, shows similar activity profiles in both ways of application, i.e. prompt action with similar duration. It should be noted that the dose of peptide was about three times higher than that of hirudin by weight.

The antithrombotic effect of the peptide was studied in various thrombosis models[30]. Results obtained in an arteriovenous shunt model[31] applied to rabbits[10,11] are shown in Table 4 and Table 5. The wet weight of thrombus formed before and after treatment during a 20 min test period were measured. In experiment A of Table 4, 1 mg/kg of peptide was given by iv infusion for 60 min. In the last third of infusion the wet weight of thrombus indicated a 74% inhibition, and 10 to 30 min after the infusion, nearly 84%. When the same dose was applied sc, a similar reduction was observed in 30-50 min after the injection. Results obtained in this model after oral application are shown in Table 5. A substantial inhibition thrombus formation could be detected at a dose of 10 mg/kg, inhibition was nearly 95% 4 h after administration. At a 20 mg/kg dose, inhibition persisted for more than 6 hours.

Inhibitory effects of the peptide after sc, im and oral administration were also studied in a venous thrombosis model with stasis based on vascular lesion in rats[32] (Table 6). Even low doses, 1 mg/kg given sc or im and the oral dose of 5 mg/kg, produced significant reduction of thrombus weight.

In an experimental arterial thrombosis model in rats[33] (Table 7), the inhibitory effect of orally applied peptide could also be demonstrated. Thrombotic occlusion as recorded by the decrease of vessel-surface temperature, was inhibited by about 58 and 78% depending on the dose applied.

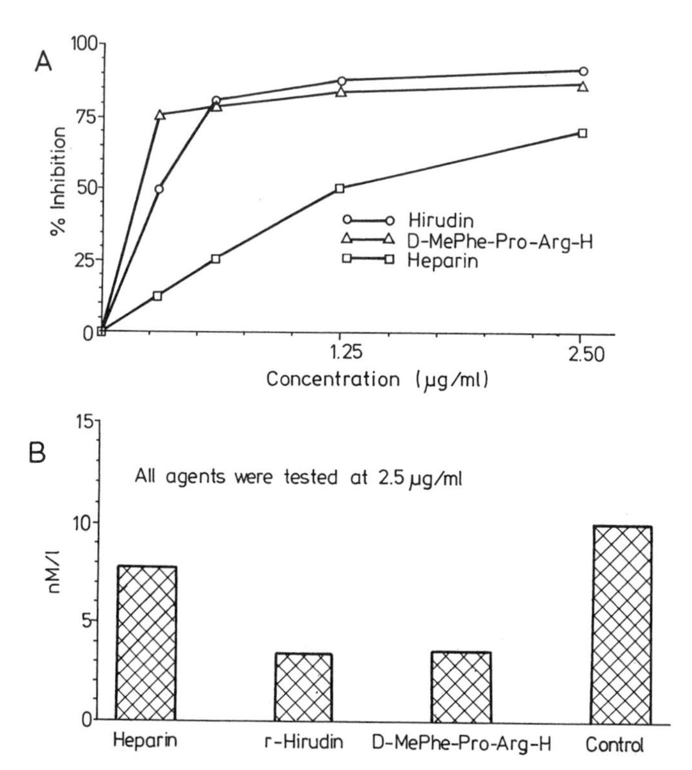

Figure 5 Comparison of r-hirudin (Knoll), D-MePhe-Pro-Arg-H and heparin. **A**, anti-IIa activities in normal human plasma. **B**, inhibition of prothrombin fragmentF_{1+2} generation.

The effect of D-MePhe-Pro-Arg-H along with heparin and two thrombolytics was also studied in a fibrin induced coronary embolization model in pigs[34] (Table 8). Fibrin clots (⌀ = 400 μm) injected into the left coronary artery produced microinfarction of the heart muscle with typical haemodynamic parameters and killed control animals in about an hour. Infusion of the peptide in a dose of 17 μg/kg/min for 180 min protected all the animals.

Table 4 Inhibitory effect of D-MePhe-Pro-Arg-H on thrombus formation in an arteriovenous shunt model[a] in rabbits at a dose of 1 mg/kg upon intravenous infusion in 60 min (A1 and A2) and subcutaneous injection (B) (n=5).

Period for thrombus formation	Wet weight of thrombus, mg	Inhibition %
A1, Control[b]	93.1 ± 8.4	-
40-60 min (during infusion)	23.8 ± 3.7[c]	74.4
A2, Control[b]	86.1 ± 9.4	-
70-90 min (10 to 30 min after infusion)	13.9 ± 3.8[c]	83.8
B, Control[b]	89.28 ± 4.93	-
30-to 50 min (after injection)	21.88 ± 2.08[c]	75.5

Table 5 Inhibitory effect of D-MePhe-Pro-Arg-H on thrombus formation in an arteriovenous shunt model[a] in rabbits after single doses of 10, 15 and 20 mg/kg (n=5-12 each dose).

Dose mg/kg	Period for thrombus formation, min	Wet weight of thrombus, mg	Inhibition %
0	Control[b]	106.1 ± 14.2	-
10	40 to 60	34.6 ± 4.7[c]	67.4
	220 to 240	5.7 ± 2.3[c]	94.6
15	40 to 60	24.8 ± 7.0[c]	76.6
	220 to 240	2.6 ± 1.3[c]	97.6
20	40 to 60	19.6 ± 7.0[c]	81.5
	220 to 240	4.2 ± 1.5[c]	96.0
	340 to 360	1.1 ± 0.6[c]	99.0

[a-c] See Table 3
See Table 4

Table 6. Inhibitory effect of D-MePhe-Pro-Arg-H in a venous thrombosis model after iv (A), sc (B), im (C) and oral (D1 and D2) administration to rats (n=7-9 each dose).

Doses mg/kg		Wet weight of thrombus, mg[b]	Inhibition %	Dry weight of thrombus, mg[b]	Inhibition %
A	0	20.68±0.23[c]	-	7.22±0.41[c]	-
	1	4.72±0.13	77.18	1.14±0.07[d]	84.21
B	0	20.74±0.26[c]	-	8.25±0.20[c]	-
	1	9.24±0.41	55.45	3.38±0.30[d]	59.03
	3	7.54±0.56	63.64	2.74±0.36[d]	66.79
C	0	20.74±0.26[c]	-	8.25±0.20[c]	-
	1	8.60±0.58	58.53	3.19±0.24[d]	61.33
	3	5.34±0.79	74.25	1.48±0.14[d]	82.06
D1	0	20.68±0.23[c]	-	7.22±0.41[c]	-
	5	13.94±1.21	32.59	4.46±0.55[d]	38.23
D2	0	22.60±1.01[c]	-	6.5±0.32[c]	-
	10	5.20±1.1	76.99	1.63±0.27[d]	74.92
	15	3.84±0.39	83.01	1.44±0.13[d]	77.85

[a]Pescador et al's model [31]. [b]Thrombus formed 150 min after administration of the inhibitor was placed on a saline-soaked filter paper to remove adhered blood. After weighing (wet weight) it was dried at 80°C overnight (dry weight). [c]Control values. [d]P<0.001 compared with control value.

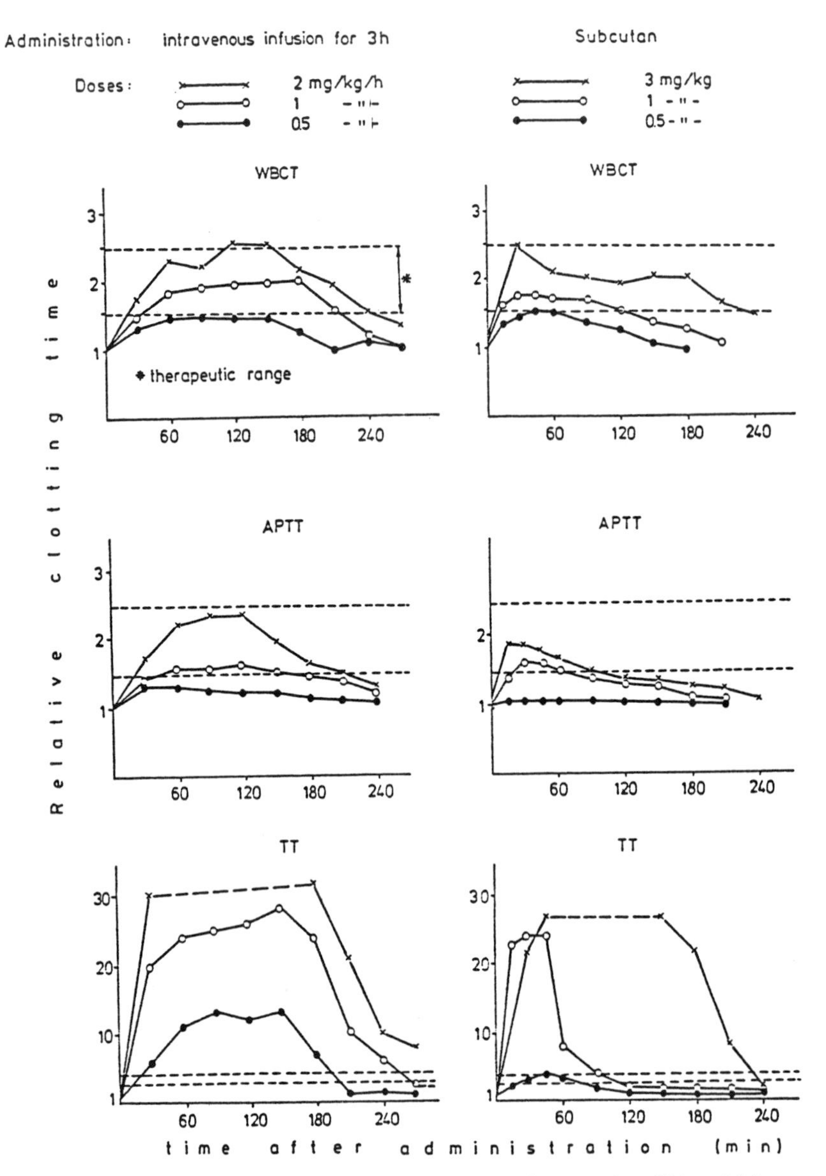

Figure 6 *In vivo* anticoagulant activities of D-MePhe-Pro-Arg-H in rabbits upon i.v. infusion, sc, im and oral administration as measured by the WBCT, APTT and TT assays.

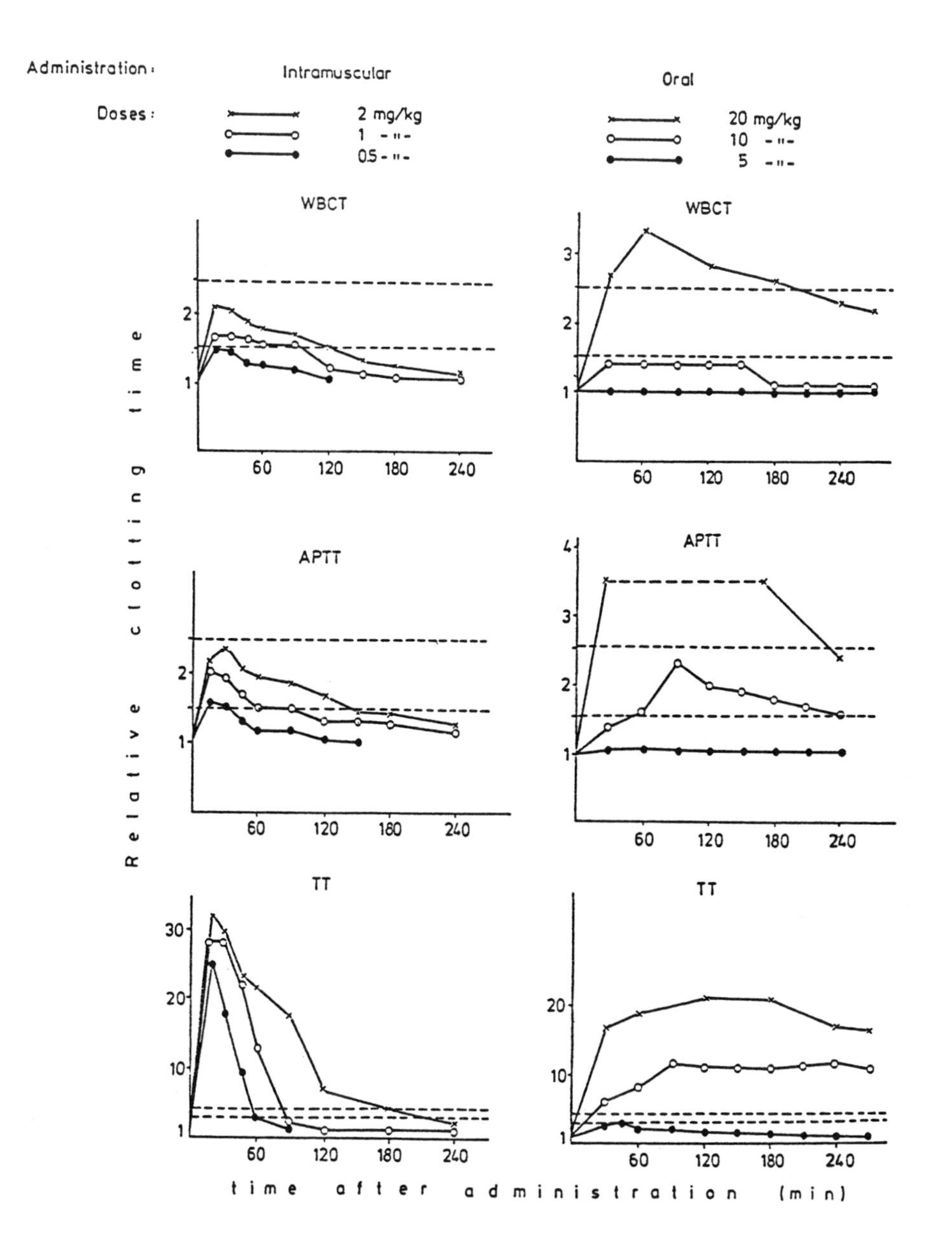

Administration:
Intramuscular
Oral
Doses:
2 mg/kg
1 -"-
0.5 -"-
20 mg/kg
10 -"-
5 -"-
WBCT
APTT
TT
Relative clotting time
time after administration (min)

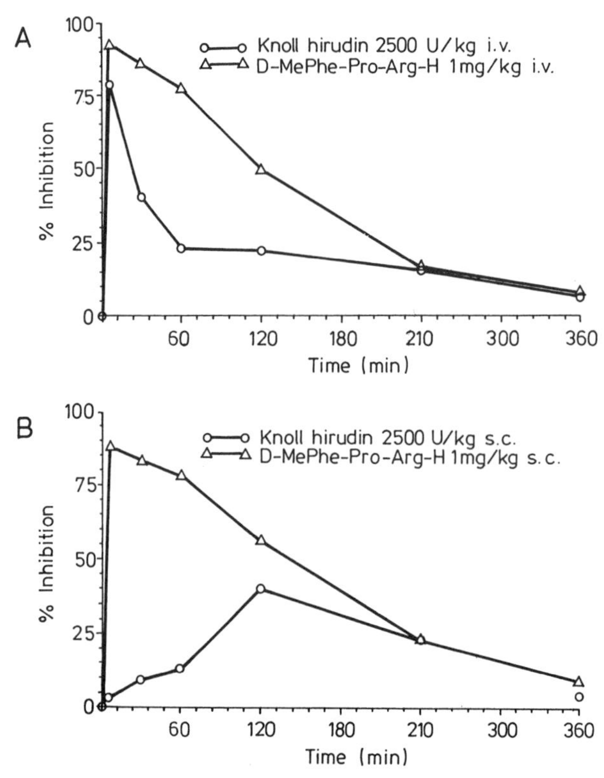

Figure 7 Comparison of the pharmacodynamics of r-hirudin (Knoll) and D-MePhe-Pro-Arg-H given sc and iv to monkeys.

Table 7 Inhibitory effect of orally administered D-MePhe-Pro-Arg-H on an experimental arterial thrombosis model[a] in rats (n=11-13 each dose).

Doses mg/kg	Vessel wall temperature Before provocation (°C)	After provocation (°C)	Average decrease in temperature °C	Inhibition %
0	33.72 ± 0.18	32.16 ± 0.19[b]	1.56	-
10	33.84 ± 0.22	33.19 ± 0.21[b]	0.66	57.7
15	33.93 ± 0.10	33.58 ± 0.10[b]	0.35	77.6

[a]Hladovecz's model (32) with modification (29). Thrombus formation in the carotid artery was induced by mechanical constriction for 5 min and allowed to proceed for 30 min. Inhibitor was given 30 min before mechanical constriction.
[b]$p < 0.001$ compared with control value.

Table 8 Effect of different antithrombotic/thrombolytic compounds on fibrin-induced coronary embolization in pigs as characterized by the increase of LVEDP, PCWP and HR and by survival time[ab].

Compound	μg,IU/kg/min infused for 180 min	Relative increase of LVEDP	PCWP	HR	Survival time min
Control	-	100	100	100	56±12
GYKI-14766	17 μg	37.6	17.9	6.1	>180
Heparin	17 IU	52.9	24.3	14.4	115 for 2/3
Saruplase[c]	10 μg	39.9	16.2	14.4	>180
Streptokinase	11 IU	8.0	7.3	2.2	>180

[a][33]. [b]LVDP, left ventricular end diastolic pressure;
PCWP, pulmonary capillary wedge pressure (PCWP); HR, heart rate;
[c]Recombinant single chain urokinase-PA.

Table 9 Changes in the bleeding time vs clotting times induced by D-MePhe-Pro-Arg-H administered to New Zealand white rabbits[a].

Route and dose of administration(n)	Assay time	Bleeding time, min	Clotting time WBCT, min	TT, s
IV bolus injection				
2 mg/kg (n=5)	0 min	2.1±0.1	12.6± 1.5	21.6±1.7
	5 min	2.0±0.3	37.2± 5.5	>600
10 mg/kg (n=5)	0 min	1.8±0.35	10.8± 0.38	19.0±0.77
	5 min	2.2±0.40	65.2±14.3	>600
	15 min	2.0±0.40	55.0±14.1	>600
IV infusion for 3h				
25 mg/kg per hour (n=6)	0 min	1.4±0.09	13.0± 0.89	24.0±2.09
	3 h	2.1±0.38	>120	>600
SC injection				
2 mg/kg (n=6)	0 min	2.58±0.3	14.0± 2.1	19.6±1
	15 min	2.42±0.4	29.7± 3.7	>600
50 mg/kg (n=4)	0 min	1.98±0.4	14.0± 2.1	23.5±2.2
	6 h	1.98±0.5	53.0±22.4	>600
Oral				
2x15 mg/kg per day for 5 days (n=3)	0 min	1.83±0.3	11.0± 0.58	19.8±0.47
	1 h	2.08±0.1	22.7± 4.38	>600
	121 h	1.67±0.3	30.0± 5.30	>600

[a]Markwardt et al's method [34] with modification, i.e. standardized incision on depilated rabbit ear (6mm long, 1mm deep) made with Medicut (Medicor, Budapest, Hungary); blood was blotted with a special filter paper wetted by saline at 15 s intervals till no blood on the paper appeared.

Streptokinase in 11 IU/kg/min and Saruplase in 10 μg/kg/min also saved the animals, all pigs survived the study period, while on heparin treatment with a dose of 17 IU/kg/min, two of three pigs died after less than 3 h. The three parameters measured indicate a potency order of Streptokinase > peptide > Saruplase > heparin.

Bleeding is known to be one of the most serious side reaction that frequently occurs during anticoagulation. Such an effect of D-MePhe-Pro-Arg-H was also studied after parenteral and oral application to rabbits. As indicated in Table 9 the bleeding time was measured[35] before starting the experiment and at the maximum of the anticoagulant effect. Thus WBCT and TT were also recorded. These data show that the peptide did not cause bleeding. Even at a provocative dose of 25 mg/kg/h infused for 3 hours, a slight increase of the bleeding time could only be observed. It is in concert with another finding that this peptide is completely harmless to platelets[8,25-27,29,30].

Acute toxicity of the peptide was determined in mice and rabbits after iv (bolus), sc and oral administration, the respective LD50 values were 40-45, >250 and >1000 mg/kg.

In conclusion, D-MePhe-Pro-Arg-H (GYKI-14766)

- is a stable analog of D-Phe-Pro- Arg-H (GYKI-14166), our initial tripeptide anticoagulant,
- acts directly on thrombin,
- exerts its effect shortly after parenteral or oral application,
- assists fibrinolysis/thrombolysis,
- does not cause bleeding.

It appears from these findings that GYKI-14766 is an excellent candidate for development as an anticoagulant/antithrombotic agent.

REFERENCES

1. S. Bajusz, E. Balla, E. Szell, and D. Bagdy, Inhibition of blood clotting by synthetic peptides, Abstracts, 9th FEBS Meeting, Budapest, p. 372, (1974) Hungarian Biochemical Society, Budapest.
2. S Bajusz, E. Barabas, E. Szell, E, and D. Bagdy, D, Peptide aldehyde inhibitors of the fibrinogen-thrombin reaction, in: Peptides, Proceedings of the Fourth American Peptide Symposium (R. Walter and J. Meienhofer, eds.), pp. 603-608, (1975) Ann Arbor Science Publ., Ann Arbor, MI, USA.
3. B. Blomback, M. Blomback, P. Olsson, P, Svendsen, L, and G. Aberg, 1969, Synthetic peptides with anticoagulant and vasodilating activity, Scand. J. Clin. Lab. Invest., Suppl. 24 (Suppl. 107), 59-64 (1969).
4. S. Bajusz, E. Barabas, P. Tolnay, E. Szell, and D. Bagdy, 1978, Inhibition of thrombin and trypsin by tripeptide aldehydes, Int. J. Peptide, Protein Res. 12:217-221 (1978).
5. S. Bajusz, E. Szell, E. Barabas, and D. Bagdy, Structure-activity relationships among tripeptide aldehyde inhibitors of plasmin and thrombin, in: Peptides, Proceedings of the Seventh American Peptide Symposium (D.H. Rich and E. Gross, eds.), pp. 417-420 (1981), Pierce Chem. Co., Rockford, IL, USA.
6. D. Bagdy, E. Barabas, E. Szell, E, and S. Bajusz, S, Über die biochemische Pharmakologie einiger Tripeptide aldehyde, Folia Haematol., Leipzig 109:22-32 (1982).
7. D. Bagdy, E. Barabas, M. Dioszegi, L. Sebestyen, S. Bajusz, E. Szell, and M. Antmann, M, In vitro and ex vivo effect of tripeptide aldehydes on blood

coagulation, Kiserletes Orvostudomany 35:650-661 (Hung); Chem. Abs.1984, 100:114752k (1983).
8. D. Bagdy, M. Dioszegi, L. Sebestyen, E. Barabas, S. Bajusz, and E. Szell, E, Anticoagulant and thrombocyte function-inhibiting effect of D-Phe-Pro-Arg-H(GYKI-14166) *in vivo,* Kiserletes Orvostudomany 36, 24-46 (Hung); Chem. Abs 1984, 100:185519k (1984).
9. D. Bagdy, E. Barabas, Z. Fittler, E. Orban, G. Rabloczky, S. Bajusz, E. Szell, and F. Jozsa, Anticoagulant activity of orally administered D-PhePro-Arg-H (GYKI-14166), Kiserletes Orvostudomany 37, 528-534 (Hung); Chem. Abs. 1986, 104:61729g (1985).
10. D. Bagdy, Z. Fittler, and S. Bajusz, Effectiveness of a synthetic thrombin antagonist (D-PhePro-Arg-H) in vivo in experiments with model thrombosis, Kiserletes Orvostudomany 37:642-648 (Hung); Chem. Abs. 1986, 104:122813d (1985).
11. D. Bagdy, S. Bajusz, and G. Rabloczky, RGH-2958, D-Phenylalanyl -L-prolyl -L-arginine aldehyde sulfate, Drugs of the Future 10:829-834 (1985).
12. D. Bagdy, E. Barabas, Z. Fittler, E. Orban, G. Rabloczky, S. Bajusz, and E. Szell, Experimental oral anticoagulation by a direct acting thrombin inhibitor (RGH-2958), Folia Haematol, Leipzig 115:136-140 (1988).
13. J.I. Witting, C. Pouliott, J.L. Catalfamo, J. Fareed, and J.W. Fenton II, Thrombin inhibition with dipeptidyl arginals, Thromb. Res. 50:461-468 (1988).
14. W. Bode, I. Mayr, U. Baumann, R. Huber, S.R. Stone, and J. Hofsteenge, The refined 1.9 A crystal structure of human α-thrombin:interaction with D-Phe-Pro-Arg chloromethylketone and significance of the Tyr-Pro-Pro-Trp insertion segment, EMBO J.8:3467 (1989).
15. C. Kettner, and E. Shaw, D-Phe-Pro-Arg-CH2Cl - A selective affinity label for thrombin, Thromb. Res. 14:969-973 (1979).
16. W. Stüber, H. Kosina, and H. Heimburger, Synthesis of a tripeptide with C-terminal nitrile moiety and the inhibition of proteinases, Int. J. Peptide Protein Res. 31:63-70 (1988).
17. S. Bajusz, E. Barabas, E. Szell, and D. Bagdy, Inhibition of thrombin with H- and Boc-D-Phe-Pro-Agm, in: Peptides 1982 (K. Blaha and P. Melon, eds), pp. 643-647 (1988), deGruyter, Berlin-New York.
18. M. Szelke, and D.M. Jones, Peptide analogs and their use in enzyme inhibition, European Patent Application, EP 118:280; Chem. Abs. 1985, 102:185498j (1984).
19. C. Kettner, L. Mersinger, and R. Knabb, The selective inhibition of thrombin by peptides of boroarginine, J. Biol. Chem. 265:18289-18297 (1990).
20. J.M. Maraganore, P. Bourdon, J. Jablonski, K.L. Ramachandran, and J.W. Fenton II, Design and characterization of hirulogs: a novel class of bivalent peptide inhibitors of thrombin, Biochemistry 29:7095 (1990).
21. J. DiMaio, B. Gibbs, D. Munn, J. Lefevre, Ni, Feng, and Y. Konishi, Bifunctional thrombin inhibitors based on the sequence of hirudin45-65, J. Biol. Chem. 265:21698-21703 (1990).
22. J. DiMaio, B. Gibbs, D. Munn, J. Lefevre, Ni, Feng, and Y. Konishi, Design of bifunctional thrombin inhibitors based on the sequence of hirudin45-65, in: Peptides 1990 (E. Giralt and D. Andreu, eds.), ESCOM, Leiden, The Netherlands (1991).
23. S. Bajusz, E. Szell nee Hasenorl, E. Barabas, and D. Bagdy, D-Phenylalanyl-L-prolyl-L-arginine aldehyde sulfate and process for the preparation thereof, US Patent, US 4,399 036 (1983).
24. S. Bajusz, E. Szell nee Hasenorl, D. Bagdy, E. Barabas, M. Dioszegi, Z. Fittler, Zs, F. Jozsa, G. Horvath, and E. Tomori nee Jozst, Peptide-aldehydes, process

for the preparation thereof and pharmaceutical compositions containing the same, US Patent, US 4,703 036 (1987).

25. S. Bajusz, E. Szell, D. Bagdy, E. Barabas, G. Horvath, M. Dioszegi, Z. Fittler, Zs, G. Szabo, A. Juhasz, A, E. Tomori, and G. Szilagyi, Highly active and selective anticoagulants: D-Phe-Pro-Arg-H, a free tripeptide aldehyde prone to spontaneous inactivation, and its stable N-methyl derivative, D-MePhe-Pro-Arg-H, J. Med. Chem. 33:1729-1735 (1990).
26. D. Bagdy, E. Barabas, G. Szabo, E. Szell, and S. Bajusz, Pharmacodynamic effects and pharmacokinetics of D-MePhe-Pro-Arg-H.H2S04 (GYKI-14766), a novel anticoagulant with antithrombotic properties, Eur. J. Pharmacol. 183:1844-1845 (1990).
27. D. Bagdy, E. Barabas, S. Bajusz, and E. Szell, In vitro inhibition of blood coagulation by tripeptide aldehydes. A retrospective screening study focused on the stable D-MePhe-Pro-Arg-H. H2SO4, Thromb. Haemostas. 67:325 (1992).
28. J. Fareed, Unpublished results (1991).
29. D. Bagdy, E. Barabas, G. Szabo, G, S. Bajusz, and E. Szell, *In vivo* anticoagulant and antiplatelet effects of D-Phe-Pro-Arg-H and D-MePhe-Pro-Arg-H, Thromb. Haemostas. 67:357 (1992).
30. D. Bagdy, G. Szabo, E. Barabas, S. Bajusz, and E. Szell, Inhibition by D-MePhe-Pro-Arg-H (GYKI-14766) of thrombus growth in experimental models of thrombosis, Thromb. Haemostas. 68:125 (1992).
31. J.R. Smith, and A.M. White, Fibrin, red cell and platelet interactions in an experimental model of thrombosis, Br. J. Pharmacol. 77:29-38 (1982).
32. R. Pescador, R. Port, A. Conz, and M. Mantovani, A quantitative venous thrombosis model with stasis based on vascular lesion, Thromb. Res. 53:197-201 (1989).
33. J. Hladovec, Experimental arterial thrombosis in rats with continuous registration, Thrombos. Diathes. Haemorrh. 26:407-410 (1971).
34. I. Szelenyi, Unpublished results (1991).
35. F. Markwardt, H.P. Klocking, and G. Nowak, Antithrombin und Antiplasmin-Wirkung von 4-amidinophenyl-brenztraubsaure (APPA) *in vivo*, Thromb. Diathes. Haemorrh. 24:240-247 (1970).

PEPTIDE BORONIC ACID INHIBITORS OF THROMBIN

Charles Kettner and Robert M. Knabb

The DuPont Merck Pharmaceutical Company
Wilmington, DE 19880-328, USA

Fibrin polymerization is the major component in venous thrombosis while both fibrin deposition and platelet aggregation are important in arterial thrombosis. Thrombin, the last protease in the blood coagulation cascade, hydrolyzes two bonds of fibrinogen to convert it into insoluble fibrin. Thrombin also catalyzes platelet aggregation. However, the predominance of thrombin catalysis in the latter reaction and in arterial thrombosis was not recognized until recently[1]. From earlier observations, it was known that the high molecular weight inhibitor, heparin-antithrombin III, was effective in blocking venous thrombosis, but was ineffective in blocking arterial thrombosis. This suggested that the thrombin catalyzed reaction was of limited importance in platelet aggregation. More recently, studies of Hanson and Harker[1] demonstrated that the small peptide thrombin inhibitor, H-(D)Phe-Pro-ArgCH_2Cl, was highly effective in blocking arterial thrombosis in animal models. The differences in the behavior of the two thrombin inhibitors in blocking arterial thrombosis was attributed to the limited accessibility of the high molecular weight inhibitor to thrombin on the platelet surface. This result clearly suggests that the use of low molecular weight thrombin inhibitors in the control of arterial thrombosis will be advantageous.

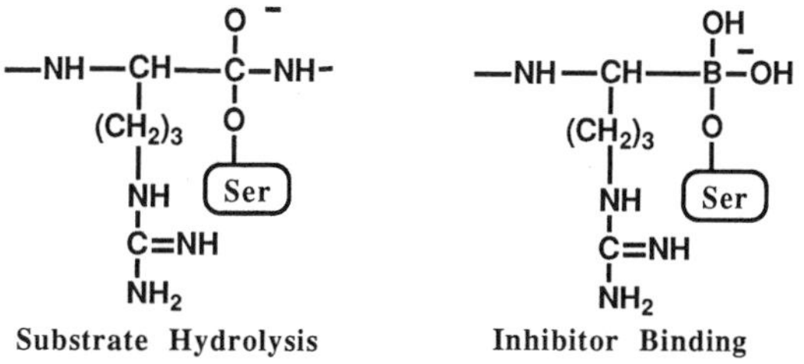

Substrate Hydrolysis Inhibitor Binding

Scheme 1

The Design of Synthetic Inhibitors of Thrombin
Edited by G. Claeson, *et al.*, Plenum Press, New York, 1993

Peptide boronic acids[3], which function as reaction intermediate analogs[4,5], are highly effective reversible inhibitors of serine proteases. As illustrated in Scheme 1, the trigonal boronic acid [$-B(OH)_2$] forms a tetrahedral complex with the active site serine. This mimics the tetrahedral complex obtained during normal substrate hydrolysis. A synthetic protocol has been developed for the preparation of peptides containing the boronic acid analog of arginine, boroArg-OH, and it has been applied to the synthesis of inhibitors of thrombin using the -(D)Phe-Pro-Arg- sequence. In this communication, we summarize the properties of these inhibitors both *in vitro*[6] and *in vivo*[7].

Reactivity of Peptide Boronic Acids with Thrombin *In Vitro*

Binding constants for thrombin. All inhibitors in Table 1 are slow binding inhibitors and exhibit kinetic properties consistent with the model described by Williams and Morrison[8] (Equ 1-3). Similar properties have been observed for peptide boronic acid inhibitors with other serine proteases[3,9].

$$E + I \underset{k_2}{\overset{k_1}{\rightleftharpoons}} EI \underset{k_4}{\overset{k_3}{\rightleftharpoons}} EI^*$$

$$K_i\,(\text{initial}) = \frac{k_2}{k_1} \quad (1)$$

$$\frac{K_i\,(\text{initial})}{K_i\,(\text{final})} = \frac{k_3}{k_4} + 1 \quad (2)$$

$$k' = k_4 + k_3 \left[\frac{I/K_i\,(\text{initial})}{I/K_i\,(\text{initial}) + 1 + S/K_m} \right] \quad (3)$$

k' is the apparent first-order rate constant and k_3 and k_4 are the first-order rate constants for the slow binding reactions. K_i(initial) and K_i(final) are the dissociation constants for the El and El* complexes, l is the concentration of inhibitor, S is the concentration of substrate, and K_m is the Michaelis constant.

Comparison of the effects of the N-terminal protecting groups of inhibitors with the -(D)Phe-Pro-boroArg- sequence indicates that the binding of the final complex, EI*, is dependent on these groups. Ac-(D)Phe-Pro-boroArg-OH has final K_i of 41 pM while substitution of a Boc- group for Ac- results in a 10-fold increase in final binding shown by the decrease in K_i to 3.6 pM. Inhibitors with a free amino group were the most effective inhibitors binding with a final K_i of <3.6 pM. On the other hand, values of K_i(initial) and k_3, the rate limiting step for the formation of the EI* complex, are similar for all thrombin inhibitors. An initial complex is formed with a K_i(initial) of 720-1200 pM and k_3 values for conversion to a final complex are 4.3×10^{-3} to $9.7\times10^{-3}S^{-1}$. Association rate constants, k_3/K_i(initial), range from 5×10^6 to 9×10^6 M^{-1} s^{-1}. Significant differences observed in K_i(final) are due to differences in the values of k_4, the limiting rate constant for the dissociation of the El* complex. For Ac-(D)Phe-Pro-boroArg-OH and the corresponding Boc compound, values measured for k_4 are 2.3×10^{-4} and 1.8×10^{-5}

Table 1 Kinetic constants for the binding of peptide boronic acids to thrombin[a,c].

	Inhibitor	K_i (initial) pM	K_i (final) pM	$10^3 \cdot k_3$ s^{-1}	$10^4 \cdot k_4$ s^{-1}
I.	Ac-(D)Phe-Pro-boroArg-OH	790±170	41±2	4.3±0.3	2.3
II.	Boc-(D)Phe-Pro-boroArg-$C_{10}H_{16}$[b]	720±120	3.6±0.6	6.7±0.2	0.17
III.	H-(D)Phe-Pro-boroArg-OH	1200±150	< 3.6	9.7±1.5	< 0.17
IV.	H-(D)Phe-Pro-boroArg-$C_{10}H_{16}$	1200±60	< 3.6	7.3±0.3	< 0.17

[a] All kinetic constants were measured at 25°C in phosphate buffer, pH7.5. Assays were run in the presence of inhibitor at 5 different levels (0.20-0.020 mM) of the substrate, S2238, and at a thrombin level of 0.19 nM. K_i(initial) and K_i(final) were determined by the method of Lineweaver and Burk from values of initial velocity and steady state velocity, respectively, and controls run in the absence of inhibitor over the susbtrate range of 0.20-0.020 mM. Values of the rate constants, k_3 and k_4, were determined by fitting values of k', the apparent first order rate constant, to the equation for slow binding inhibitor[8], $k' = k_4 + k_3[I/K_i(\text{initial})/(I/K_i(\text{final})+1+S/K_m)]$.
[b] "-$C_{10}H_{16}$" is the abbreviation for the pinanediol protection group of the boronic acid. Almost identical kinetic properties were obtained for the free boronic acid and the pinanediol ester at the concentration used here.
[c] Kinetic constants have been published previously[6] and a portion of the published Table is presented here with permission.

s^{-1}, respectively. In other terms, approximately 1 h and 10 h are the shortest possible times for 50% dissociation of the two respective inhibitors under optimum conditions. For H-(D)Phe-Pro-boroArg-OH, an even tighter complex with the enzyme is formed where the half life is several days although this particular compound is much less stable than other boronic acids in aqueous solution and decomposes with a half life of approximately 30 min.

Determinations of Kinetic Constants

Kinetics of the interaction of thrombin with the peptide boronic acids were determined using the substrate, H-(D)Phe-Pip-Arg-*p*NA. The sensitivity of this assay and the low K_m of this substrate allowed saturating levels of substrate and low concentration of enzyme to be used. Substrate was varied over a concentration range of 0.20 to 0.020 mM, a concentration range which is 50-5 times the K_m of the substrate ($K_m = 3.8 \pm 1.5$ μM). Kinetic constants for the binding of thrombin inhibitors were determined under pseudo-first-order conditions in which an excess of inhibitor over enzyme was maintained. Routinely, for thrombin inhibitors an inhibitor level of 1.0 nM and an enzyme level of 0.19 nM was used.

First, values of K_i(initial) were determined by "enzyme initiated reactions" where enzyme was added to premixed solutions of substrate and inhibitor using initial velocities measured in a time frame of 0-3 min after initiating the reaction. Prolonged reactions under these conditions resulted in unsatisfactory values of the steady state velocities due to substrate depletion. Values of the steady state velocities and the apparent first-order

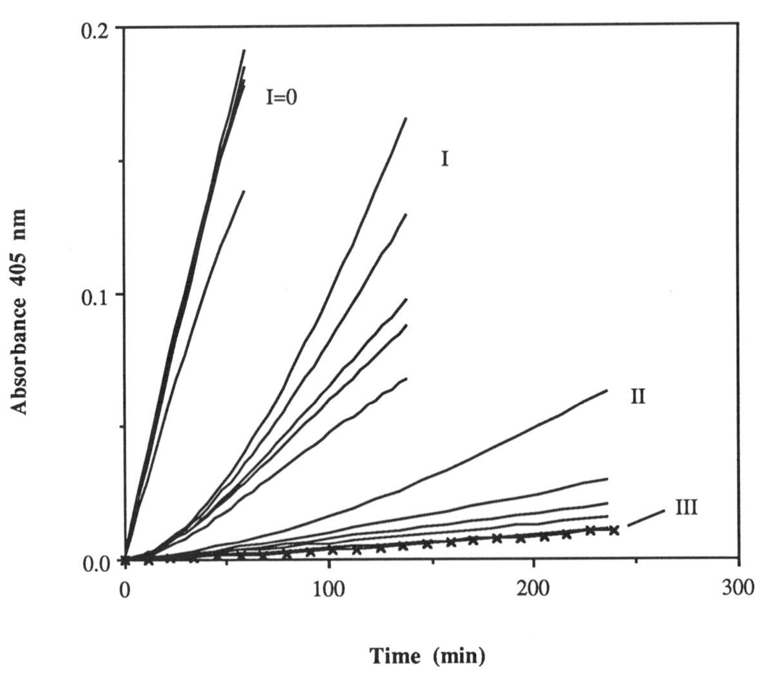

Figure 1 Effect of boroarginine inhibitors on thrombin assays. Hydrolysis of substrate, S2238, by 0.19 nM thrombin at 25°C and pH 7.5 was determined at 5 different substrate levels (0.20, 0.062, 0.036, 0.025 and 0.020 mM) by measuring the increase in absorbance at 405 nm. "I" are assays run in the absence of inhibitor. "II", "III", and "IV" are assays run in the presence of Ac(D)Phe-Pro-boroArg-OH, Boc-(D)Phe-Pro-boroArg $C_{10}H_{16}$, and H-(D)Phe-Pro-boroArg-OH, respectively, in which thrombin was incubated with the inhibitors and then diluted 100-fold into solutions of substrate to yield a final inhibitor concentration of 1.0 nM. For "II" and "III" the five substrate concentrations described above were used while for "IV" the 0.20 mM was used. These results have been published previously as "Supplemental Material"[6] and is reproduced with permission.

rate constants (k′) for I and II were measured by "substrate initiated reactions" where thrombin was preincubated with inhibitor and then diluted 100-fold into solutions of substrate yielding an inhibitor concentration of 1.0 nM. It should be noted that values of k' and the steady state velocities for "enzyme initiated reactions" and for "substrate initiated reactions" are equivalent for slow-binding inhibitors[10].

Typical data is shown in Figure 1 for Ac-(D)Phe-Pro-boroArg-OH, Boc-(D)Phe-Pro-boroArg-$C_{10}H_{16}$ and H-(D)Phe-Pro-boroArg-OH the corresponding controls run in the absence of inhibitor . The activity of thrombin was monitored by measuring the increase in absorbance at 405 nm with time. Controls were run by first incubating thrombin in the

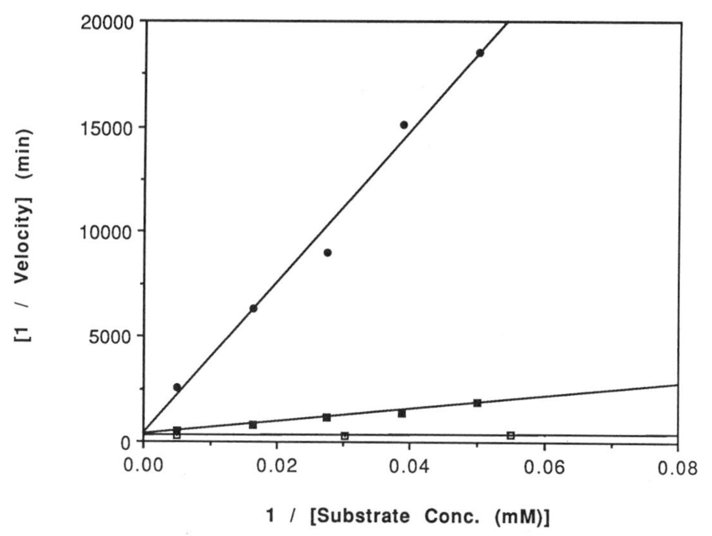

Figure 2 Lineweaver Burk plots of steady state velocities. Double reciprocal plots of steady state velocities *vs* substrate concentration were made for data in Figure 1. Velocities for 1.0 nM Boc-(D)Phe-Pro-boroArg-$C_{10}H_{16}$ (-●-) and Ac-(D)Phe-Pro-boroArg-OH (-■-) assayed over a substrate range of 0.20-0.020 mM were compared with controls (-□-) run in the absence of inhibitor using a substrate range of 0.20-0.010 mM. The ratio of slopes of the lines are 440:36:1. This Figure is being reproduced with permission from ref.6.

assay buffer for 1 h and then initiating the reaction by adding substrate. This step was added to account for the small loss of thrombin (approximately 20%) obtained on the initial incubation of thrombin at this low dilution. Subsequent incubation of thrombin for 2 and 3 h in the absence of susbtrate did not result in the further loss of activity. Values of the steady state velocities for the former two inhibitors and those of controls, measured over a substrate concentration range of 0.20-0.010 mM, were used to determine K_i(final) values by the method of Lineweaver and Burk (Figure 2).

Examination of the data in Figure 1 demonstrates that in the absence of inhibitor the 5 to 50-fold K_m substrate concentrations are saturating levels for the enzyme. For this set of data, Ac-(D)Phe-Pro-boroArg-OH at 1.0 nM with a substrate level 50 times K_m inhibits the enzyme 45% after equilibrium has been reached. This result is consistent with the K_i(final) of 41 pM we report in Table 1. Similarly, the Boc-(D)Phe-Pro-boroArg-$C_{10}H_{16}$ at 1.0 nM inhibits the enzyme by 88% at this level of substrate. After preincubation of H-(D)Phe-Pro-boroArg-OH with thrombin and then diluting into substrate at a level of $50xK_m$, a low level of activity ($\approx$2%) is recovered after incubation for 3 h. We estimate values of 0.3 pM and $2x10^{-6}s^{-1}$ for K_i(final) and k_4. This complex is extremely tight considering that the molecule is unstable at physiological pH as described previously.

Selectivity

Ac-(D)Phe-Pro-boroArg-OH (DuP714) was chosen for *in vivo* studies. The reactivity of the compounds in Table 1 were determined for additional trypsin-like proteases[6] and results are shown in Table 2 for DuP714. Comparison of the values of K_i(final) indicate a 40-fold greater affinity for thrombin than plasma kallikrein. For the remaining proteases, differences are at least 2-orders in magnitude. DuP714 is a slow binding inhibitor of thrombin, factor Xa, and tissue plasminogen activator. Comparison of values of k_{assoc}, the association rate constant, for these inhibitors again indicate differences greater than 2-orders in magnitude.

***In vivo* Properties of Ac-(D)Phe-Pro-boroArg-OH (DuP714)**

Plasma levels of DuP714 The effectiveness of DuP714 in the *in vivo* inhibition of thrombin was determined. In Figure 3, the effect of a single, i.v. (intravenous) dose of DuP714 (0.025 to 0.075 mg/kg) in conscious rabbits is shown. A rapid, dose-dependent increase in clotting times of plasma, measured by the APTT assay, was observed. Normal clotting was observed within 10-15 min indicating that DuP714 is rapidly cleared from circulation. Continuous i.v. infusion of DuP714 (0.05-0.3 mg/kg/hr) in rabbits results in a sustained blood level of inhibitor (Figure 4). Subcutaneous dosing is also effective in obtaining sustained blood levels[6]. A single dose of 0.25-0.5 mg/kg elevates APTT for 3 h.

The level of inhibitor that has been obtained by i.v. administration is effective in blocking both venous and arterial thrombosis as shown in the following animal models.

Venous thrombosis The effectiveness of DuP714 in blocking venous thrombosis was demonstrated in the Wessler model[11]. In this model, restriction of blood flow in the jugular vein of rabbits and injection of a thrombogenic stimulus (canine serum) results in thrombus formation. As shown in Figure 5, a 0.50 mg/kg bolus, i.v. injection of DuP714 (0.5 mg/kg) completely inhibits thrombosis in 70% of the animals, while in the remaining animals clot formation was slight[7]. This differs from controls where 100% thrombosis was observed and the majority of the clots were severe.

Arterial thrombosis DuP714 was tested in a rabbit high flow model of arterial thrombosis[7]. In this model, the carotid artery and the jugular vein are connected by a shunt consisting of small diameter polyethylene tubing. Occlusion occurred within 30 min in 72% of the control animals and in only 11% of the rabbits treated with 0.10 mg/kg/h i.v. DuP714.

Table 2 Selectivity of DuP714 in the inhibition of thrombin[a]

Protease	K_i(final) (pM)	k_{assoc} $s^{-1}M^{-1}$
Thrombin	41	$5.5x10^6$
Plasma Kallikrein	1,900	NSB[b]
Factor Xa	9,000	$5.1x10^4$
Plasmin	5,100	NSB
Tissue Plasminogen Activator	5,700	$3.5x10^4$

[a]All inhibition constants were measured a pH 7.5 except for tissue plasminogen activator, which was measured at pH 7.0, according to the methods described previously[6]. Reported values of K_i(final) are the average of at least two separated determination. For thrombin, values of k_{assoc} were calculated from individual constants reported in Table I using the relationship $k_{assoc} = k_3 / K_i(initial)$; for the remaining enzymes, k_{assoc} was determined from the relationship $k' = k_4 + k_{assoc} [I / (1+ S / K_m)]$ by plotting values of k' *vs* the terms in brackets.

[b]NSB indicates that "non-slow binding inhibition" was observed.

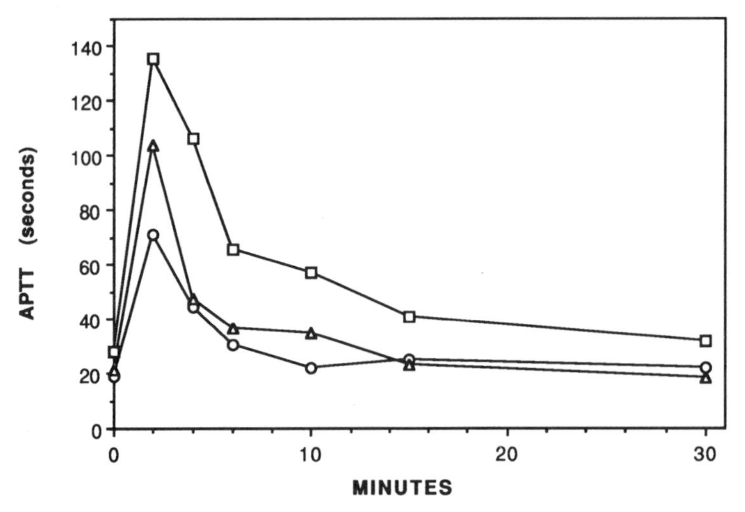

Figure 3 Effect of bolus i.v. injection of DuP714 on APTT (activated partial thromboplastin times) in rabbits. Animals received the following doses: 0, 0.025 mg/kg; ▽, 0.05 mg/mg; □, 0.075 mg/kg. Data represent mean values for four rabbits at each dose. When values of APTT exceeded 160 s, they were treated as 150 s for analysis. This Figure is reproduced from ref. 7 with permission.

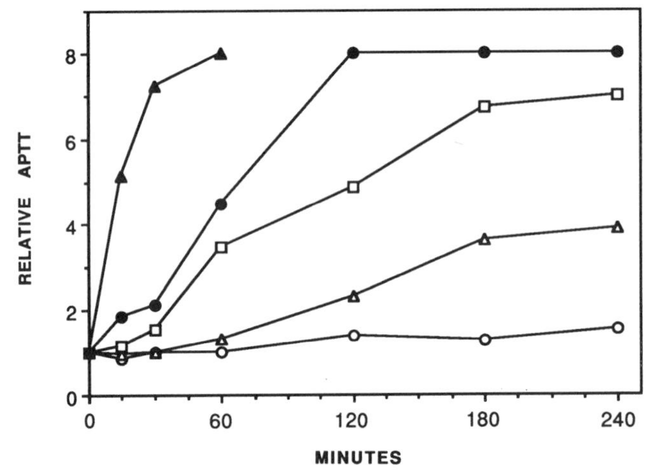

Figure 4 Effect of constant rate intravenous infusion of DuP714 in conscious rabbits. Mean values (n=4) of relative APTT of animal receiving DuP714 were compared with controls receiving normal saline at various time points during infusion. Data is shown for the following groups of animals: ○, control; △, 0.05 mg/kg/h; □, 0.1 mg/kg/h; ●, 0.2 mg/hg/h; ▲, 0.3 mg/kg/h. When APTT was greater than 8 times the baseline value, it was expressed as 8. This Figure is reproduced from ref.7 with permission.

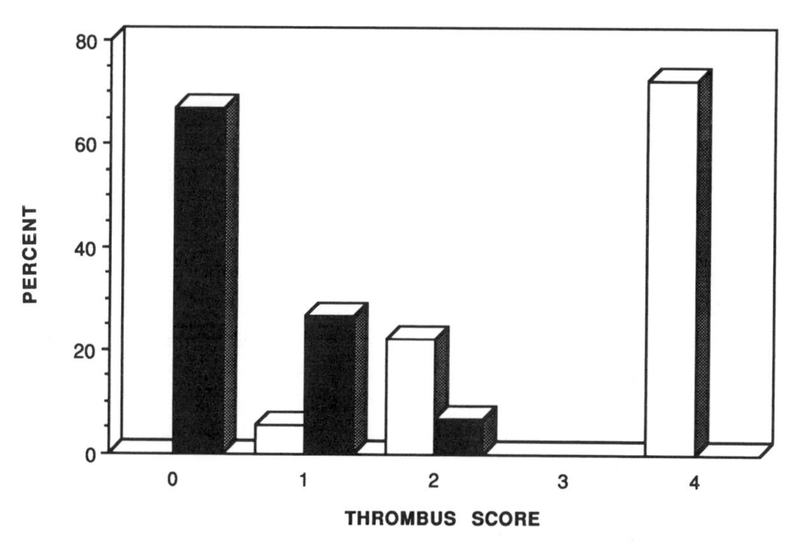

Figure 5 Effect of treatment with DuP714 on venous thrombosis. Solid bars represent results in rabbits (n=15) which received 0.5 mg/kg DuP714. Open bars represent results in control animals (n=18). Thrombus scores were graded visually where 0 represents no visible clot; 1: a small particulate clots; 2: larger strands of clots; 3: a solid clot with residual blood; 4: a fully formed clot resembling a cast of the vessel segment. This Figure is reproduced from ref. 7 with permission.

CONCLUSIONS

Ac-(D)Phe-Pro-boroArg-OH (DuP714) is a highly effective, selective inhibitor of thrombin binding with a K_i of 40 pM. A single, bolus i.v. dose results in a rapid elevation of anticoagulant activity which is rapidly cleared. Administration of a single subcutaneous dose (0.2-0.5 mg/kg) or continuous i.v. infusion (0.05-0.30 mg/kg/h) results in sustained anticoagulant activity. Our studies show that DuP714 is effective in preventing clotting in both venous and arterial rabbit models of thrombosis. Furthermore, DuP714 has been shown to be more effective than heparin in reducing clot accretion on preformed jugular vein thrombi in rabbits[12]. Low molecular weight inhibitors, like DuP714 have been shown to be more effective against clot-bound thrombin than the large heparin-antithrombin III complex[13]. This is expected to be valuable in the clinical treatment of established venous thrombosis.

Similarly, the poor efficacy of heparin in blocking arterial thrombosis is attributable to the inaccessibility of the bulky heparin-antithrombin III complex to thrombin on the platelet surface. In contrast, DuP714 and other low molecular weight thrombin inhibitors are highly effective in blocking platelet deposition and thrombosis as shown in our studies in rabbit[7] and in studies in baboons[14].

Our data indicates that DuP71 4 has the potential to offer significant advantages over heparin in the treatment of both venous and arterial thrombosis. Further studies to explore its clinical use are underway.

ACKNOWLEDGEMENTS

The technical assistance of Lawrence Mersinger, Joseph M. Luettgen, Anne C. Dobies, Margaret Buchan, and Joseph Shaw is gratefully acknowledged. We thank John V. Schloss, Stephen Brenner, Thomas Reilly, and Pieter B. Timmermans for their encouragement in these studies.

REFERENCES

1. S.R. Hanson, and L.A. Harker, L.A, Interruption of acute platelet-dependent thrombosis by the synthetic antithrombin D-phenylalanyl-L-prolyl-L-arginyl chloromethyl ketone, Proc. Natl. Acad. Sci. USA 85:3184 (1988).
2. C. Kettner, and E. Shaw, D-Phe-Pro-ArgCH$_2$Cl - A selective affinity label for thrombin, Thromb. Res. 14:969 (1979).
3. C. Kettner, and A.B. Shenvi, Inhibition of the serine proteases, leukocyte elastase, pancreatic elastase, cathepsin G, and chymotrypsin, by peptide boronic acids, J. Biol. Chem. 259:15106 (1984).
4. W.W. Bachovchin, B. Wyl-Wong, S. Farr-Jones, A.B. Shenvi, and C. Kettner, ^{15}N NMR spectrometry of the active-site histidyl residue of serine proteases in complexes formed with peptide boronic acids inhibitors, Biochemistry 27:7689 (1988).
5. R. Bone, A.B. Shenvi, C.A. Kettner, and D.A. Agard, Serine protease mechanism: Structure of an inhibitory complex of α-lytic protease and a tightly bound peptide boronic acid, Biochemistry 26:7609 (1987).
6. C.A. Kettner, L. Mersinger, and R. Knabb, Selective inhibition of thrombin by peptides of boroArginine, J. Biol. Chem. 265:18289 (1990).
7. R.M. Knabb, C.A. Kettner, P.B.M.W.M. Timmermans, and T. Reilly, *In vivo* characterization of a new synthetic thrombin inhibitor, Thromb. Haemost. 67:56 (1992).
8. J.W. Williams, and J.F. Morrison, The kinetics of reversible tight-binding inhibition, Methods Enzymol. 63:437 (1979).
9. C.A. Kettner, R. Bone, D.A. Agard, and W. Bachovchin, Kinetic properties of the binding of peptide boronic acids to α-lytic protease, Biochemistry 27,7682 (1988).
10. J.V. Schloss, Modern aspects of enzyme inhibition with particular emphasis on reaction-intermediate analogs and other potent, reversible inhibitors, in Target Sites of Herbicide Action (Boger, P., and Sandmann, G., eds) pp. 165-245, CRC Press, Inc. Boca Raton, EL.
11. S. Wessler, and S.N. Gitel, Pharmacology of heparin and warfarin, J. Am. Coll. Cardiol. 8:10B (1986).
12. P. Brill-Edwards, J. Van Ryn-McKenna, L. Cai, F.A. Ofosu, J. Hirsh, and M.R. Buchanan, Antithrombin III independent thrombin inhibitors prevent thrombin growth more effectively than heparin: Implications for antithrombotic therapy. Thromb. Haemost. 66:829 (1991).
13. J.I. Weitz, M. Hudoba, D. Massel, J. Maraganore, and J. Hirsh, Clot-bound thrombin is protected from inhibition by heparin-antithrombin III but is susceptible to inactivation by antithrombin III-independent inhibitors, J. Clin. Invest. 86:385 (1990).
14. A.B. Kelly, S.R. Hanson, R. Knabb, T.M. Reilly, and L.A. Harker, Relative antithrombotic potencies and hemostatic risks of reversible D-Phe-Pro-Arg (D-FPR) antithrombin derivatives, Thromb. Haemost. 66:737 (1991).

IN VITRO AND *IN VIVO* PROPERTIES OF SYNTHETIC INHIBITORS OF THROMBIN: RECENT ADVANCES

S. Okamoto, K. Wanaka and A. Hijikata-Okunomiya*

Kobe Research Projects on Thrombosis and Haemostasis
Asahigaoka 3-15-18
Tarumi-ku, Kobe 655
*School of Allied Medical Sciences
Kobe University
Tomogaoka 7-10-2, Suma-ku
Kobe 654-01, Japan

The purpose of our study of synthetic inhibitors of thrombin was to find out novel compounds which inhibit thrombin greatly and selectively and to establish their structural features. In the process of the research, we recognized that synthetic inhibitors of thrombin should have additional properties, such that the acute and subacute toxicities are low, and that the Vd (volume of distribution) is small. We considered that our thrombin inhibitor should meet these requirements, in perspective of using it in animal experiments and further even in clinical trials.

Chemical Modification of TAMe

Nα-tosyl-L-arginine methyl ester (TAMe)[1] was reported by Sherry et al to be a substrate of and, at the same time, an inhibitor of thrombin, although TAMe had a very low antithrombin effect and high susceptibility to hydrolysis by thrombin. On the other hand, Nα-4-methoxy-benzene-sulfonyl-L-arginine methyl ester[2] and Nα-tosyl-L-arginine sarcosine methyl ester[3], which are structurally analogous to TAMe, had inhibitory effects similar to that of TAMe on thrombin but markedly reduced susceptibility to hydrolysis by thrombin (Figure 1). These results suggested that thrombin inhibitors which are undegradable with thrombin can be obtained by chemically modifying the alpha-NH_2 and alpha-COOH groups of the arginine.

TAMe, as a basic structure, was modified systemically at alpha-NH_2 and alpha-COOH groups of the arginine (Figure 2)[4-6]. Substituting dansyl group for tosyl group of TAMe markedly elevated the antithrombin activity (I_{50}, $2x10^{-5}$M). Moreover, when n-butylamide[5] was substituted for its methyl ester, the antithrombin activity was further elevated (I_{50}, $2x10^{-6}$M). This n-butylamide was replaced with 4-ethylpiperidine, and the

The Design of Synthetic Inhibitors of Thrombin
Edited by G. Claeson, *et al.*, Plenum Press, New York, 1993

cyclic structure resulted in 20 times more potent antithrombin activity (I_{50}, $1x10^{-7}M$; Ki, $3.7x10^{-8}M$). The compound, designated as No.205: Nα-dansyl-L-arginine 4-ethyl-piperidine amide, inhibited other analogous enzymes (plasmin, trypsin, Xa and urokinase) relatively weakly (Ki value, 10^{-4}~$10^{-5}M$), which showed its high selectivity[7]. No.205 has the structural features of thrombin inhibitor, which was characterized as a tripod structure. It has (i) a positive charge of L-arginine (D-arginine has no antithrombin activity) at its centre, (ii) an aromatic structure at its N-terminus, and (iii) a hydrophobic structure of a certain size at its C-terminus.

Reduction in Acute and Subacute Toxicities and in Volume of Distribution

No.205 had a potent antithrombin activity and high selectivity. However, this compound showed high toxicity (LD_{50}, 80mg/kg IP in mice) and high Vd (volume of distribution in the body). These properties of this compound required a deliberate protocol when it was used in animal experiments.

In the studies of No.205 and homologous compounds, we found that there was a correlation between the acute toxicity and the anti-pseudocholinesterase action. To exclude this anti-pseudocholinesterase action, $-CH_2COOH$ group was introduced into the carboxyl-

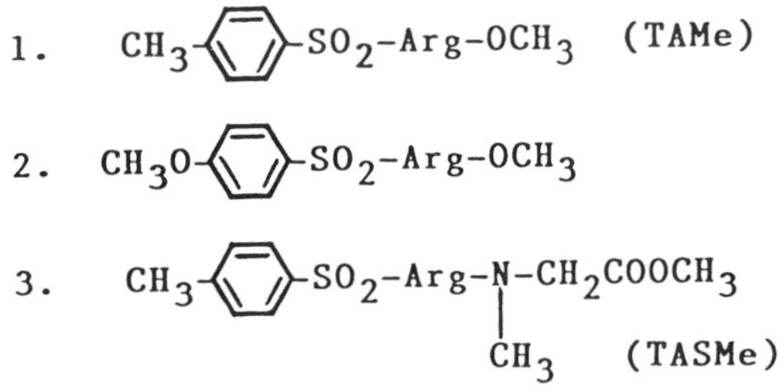

1. CH_3-C₆H₄-SO_2-Arg-OCH_3 (TAMe)

2. CH_3O-C₆H₄-SO_2-Arg-OCH_3

3. CH_3-C₆H₄-SO_2-Arg-N(CH_3)-CH_2COOCH_3 (TASMe)

Figure 1 TAMe and its homologues.

protecting group, and No.407, Nα-6,7-dimethoxynapthalene-2-sulfonyl-L-arginyl-N-methoxy-ethylglycine[7] was synthesized (Figure 2). In the step of chemical modification, we recognized that, in No.407, its anti-pseudocholinesterase activity was so weak as to be negligible, although that antithrombin activity was maintained high. The acute toxicity was greatly reduced and Vd was also reduced[8]. Replacement of the N-methoxyethylglycine of No.407 with 4-methylpiperidine with -COOH at position 2 yielded a compound with almost the same properties. Thus the introduction of a carboxyl group in the carboxyl-protecting group of arginine was important in reducing toxicity. After a series of pharmacological and adverse effect studies of synthetic inhibitors, we finally obtained No.805: (2R,4R)-4-methyl-1-[N^2-[(3-methyl-1,2,3,4-tetrahydro-8-quinolinyl)sulfonyl]-L-arginyl)]-2-piperidinecarboxylic acid, which had a potent antithrombin activity, high selectivity and safety (Figure 3)[9,10]. This compound was then designated as argatroban (previously reported as No.805, MD-805 or MCI-9038).

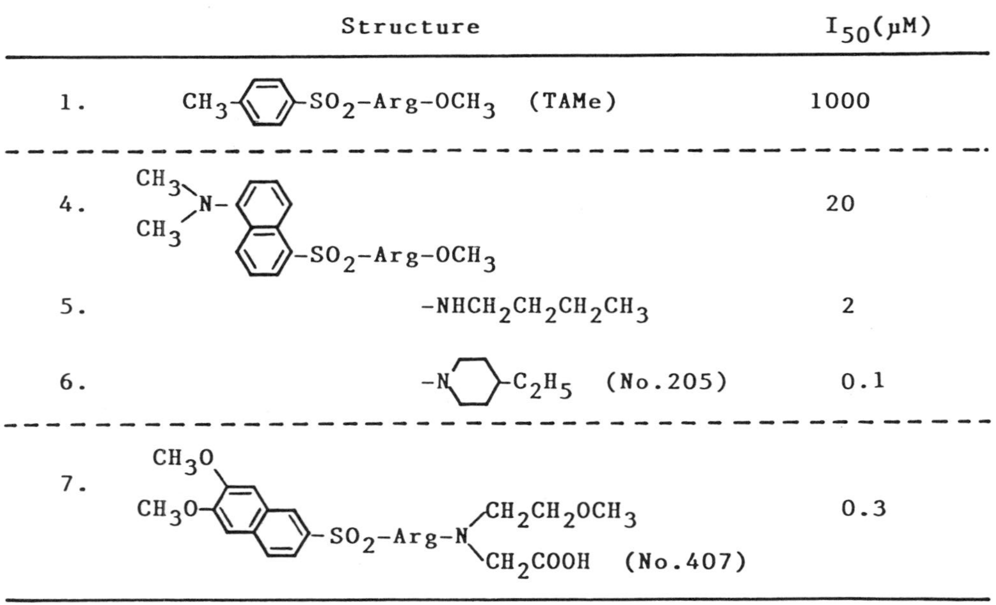

	Structure	I_{50}(μM)
1.	CH_3- -SO_2-Arg-OCH_3 (TAMe)	1000
4.	CH_3 CH_3 >N- -SO_2-Arg-OCH_3	20
5.	-$NHCH_2CH_2CH_2CH_3$	2
6.	-N -C_2H_5 (No.205)	0.1
7.	CH_3O CH_3O- -SO_2-Arg-N< $CH_2CH_2OCH_3$ CH_2COOH (No.407)	0.3

Figure 2 Structure-function relation of arginine derivatives.

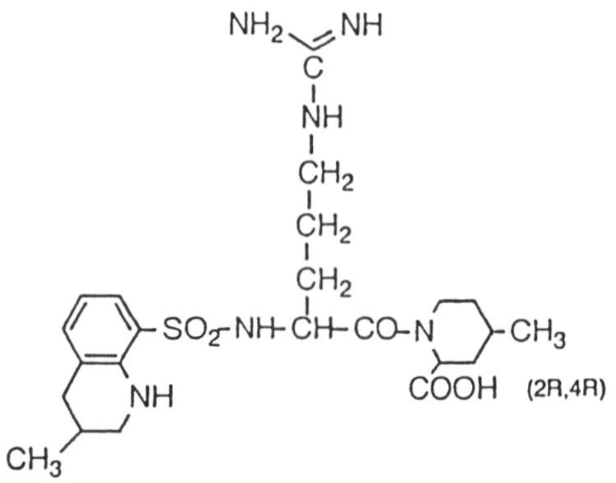

K_i (μM)				
Thrombin	Trypsin	Plasmin	Factor Xa	Plasma Kallikrein
0.019 (Bovine) 0.038 (Human)	5.0	800	210	1,500

Figure 3 Nature of argatroban.

Decoding Mechanism of Thrombin, and Thrombin Inhibitor as Noise

Thrombin hydrolyzes the fibrinogen Aα chain at the site of Arg^{16}-Gly^{17} and fibrinogen Bß at Arg^{14}-Gly^{15}, with the release of fibrinopeptides A and B, forming fibrin. In view of the fact that various kinds of animals have the similar amino acid sequence of fibrinopeptide A, Blombäck et al synthesized thrombin inhibitors which consisted of peptides, and demonstrated that Arg and the Phe at its N-terminus play an important role in the interaction between thrombin and fibrinogen[11]. The above suggested that in its active centre thrombin has a portion that has a mechanism of recognizing Arg and the aromatic structure bound to its N-terminus. Our studies also showed that Arg and the aromatic structure at its N-terminus are important in the interaction between thrombin and its inhibitors.

The 4-methyl-2-piperidinecarboxylic acid portion includes 4 types of stereoisomers[10]. When the 4-methyl-2-piperidinecarboxylic acid took the configuration of 2R and 4R, the most potent antithrombin activity (Ki, 0.019μM) was exerted. When this portion took the configuration of 2S and 4S, in which the lowest antithrombin activity was exerted, Ki value was 280 μM. The former was Ca. 10,000 times more potent than the latter (Figure 4). These results show that the active site is able to distinguish precisely the stereoisomers at the C-terminus of arginine.

Recent X-ray analysis of the complex of argatroban with thrombin revealed the mode of binding of argatroban to the active site of thrombin[12]. It was found that two hydrophobic pockets (P-pocket and D-pocket) are present near the active centre of

	Ki for thrombin (μM)
2R , 4R	0.019
2R , 4S	0.24
2S , 4R	1.9
2S , 4S	280

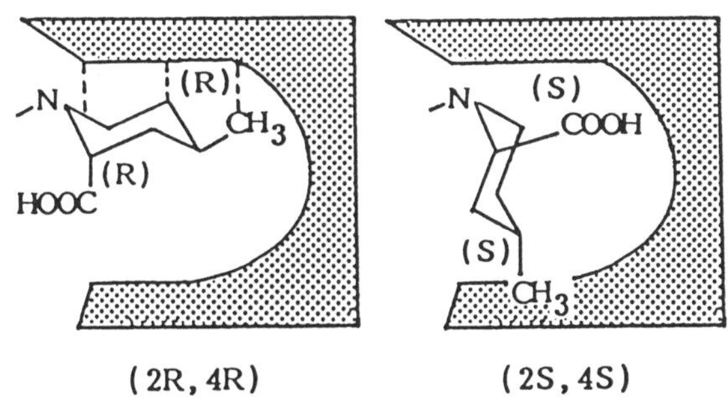

Figure 4 Stereo-isomers of piperidine derivatives and Ki values for thrombin.

thrombin and that a portion of 4-methylpiperidine of argatroban fits into the shallow pocket (P-pocket), whereas R-COOH at position 2 is located externally, thus preventing steric hindrance. The positively charged guanidino group of arginine binds to negatively charged Asp[199] in the active centre and 1,2,3,4-tetrahydroquinoline ring at the N-terminus of arginine fits into the D-pocket that contains of Trp[227], thus inducing hydrophobic interaction. This confirmed that argatroban binds directly to the active centre of the enzyme, and the finding from the structure-activity relationship studies contributed to determining the stereostructure of the active centre of the enzyme.

Antithrombotic Action of Argatroban

Thrombin plays the leading part of blood coagulation. Argatroban exerts an antithrombosis action by strongly suppressing thrombin action: i.e. (1) conversion of fibrinogen to fibrin[10], (2) activation of factor XIII[13] and (3) platelet aggregation[14]. Unlike heparin, (4) argatroban requires no cofactor, i.e. antithrombin III (ATIII), for its action[15]. Therefore, argatroban is expected to exhibit less individual variation in its efficacy in the patient. In fact, argatroban reveals an antithrombosis action regardless of the amount of ATIII, providing a stable pharmacological effect[16].

(A) Chronic Arterial Occlusion

Argatroban was approved as a new drug by the Japanese Government in 1990. It is now clinically used as a remedy for chronic arterial occlusion. This disease is a chronic condition that is often associated with ulcer or necrosis in the lower extremities. The characteristics of the disease are intravascular thrombus formation, and repeated aggravation and recurrences. Argatroban inhibited the progress of the disease significantly in rat model for peripheral arterial occlusion[17]. Clinical trials of argatroban were conducted in patients with chronic arterial occlusion. The administration of argatroban (20 mg daily in two divided doses by continuous intravenous infusion for 4 weeks) markedly improved ischaemic ulcer, pain and cold sensation, and the drug was confirmed to be useful clinically[18,19]. These results suggest that thrombin is involved in the mechanism of aggravation of chronic arterial occlusion.

(B) Cerebral Thrombosis

It has been reported that increased fibrinopeptide A (FPA) and thrombin-antithrombin III complex (TAT) occur in acute-phase cerebral thrombosis[20,21]. The increased levels of FPA are maintained over several weeks after onset of the disease. This suggests that thrombin continues to be active after onset of the disease, thus providing a state of accelerated blood coagulation. Since argatroban is powerful in antithrombin properties *in vitro* and demonstrated an antithrombosis action in animal experiments, its efficacy on thrombosis such as cerebral thrombosis is now being evaluated in clinical trials. Results obtained are distinctly promising. Tanaka et al reported that FPA was significantly reduced by administration of argatroban in patients with acute-phase cerebral thrombosis[22], and Imiya et al reported that TAT was reduced by administration of argatroban in patients with subacute cerebral thrombosis[23]. Kobayashi et al and Yonekawa et al also reported that this drug was useful in acute-phase cerebral thrombosis[24,25], which indicated that thrombin continues to exert an active role in the progress of acute-phase cerebral thrombosis. Thrombin has the action of vasoconstriction in the basilar artery[26] and stimulates endotheline release from vascular endothelial cells[27]. Argatroban greatly inhibits these actions of thrombin. This fact has aroused the keen interest of a number of investigators.

(C) Myocardial Infarction

Thrombolytic therapy has recently been used extensively as one of the therapeutic approaches to myocardial infarction. Although this treatment provides high rates of restored patency of arteries in patients with arterial occlusion, reocclusion is a new problem of this approach. Gulba et al reported apparently increased TAT level in patients with myocardial infarction who had reocclusion after receiving thrombolytic therapy[28], and Owen et al also reported increased FPA levels in patients who underwent thrombolytic therapy[29]. Argatroban apparently inhibited reocclusion occurring after thrombolytic treatment in an experimental arterial thrombosis model[30,31]. These results suggested strongly that thrombin continued to be formed or released during thrombolytic treatment and that thrombin is involved in the mechanism of reocclusion.

Argatroban was reported to have another action, by which it accelerates thrombolysis due to thrombolytic drugs (u-PA and t-PA). In Table 1, heparin and argatroban, each amount of which was adjusted to exert a similar anticoagulant action, were compared for their effects on the time of thrombolysis induced with t-PA. As a result, argatroban clearly shortened the time of thrombolysis, whereas heparin had no such effect[13]. Argatroban accelerated thrombolysis more markedly than heparin in the model for experimental arterial thrombosis[13,30,31].

The fact that argatroban accelerated the thrombolysis induced with the thrombolytic drugs and that it inhibited reocclusion after thrombolytic treatment suggest that this drug is useful in establishing a new thrombolytic therapy.

Table 1 Effect of argatroban on clot lysis by t-PA.

Compound	Conc.	Clot lysis induced by t-PA	
		Lysis time (min)	Clotting time (min)
Argatroban	0 μM	78.3±13.8	3.8±0.2
	0.01	57.2± 8.8	4.4±0.2*
	0.03	46.4± 7.5	4.9±0.3*
	0.10	33.1± 6.5*	6.7±0.4***
	0.30	25.1± 2.4**	9.0±0.4***
Heparin	0 IU/ml	84.9± 9.4	3.5±0.2
	0.01	76.8±10.8	4.6±0.3*
	0.02	56.5± 4.2*	6.8±0.4***
	0.03	65.8± 5.7	9.0±0.6***

*($p<0.05$), **($p<0.01$), ***($p<0.001$)

Development of Other Proteinase-Selective Inhibitors

We searched for the agent that has high affinity and selectivity to thrombin, and developed successfully a small molecular inhibitor of thrombin, argatroban. The chemical structure of this inhibitor suggested that thrombin has the portion within about 15 Å in diameter in its active centre, which controls the specificity of the enzyme.

On the assumption that each of the other related enzymes also has its own decoding mechanism which is possibly inhibited by noise, i.e. chemical inhibitor[32], we searched for a selective inhibitor against each of the enzymes. Very fortunately, we were able to find a series of synthetic selective inhibitors against pseudocholinesterase, plasmin and plasma kallikrein, respectively (Figure 5).

Pseudocholinesterase inhibitor

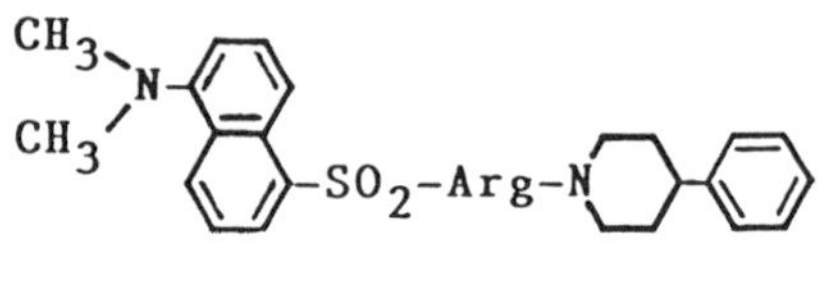

CH_3, CH_3 N – SO_2-Arg-N

Plasmin inhibitor (OS-535)

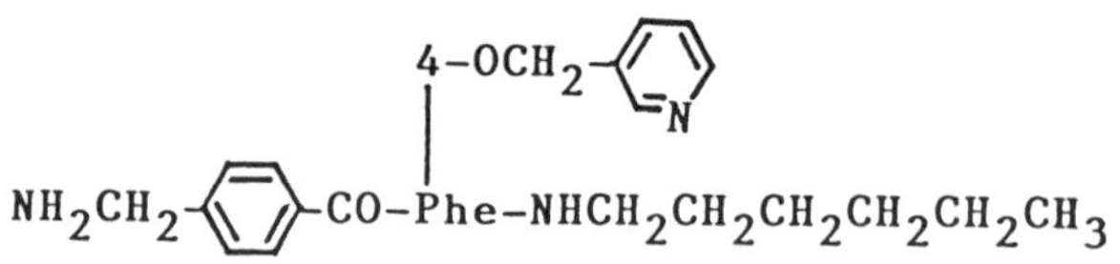

4-OCH_2 – N

NH_2CH_2 – CO-Phe-$NHCH_2CH_2CH_2CH_2CH_2CH_3$

Plasma kallikrein inhibitor (PKSI-527)

NH_2CH_2 – H – CO-Phe-NH – CH_2COOH

Figure 5 Synthetic selective inhibitors on pseudocholinesterase, plasmin and plasma kallikrein.

(A) Pseudocholinesterase

The search for thrombin inhibitors also gave us a series of anti-pseudocholinesterase. No.205 exhibited a strong antithrombin activity and anti-pseudocholinesterase activity as well. Antithrombin activity was removed by introduction of phenyl group into piperidine ring of No.205 without any decrease of anti-pseudocholinesterase activity. Thus a selective anti-pseudocholinesterase agent, Nα-dansyl-L-arginine-4-phenylpiperidine amide (Ki, 1.6×10^{-8}M) was successfully obtained[33].

(B) Plasmin

With respect to antiplasmin agents, epsilon-aminocaproic acid (EACA) and tranexamic acid (t-AMCHA) are known. However, these antiplasmin agents bind to lysine binding site apart from the active centre of plasmin to exert its antiplasmin action, which is different from argatroban that acts on the active centre of thrombin. Therefore, these antiplasmin agents inhibited only fibrin degradation due to plasmin (I_{50}, $6x10^{-5}$M) and very rarely inhibited hydrolysis of synthetic substrates and fibrinogen and kinin formation induced with plasmin[34].

After the search for antiplasmin agents that act on the active centre and selectively inhibit plasmin, we finally found out OS-535: Nα-(4-aminomethylbenzoyl)-4-(3-picolyloxy)-L-phenylalanine-n-hexylamide[35]. OS-535 inhibited not only degradation of fibrin by plasmin (I_{50}, $2.9x10^{-6}$M) but also other actions of plasmin, which could not be inhibited with t-AMCHA and other plasmin inhibitors: OS-535 at a dose of approximately 1/40 of that of t-AMCHA inhibited experimental ascites tumour growth, which was accompanied with increased plasmin activity in ascites[36]. Results obtained indicated that OS-535 is considered promising as an antiplasmin agent.

(C) Plasma Kallikrein

Aprotinin is well known as an antikallikrein agent. This drug, however, more strongly inhibits glandular kallikrein and plasmin rather than plasma kallikrein[37]. Our study of potent and selective inhibitor of plasma kallikrein finally reached to PKSI-527: Nα-(trans-4-aminomethyl-cyclohexylcarbonyl)-L-phenylalanine 4-carboxymethylanilide. PKSI-527 greatly inhibited plasma kallikrein (Ki, $8.6x10^{-7}$M) but only weakly inhibited other related enzymes in the blood, and thereby this compound was considered to be a selective inhibitor of plasma kallikrein[38]. PKSI-527 inhibited blood coagulation, fibrinolysis and kinin formation through activation of factor XII, thus indicating the important role of plasma kallikrein in the mechanism involving factor XII[39]. How far plasma kallikrein is implicated in various diseases is a problem remaining unsolved, but PSKI-527 may be useful in providing insights into the problem.

CONCLUSION

We studied compounds, each of which inhibits a certain enzyme greatly and selectively, and determined their structural formulas. The structural formula of thrombin inhibitor suggested that a proteinase has a portion within 15 Å of its active centre, where structural information is precisely decoded, although its mechanism and the structure of the portion probably differ from enzyme to enzyme, and also that a small molecular selective inhibitor against each enzyme can be present. Moreover, these results suggested that such synthetic selective inhibitors of enzymes are useful in elucidating the physiological and pathological roles of each enzyme and providing clinically useful remedies.

REFERENCES

1. S. Sherry, and W. Troll, The action of thrombin on synthetic substrates, J. Biol. Chem. 208:95 (1954).
2. S. Okamoto, A. Hijikata, K. Ikezawa, E. Mori, R. Kikumoto, Y. Tamao, K. Ohkubo, T. Tezuka, and S. Tonomura, Structure-activity relationship in a

series of synthetic thrombin inhibitors: No.205, No.407 and No.700, Blood & Vessel 11:230 (1980).
3. M.J. Weinstein, and R.F. Doolittle, Differential specificities of thrombin, plasmin and trypsin with regard to synthetic and natural substrates and inhibitors, Biochim. Biophys. Acta. 258:577 (1972).
4. S. Okamoto, K. Kinjo, A. Hijikata, R. Kikumoto, Y. Tamao, K. Ohkubo, and S. Tonomura, Thrombin inhibitors. 1. Ester derivatives of Nα-(arylsulfonyl)-L-arginine, J. Med. Chem. 23:827 (1980).
5. R. Kikumoto, Y. Tamao, K. Ohkubo, T. Tezuka, S. Tonomura, S. Okamoto, Y. Funahara, and A. Hijikata, Thrombin inhibitors. 2. Amide derivatives of Nα-substituted L-arginine, J. Med. Chem. 23:830 (1980).
6. R. Kikumoto, Y. Tamao, K. Ohkuba, T. Tezuka, S. Tonomura, S. Okamoto, and A. Hijikata, Thrombin inhibitors. 3. Carboxyl-containing amide derivatives of Nα-substituted L-arginine, J. Med. Chem. 23:1293 (1980).
7. A. Hijikata, S. Okamoto, R. Kikumoto, and Y. Tamao, Kinetic studies on the selectivity of a synthetic thrombin-inhibitor using synthetic peptide substrates, Thromb. Haemostas. 42:1039 (1979).
8. K. Oda, K. Ohtsu, Y. Tamao, R. Kikumoto, A. Hijikata, K. Kinjo, and S. Okamoto, Comparison of plasma levels and excretory routes between No.189 and No.407, potent thrombin inhibitors, Kobe J. Med. Sci. 26:11 (1980).
9. S. Okamoto, A. Hijikata, R. Kikumoto, S. Tonomura, H. Hara, K. Ninomiya, A. Maruyama, M. Sugano, and Y. Tamao, Potent inhibition of thrombin by the newly synthesized arginine derivative No.805, The importance of stereo-structure of its hydrophobic carboxamide portion, Biochem. Biophys. Res. Commun. 101:440 (1981).
10. R. Kikumoto, Y. Tamao, T. Tezuka, S. Tonomura, H. Hara, K. Ninomiya, A. Hijikata, and S. Okamoto, Selective inhibition of thrombin by (2R,4R)-4-methyl-1-[N^2-[3-methyl-1,2,3,4-tetrahydro-8-quinolinyl)sulfonyl]-L-arginyl)]-2-piperidinecarboxylic acid, Biochemistry 23:85 (1984).
11. B. Blombäck, M. Blombäck, P. Olsson, L. Svendsen, and G. Åberg, Synthetic peptides with anticoagulant and vasodilating activity, Scand. J. Clin. Lab. Invest. (Suppl.) 107:59 (1969).
12. D.W. Banner, and P. Hadvary, Crystallographic analysis at 3.0-Å resolution of the binding to human thrombin of four active site-directed inhibitors, J. Biol. Chem. 266:20085 (1991).
13. Y. Tamao, T. Yamamoto, R. Kikumoto, H. Hara, J. Itoh, T. Hirata, K. Mineo, and S. Okamoto, Effect of a selective thrombin inhibitor MCI-9038 on fibrinolysis *in vitro* and *in vivo*, Thromb. Haemostas. 56:28 (1986).
14. H. Hara, Y. Tamao, and R. Kikumoto, Effect of argipidine (MD-805) on platelet function, Jpn. Pharmacol. Ther. (Suppl.5) 14:875 (1986).
15. T. Kumada, and Y. Abiko, Comparative study on heparin and a synthetic thrombin inhibitor No.805 (MD-805) in experimental antithrombin III-deficient animals, Thromb. Res. 24:285 (1981).
16. K. Fukutake, M. Tateyama, K. Amano, M. Ueda, and M. Fujimaki, Clinical pharmacology of argatroban - selective antithrombin agent - Comparative study with heparin. Cardioangiology 29:190 (1991).
17. M. Iwaoto, T. Hara, H. Ogawa, and M. Tomikawa, Prophylactic effect of argipidine on development of lesions in rat peripheral arterial occlusion model, Jpn. Pharmacol. Ther. (Suppl.5) 14:903 (1986).
18. T. Tanabe, Y. Mishima, K. Furukawa, S. Sakaguchi, K. Kamiya, S. Shionoya, T. Katsumura, and A. Kusaba, Clinical results of MD-805, antithrombin agent, on chronic arterial occlusion. A multicentre cooperative study, J. Clin. Thr. Med. 2:1645 (1986).

19. T. Tanabe, K. Yasuda, M. Sakuma, H. Kubota, N. Kiyota, T. Imai, M. Kawabata, M. Matsuyama, M. Kawabata, S. Kuroshima, T. Shiono, T. Takahashi, Y. Marda, T. Kawakami, N. Kashimura, and S. Machida, Clinical experience of MD-805, antithrombin agent, on chronic arterial occlusion, J. Clin. Thr. Med. 2:1635 (1986).
20. W.M. Feinberg, D.C. Bruck, M.E. Ring, and J.J. Corrigan Jr, Hemostatic markers in acute stroke, Stroke 20:592 (1989).
21. K. Iijima, Activated coagulation and fibrinolytic factors. Detection of protease/inhibitor complex, Jpn. J. Clin. Pathol. 36:623 (1988).
22. T. Tanaka, N, Kawahata, L. Shin, K. Takasuka, Y. Nishimura, N. Yoshikawa, and T. Komatsu, Therapeutic effect of argipidine (MD-805), antithrombin agent, in the acute stage of cerebral thrombosis, J. Clin. Thr. Med. 3:133 (1987).
23. M. Imiya, and T. Matsuo, *In vitro* and *in vivo* inhibition of platelet aggregation by argatroban (MD-805) in a patient with subacute cerebral thrombosis, Med. Consult. New Remed. 29:911 (1992).
24. Y. Yonekawa, H. Handa, S. Okamoto, Y. Kamijo, Y. Oda, J. Ishikawa, H. Tsuda, Y. Shimizu, M. Satoh, T. Yamagami, I. Yano, Y. Horikawa, and E. Tsuda, Treatment of cerebral infarction in the acute stage with synthetic antithrombin MD -805: Clinical study among multiple institutions, Arch. Jpn. Chir. 55:711 (1986).
25. S. Kobayashi, M. Kitani, S. Yamaguchi, T. Suzuki, K. Okada, and T. Tsunematsu, Effects of an antithrombotic agent (MD-805) on progressing cerebral thrombosis, Thromb. Res. 53:305 (1989).
26. K. Nakamura, Y. Hatano, and K. Mori, Thrombin-induced vasoconstriction in isolated cerebral arteries and the influence of a synthetic thrombin inhibitor, Thromb. Res. 40:715 (1985).
27. I. Maruyama, Synthetic anticoagulant, Jpn. J. Clin. Hematol. 31:776 (1990).
28. D.C. Gulba, M. Barthels, G-H. Reil, and P.R. Lichtlen, Thrombin/antithrombin-III complex level as early indicator of reocclusion after successful thrombolysis, Lancet 9:97 (1988).
29. J. Owen, K.D. Friedman, B.A. Grossman, C. Wilkins, A.D. Berke, and E.R. Powers, Thrombolytic therapy with tissue plasminogen activator or streptokinase induces transient thrombin activity, Blood 72:616 (1988).
30. I-K. Jang, H.K. Gold, R.C. Leinbach, J.T. Fallon, and D. Collen, *In vivo* thrombin inhibition enhances and sustains arterial recanalization with recombinant tossie-type plasminogen activator, Circulation Res. 67:1552 (1990).
31. J. Schnieder, Heparin and the thrombin inhibitor argatroban enhance fibrinolysis by infused or bolus-injected saruplase (r-scu-PA) in rabbit femoral artery thrombosis, Thromb. Res. 64:677 (1991).
32. S. Okamoto, and A. Hijikata, Drug design VI. In: Rational approach to proteinase inhibitors, E.J. Ariens ed. Academic Press, New York, pp.143 (1975).
33. A. Hijikata-Okunomiya, S. Okamoto, Y. Tamao, and R. Kikumoto, Nα-Dansyl-L-arginine 4-phenylpiperidine amide, J. Biol. Chem. 263:11269 (1988).
34. K. Wanaka, S. Okamoto, A. Okunomiya, M. Bohgaki, T. Naito, and Y. Okada, Trial to analyse the substrate-multiplicity of plasmin by using a novel synthetic inhibitor; with reference to amidolysis, fibrino(geno)lysis and bradykinin formation, Blood and Vessel 19:368 (1988).
35. S. Okamoto, K. Wanaka, U. Okamoto, Y. Okada, N. Horie, and A. Hijikata-Okunomiya, Further studies on a newly synthetized active centre-directed plasmin inhibitor, Thromb. Haemostas. 65:886 (1991).
36. S. Okamoto, N. Horie, K. Wanaka, U. Okamoto, M. Bohgaki, and A. Hijikata-Okunomiya, A novel synthetic potent plasmin-inhibitor inhibits tumor growth in Sarcoma-180 bearing mice, Thromb. Haemostas. 62:40 (1989).

37. H. Fritz, and G. Wunderer, Biochemistry and application of aprotinin, the kallikrein inhibitor from bovine organs, Drug Res. 33:479 (1983).
38. S. Okamoto, U. Okamoto, K. Wanaka, A. Hijikata-Okunomiya, M. Bohgaki, T. Naito, N. Horie, and Y. Okada, Kinins V (part B), in: Highly selective synthetic inhibitors with regard to plasma kallikrein activities, K. Abe et al. eds. Plenum Publishing Co., New York, pp.29 (1989).
39. K. Wanaka, S. Okamoto, M. Bohgaki, A. Hijikata-Okunomiya, T. Naito, and Y. Okada, Effect of a highly selective plasma kallikrein synthetic inhibitor on contact activation relating to kinin generation, coagulation and fibrinolysis, Thromb. Res. 57:889 (1990).

THE USE OF ISOSTERIC BONDS IN THE DESIGN OF THROMBIN INHIBITORS

M.F. Scully, J. Deadman, L. Cheng, C.A. Goodwin, V. Ellis, V.V. Kakkar and G. Claeson

Thrombosis Research Institute
Manresa Road
London SW3 6LR, UK

INTRODUCTION

Pseudopeptide bonds have been extensively investigated[1] to give increased metabolic stability of peptide-based inhibitors while capitalising on the binding energy of the P and P' peptide chains to the host enzyme. Reduced amide bonds (CH_2NH) are the simplest modification and can be formed by reduction of the carbonyl of the parent peptide or incorporated into the growing peptide chain by reductive alkylation in solution or on solid phase[2]. Recent studies have shown that the increased flexibility in comparison to the amide backbone is such that the residue adopts mainly unnatural conformations, with a tendency to stabilised ß-turns[3]. Also the loss of the intraresidue H-bonding, possible for the amide carbonyl, may result in loss of binding energy[4]. The analogous retroamide bonds ($NHCH_2$) were found to favour ß-sheet formation, while ß-turns were particularly destabilised. Replacement of the scissile bond by an E-alkene (CH=CH), although mimicking the amide bond angles and bond strength[5], is complicated by isomerisation of the double bond into conjugation with the carbonyl group. Dipole moment calculations for the fluoroolefins (FC=CH) showed them to be electronically almost identical to amides, however the dipeptide unit was only accessible via an elegant synthetic protocol of nine or ten steps, giving only 1% or in another case 30% yield[6]. Electron withdrawing substituents, such as fluorine, next to the reactive carbonyl, have been incorporated to enhance the electrophilicity of the carbon so as to favour attack by nucleophilic residues of the catalytic enzyme. For example, monofluoroketones (COCHF) can be synthesised by electrophilic fluorination of *tert*-butyldimethylsilyl enol ethers in nearly 30% overall yield of the tripeptide[7,8]. Difluoromethylene retroamides ($COCF_2CH_2NH$) were efficiently synthesised by a two-step Reformatsky reaction in good yields (>40%)[8]. (CH_2O) bond replacements are accessible by an intramolecular modification of the Williamson ether synthesis, giving a delta lactam intermediate, with subsequent hydrolysis of the cis-lactam bond to give the required pseudodipeptide[8,9]. Recently aminonitrile pseudopeptides

The Design of Synthetic Inhibitors of Thrombin
Edited by G. Claeson, *et al.*, Plenum Press, New York, 1993

(CH(CN)NH) were synthesised by an assymetric Strecker synthesis, with trimethylsilyl-cyanide as the source of nitrile, in good yield without racemisation[10]. Here the nitrile functionality retains some of the hydrogen bonding properties of the substrate.

Synthesis of Peptide Inhibitors Containing a Ketomethylene Isosteric Bond

In our studies of inhibitors of serine proteases at the Thrombosis Research Institute over the last decade, we have studied the ketomethylene ($COCH_2$) isostere, which are good substrate analogues which, on interaction with the active site serine, reversibly form hemiacetal structures, with the anion presumably stabilised by hydrogen bonding in the so-called oxyanion-binding pocket of the enzyme. The Dakin-West reaction, first described in 1928, is the classical route to this isosteric bond. The first step involves formation of an oxazolinone (azolactone, 2 Scheme 1) by cyclisation of the N-acyl-mixed anhydride of the amino acid in pyridine at 100°C. In fact, formation of the oxazolinone is fast, and not rate determining, and has since been carried out under milder conditions, hence early studies by Szelke in association with this Institute[11] used DPECI-catalysed cyclisation of N α-formyl-L-(NG,NG-dibenzyl-oxycarbonyl) arginine. Addition of DMAP/pyridine causes deprotonation of C-2 and the enol can be acylated by, for instance, an ester of succinoyl chloride. Steglich and Hofle have studied this reaction by Nmr and found that 3 species are formed (3,4 and 5, Fig.1) of which the pyridyl adduct (3, Fig.1) may be a reversibly formed by-product from (4, Fig.1).

Overall, the rate determining step of O to C acyl migration gives the product of the type (5a, Fig.1), which can be decarboxylated by AcOH at 22°C for 16h to give the pseudodipeptide (6a). However, the overall yield for the pseudodipeptide (6a) was at most 27%, and in the case of an attempt to synthesise Arg-k-Ser using the acid chloride $C1OCCH_2CH_2$-)CO_2Tce)CH_2OBzl, failed completely.

McMurray and Dyckes reported[12] that the phenyl group of 2-phenyl-5-(4H)-oxazolones (2b, Fig.1) activates the intermediate carbanion, presumably by resonance delocalisation of the negative charge, such that acylation and decarboxylation are facile and occur in good yield. Such 2-phenyl oxazolones are accessible by cyclisation of the N-benzoyl amino acids, however removal of the N-benzoyl-protecting group from the resulting dipeptide prior to elongation of the peptide chain required forcing conditions, frequently leading to modification of side chains, while N-urethane protecting groups were unstable to the Dakin-West reaction conditions. Also, removal of N-Boc groups by TFA in the presence of the ketomethylene moiety frequently led to decomposition. For these reactions we decided to try if the peptide chain itself could be used to protect the C-2 position.

Hence, for access to Arg-K-Gly isosteres we performed the Dakin-West reaction with Boc-D-$(aa)_1(aa)_2$-Arg(Mtr)-OH (1 Fig.1, [R=Boc $(aa)_1(aa_2R'=(CH_2)_3$G-Mtr) and for Gpa-k-Gly derivatives we used Boc-D-(aa)-$(aa)_2$(p-NO_2)Phe-OH (1 Scheme 1, R'=Ar-NO_2 $_1$R=Boc$(aa)_1(aa)_2$]), giving yields in one pot of 80-94% and 94-97% respectively for the pseudo-tetrapeptide unit (6c, Fig.1). Also, it was reported that the powerful electron withdrawing properties of the alkyl urethane groups tend to make the ketone carbonyl susceptible to nucleophilic attack under the coupling conditions required for C-terminal elongation of the pseudo dipeptide unit, such that for R''=$CH_2CH_2CH_2CO_2H$, DMAP had to be used as catalyst to increase the rate of acylation[13]. In preference then, we carried out acylations with the fully derivatised succinate; however, attempted acylation with monopiperidyl succinoyl chloride gave a mixture of products, and the best yields were achieved by hydrolysis of the C-terminal methyl ester, activation as the succinimide ester and reaction with piperidine in dimethoxyethane, in greater than 80% yield for the three steps. Subsequent removal of the Mtr and N-terminal Boc-protecting group, with TFA and thioanisole occurred in good yield to give the target compounds. The Gpa functionality

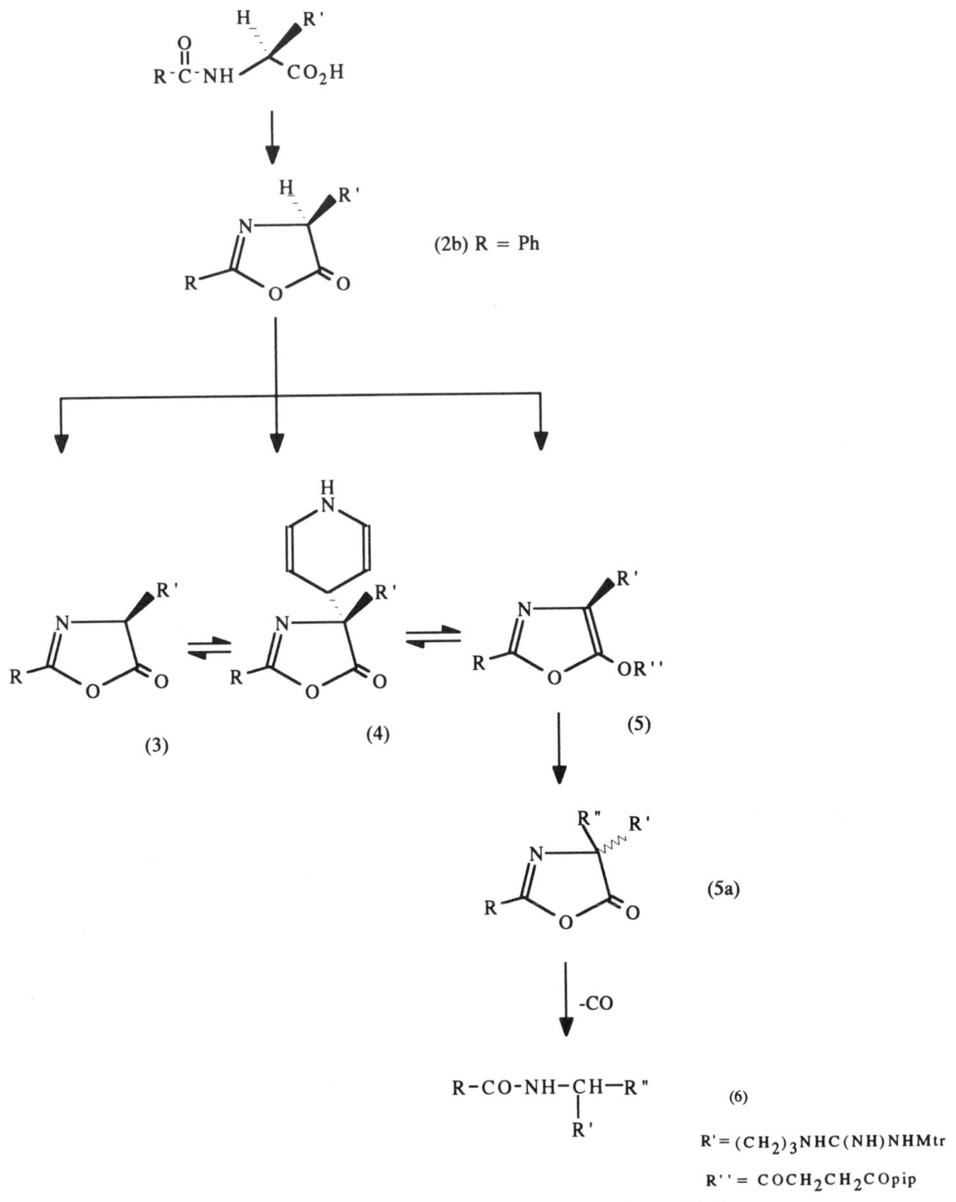

Figure 1 Approaches to the synthesis of peptide isosteres.

Table 1 Properties of various peptide isosteres in inhibiting human thrombin (HT), human factor Xa (FXa), human trypsin (Try) and prolonging the activated partial thromboplastin clotting time (APTT). Values given (μM) are the Ki (for each enzyme) or concentration required to double the clotting time. The highest concentrations tested are shown as greater than (>). Compounds with the prefix H were synthesized by M. Szelke and J. Jones at the Royal Postgraduate Medical School, Hammersmith, London.

Code	Peptide	HT		FXa	Try	APTT
H 171	Boc-DPhe-Pro-L-ArgkGly-Pro-NHEt	155		>1000	18	85
H 172	Boc-DPhe-Pro-Arg-r-Gly-Pro-OH	125		1450	65	900
H 174	H-D-Phe-Pro-Arg-Gly-Pro-OH	432		4027	291	66
H 179	H-DPhe-Pro-L-ArgkGly-Pro-NHEt	5.8		NA	17	14.6
H 180	Ac-DPhe-Pro-L-ArgkGly-Pro-NHEt	580		>1000	220	500
H 182	H-DPhe-Pro-Arg-Gly-Pro-NHEt	104		>1000	18	71
H 200	H-DPhe-Pro-L-ArgkGly-Pro-Arg-Val-NHEt	3.8		NA	NA	
H 248	H-DPhe-Pro-L-Arg(r)OHGly-Pro-NHEt	430		NA	NA	
H 250	H-DPhe-Pro-L-Arg(s)OHGly-Pro-NHEt	950		NA	NA	
H 284	Dns-Pro-ArgkGly-Pro-NHEt	73		300	NA	1000
H 333	ßNal-Gly-DL-Apa-Gly-Pro-NHEt	8.7	27.8	1.1	50	
H 335	H-D-Phe-Pro-Apa-kGly-Pro-NHEt	8.4		>100	20	30
H 337	H-DPhe-Pro-L-ArgkGly-Pyr	0.9	100	3.5	14	
H 338	H-DPhe-Pro-L-ArgkGly-Pro-Asp-OH	138	100	100	>28	
H 356	PhCO-Pro-L-ArgkGly-Pro-NHEt	>215	>100	>100	>28	
H 363	H-DPhe-Pro-L-ArgkLeu(CO_2H)OH	99	>177	>241	>151	
H 367	H-DPhe-Pro-D-ArgkGly-OH	7.8	>65	146	110	
H 372	H-DPhe-Pro-L-ArgkGly-pip	1.3	>58	27	17	
TRI 4	H-D,L-Dpa-Pro-ArgkGly-pip	0.6	NA	NA	45	
TRI 5	H-D-Phe-Pro-GpakGly-pip	100	NA	NA		
TRI 6	H-D-Dpa-Pro-ArgkGly-pip	0.2	NA	NA	8.0	
TRI 7	H-L-Dpa-Pro-ArgkGly-pip	1.7	NA	NA	100	
TRI 8	H-D,L-Fgl-Pro-ArgkGly-pip	100	NA	NA		
TRI 9	H-D,L- -Nal-Pro-ArgkGly-pip	13.5	NA	NA		
TRI14	H-D,L-ß-Nal-Pro-ArgkGly-pip	11.6	NA	NA	100	
TRI54	D-ß-Nal-Pro-Arg-k-Gly-pip	2.87		NA	NA	45

Abbreviations Apa-amidinophenylalanine; PhCO-benzoyl; pip-piperidine; pyr-pyrrolidine Dpa-diphenylalanine; Fgl-fluorenylglycine; Nal-napthylalanine; Gpa-guanidinophenylalanine; K-keto isostere; OH-hydroxy isostere.

yield to give the target compounds. The Gpa functionality was obtained by catalytic hydrogenation of the p-NO_2 then elaboration with guanyl-1,3-dimethylpyrazole (Habbeeb's reagent) in 57% yield.

While investigating the influence of the P3 position for inhibition of thrombin by serine proteases we selected Phe, α-Nal (Scheme 2) and ß-Nal, Fgl and Dpa for this position. The possible significance of such modifications were highlighted by Maraganore[14] et al who found that for a series of ketomethylene-based hirulogs the P2 and P3 positions, by comparison to peptides shortened by these residues, contribute approximately 20 Kj/Mol of binding energy to the inhibitor. McMurray and Dyckes reported that although the deprotonation at C-2 involved in the Dakin-West reaction is expected to lead to a racemic product, that no separation of the pseudotripeptide product isomers was seen on HPLC. In our study the Dpa as obtained by the Strecker synthesis is racemic and a mixture of four components of very similar retention times was seen as expected, clearly indicating the racemic nature of the ketomethylene residue.

Biological Findings with Isosteres

The biological activity of the isosteric peptides was tested as Ki for inhibition of thrombin, factor Xa and trypsin (measured by Dixon plots) the ability to prolong activated partial thromboplastin time (APTT) and thrombin time (TT).

Results are shown in Table 1 of selected compounds. In general, it was found that a substantial improvement in potency and specificity was introduced by incorporation of the isosteric peptide bond (eg. compare H182 to H179). The type of isostere was important also and again, in general, the potency of keto isosteres was greater than reduced isosteres which was considerably greater than the hydroxy isostere (compare H179 to H172, H248 and H250). The keto isostere was adopted, therefore, for future development.

Modifications of the lead compound H179 were made on the amino and carboxy side of the isostere bond. The potency of the compounds was very sensitive to derivatisation of the amino terminus of H179 (viz H-D-Phe-Pro). Addition of blocking groups (Boc-H171, Ac-H180) had a negative effect. Deviation from D-Phe also disimproved the compounds (H356, H284, TRI 7,8,9,14) except in the case of ßNal (H333 compare H335) and D-Dpa (TRI -6). On the COOH side of the isostere a small improvement was seen upon addition of residues found in the α chain of fibrinogen (H200) but not upon introduction of acidic group found in the ß chain which is thought to be important for fibrinogen binding to the anionic binding site of thrombin (H338). The introduction of a heterocyclic amine on the COOH side was helpful (H337, H372). The glycine but not the proline of H179 appeared to be important for inhibitory activity (H367, H363), in the latter position piperidine was favourable (H372).

Table 2 Specificity of isosteric peptide TRI 6 (see Table 1)

Enzyme	Ki (μM)
Thrombin	0.2
Plasmin	157
Kallikrein	199
Urokinase	492
Factor Xa	149

The specificity of the best isostere (TRI6) developed was good with Ki values for other enzymes 2 to 3 orders of magnitude higher than for the target enzyme (Table 2). From the range of inhibitors available H179 was taken forward as a lead compound in the early phase of development.

Effect of H179 on Platelet Aggregation

H179 was found to inhibit the aggregation of stirred platelet-rich plasma stimulated by 0.2 units ml^{-1} thrombin with an IC_{50} of 0.4 μM (the concentration needed to give an equal effect to a 50% reduction of agonist, when present at a suboptimal concentration). The effect of H179 on ADP-induced aggregation was also studied to determine if the inhibitor had any direct effect on platelet aggregation. With 1.5 μM ADP no effect on aggregation was seen with inhibitor concentrations of up to 50 μM. The low IC_{50} obtained for thrombin-induced platelet aggregation compared with the Ki for thrombin (0.4 μM and 5.8 μM) may be due to species differences in the thrombin inhibition. The Ki values were obtained with human thrombin and platelet aggregation was routinely performed using bovine thrombin. Preliminary experiments with H162 and H163 showed that the Ki's for these compounds were approximately 3 times lower with bovine thrombin than human thrombin. Alternatively, the lower concentration of inhibitor needed to affect thrombin-induced platelet aggregation may reflect a difference in the affinity of thrombin for fibrinogen and the platelet thrombin receptor.

As well as inhibiting the polymerization of fibrin monomers, the amino terminal tetrapeptide of the fibrin α-chain has also been shown to inhibit fibrinogen binding to platelets[15]. Therefore, H200, which contains the P'_1-P'_4 residues, was tested in the platelet aggregation system. The IC_{50} for thrombin was found to be 2.3 μM, which is 6 times higher than that for H179. The P'_1-P'_4 residues of H200 do not, therefore, appear to effect either fibrin polymerization or fibrinogen binding to platelets.

Effect of H179 *ex vivo*

The effect of H179 on the clotting times of rabbit plasma *ex vitro* was determined and found to be different to that of human plasma. Doubling of the thrombin clotting time of rabbit plasma required 10 μM H179 and doubling of the APTT required 260 μM (compared to 15μM (TT), 65 μM (APTT) plasma concentrations for human).

Prior to dosage of the rabbits with H179, but subsequent to anaesthetizing and insertion of the catheter, blood samples were removed at 5 minute intervals to ensure that the procedures did not adversely affect the clotting times. The APTT was measured in duplicate, and found to be unchanged. The APTTs and thrombin times after administration of H179 are shown in Table 3. Calculation from these clotting times indicates high bioavailability upon intravenous injection of H179. H179 was also administered at different dosages and the blood levels measured as absolute concentration from prolongation of the thrombin clotting time and also by an amidolytic method[16].

Plotting log thrombin clotting time against time (Figure 2) shows that H179 has a two-phase clearance from blood. The first phase has a half-life of approximately 2 minutes and the second phase has a half-life of approximately 60 minutes. Each of these phases is responsible for the clearance of approximately 50% of the initial activity. The time taken for the coagulation times to return to normal is dose-dependent, but at the doses employed here an antithrombin effect (expressed as a doubling of the thrombin time) is maintained for up to 1 hour. Activity was also seen upon duodenal dosing at 40 mg/kg measured as increase in thrombin clotting time and inhibition of thrombin measured by amidolytic activity (Table 4).

Table 3 Time-course of H179 concentration upon iv infusion into rabbits measured by a clotting time (TCT) and amidolytic method.

Sample time (min)	[H179] (μM) from comparison with standard curves			
		TCT		Amidolytic
	0.5mg/kg	2mg/kg	4mg/kg0.5mg/kg2mg/kg4mg/kg	
1	9	36.7	-	5.718.555
5	3.5	13.8	-	3.213.027.7
15	4.8	10.5	-	1.87.320
30	2	10	-	1.75.713.2
60	2.2	7.2	-	1.55.712.2
120	2.5	2.5	-	1-7
210	-	-	-	-610

Each set of results is an average of 2 or 3 animals.

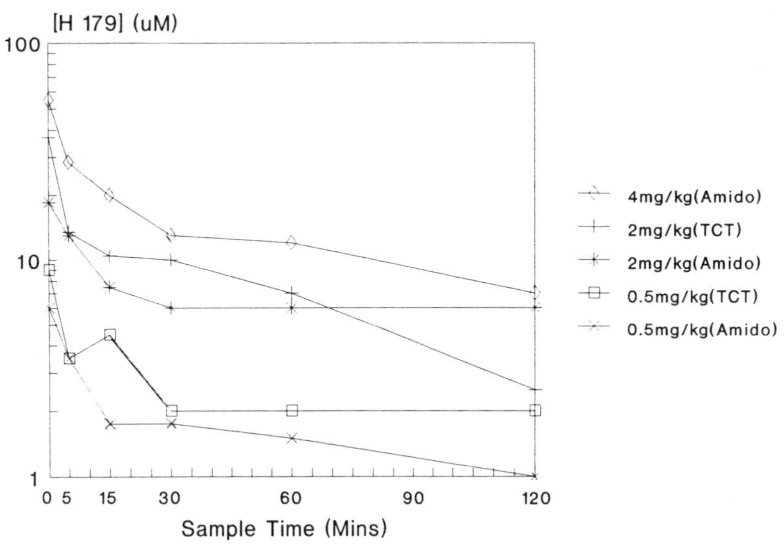

Figure 2 Clearance of H179 from blood upon iv infusion into rabbits.

Effect of H179 on the Development of the Platelet Prothrombinase Binding Site

It is known that incubation of washed platelets with a mixture of collagen plus thrombin for 10 minutes results in a large increase in platelet prothrombinase activity (reflecting the ability of the surface to support activation of prothrombin by the factor Xa/factor Va complex which is due to exposure of phosphatidylserine on the outer leaflet of the membrane). The time course of this increased activity and the effect of H179 on its development were studied.

Prostacyclin-washed platelets were incubated with 2nM α-thrombin and 10 μg ml^{-1} collagen and at timed intervals aliquots were removed into factor Xa and $CaCl_2$. The prothrombinase activity was determined by the rate of activation of added prothrombin, measured with the thrombin-specific chromogenic substrate S-2238[17]. Figure 3 shows the time course for the development of the platelet prothrombinase activity. As these experiments were performed in the absence of exogenous factor Va, the increase in activity observed is due to both development of binding sites on the platelet and the activation of released factor V. It can be seen that half the maximal activity is observed after 7 minutes incubation with collagen and thrombin. The effect of addition of varying concentrations of H179 prior to stimulation by collagen and thrombin is also shown in Figure 3. When H179 is present at approximately its Ki for thrombin (5.8 μM) there is only 15% of the maximum activity developed in 30 minutes. The concentrations of H179 required for 50% inhibition of the development of platelet prothrombinase activity is approximately 0.07 μM. This is 80 times lower than the Ki for thrombin and 6 times lower than the IC_{50} for thrombin-induced platelet aggregation.

Table 4 Inhibitory activity observed in blood upon duodenal dosing of H179 measured as increase in thrombin clotting time or reduction of amidolytic activity of thrombin.

Sample time (mins)	TCT (seconds)			Amidolytic (OD405)		
	0	10mg/kg	40mg/kg	0	10mg/kg	40mg/kg
Pre	40.3	32.2	32.4	0.210	N/A	0.210
5	47.4	33.2	45.3	0.224	N/A	0.178
15	45.5	34.4	53.0	0.217	N/A	0.173
30	43.5	34.7	59.5	0.213	N/A	0.182
60	42.5	36.6	75.6	0.205	N/A	0.163
120	38.4	36.0	96.1	0.212	N/A	0.134
205	37.3			0.196	N/A	

Assessment of Biological Activity

The pentapeptide keto isostere chosen for particular study, H179, had a Ki for thrombin of 5.8 μM which was 17.5 times less than its parent peptide H182. It also displayed much better specificity than the aromatic amine inhibitors, with a Ki for factor Xa of greater than 2,500 μM and trypsin of 17 μM. The activity of H179 in the APTT assay was also much better, with only 2.5 times the Ki for thrombin needed to double the APTT, compared to 5-100 times for the aromatic amines. The *in vivo* experiments showed that the clearance of H179 from rabbit blood was biphasic with t½s of 2 minutes and 18 minutes, each accounting for approximately 50% of the initial dose. This is in

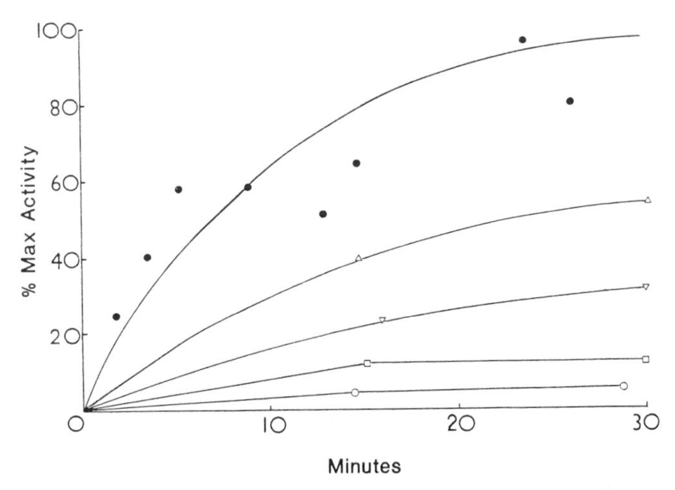

Figure 3 The effect of H179 on the development of platelet prothrombinase binding sites. Prostacyclin-washed platelets were incubated with 2nM α-thrombin and 10μg ml^{-1} collagen and at timed intervals aliquots removed for determination of prothrombinase activity (●). The incubations were also performed in the presence of varying concentrations of H179: △ 0.063μM, ▽ 0.63μM, □ 6.3μM, ○ 63μM.

contrast to the work of Collen et al[18]. with tripeptide chloromethyl ketone inhibitors who obtained a biphasic clearance with t½s of 0.7 minutes and 2.9 minutes, with the fast rate responsible for approximately 70% of the clearance. Therefore, H179 appears to be a relatively metabolically stable inhibitor, which has a high degree of specificity for thrombin and high pharmaceutical bioavailability.

The effect of H179 on thrombin-induced platelet aggregation shows that the inhibition occurs at lower concentrations of H179 than those that inhibit fibrinogen clotting. Platelet aggregation is one of the primary mechanisms in haemostasis, and it is possible therefore that the inhibition of platelet aggregation by H179 would contribute significantly to its action *in vivo*. However, it is likely that platelet activation by collagen is a more important initial stimulus for *in vivo* platelet aggregation.

Activation of platelets by thrombin alone is not sufficient to generate the maximum concentration of factor Va-Xa sites on the platelet surface. A mixture of thrombin plus collagen is needed to get this maximal stimulation. The data on the development of prothrombinase sites on the platelet after stimulation by thrombin plus collagen show that this process is relatively slow, taking 7 minutes to reach half maximal activity. This can be compared to platelet aggregation and release reaction which are essentially complete within approximately 3 minutes of stimulation. The development of this prothrombinase activity is dependent on two processes, the exposure of phosphatidylserine on the platelet surface[19] and the activation of released platelet factor V by thrombin. As the activation of factor V is relatively slow at the thrombin concentrations used, and the total platelet factor Va concentration is not sufficient to saturate all of the platelet sites, this is probably the rate limiting process. However, the exposure of phosphatidylserine by the postulated trans-bilayer "flip-flop" mechanism[19] is also likely to be a relatively slow process.

The effect of H179 on the development of platelet prothrombinase activity showed it to be a very potent inhibitor, with half-maximal effect at concentration of 0.07 μM. Thrombin-induced platelet aggregation has been shown to be inhibited by H179 with an IC_{50} of 0.4 μM. The higher potency in inhibiting the generation of prothrombinase activity is possibly due to the platelet thrombin receptor for aggregation and release being independent from, and having a different affinity to, the receptor which initiates trans-bilayer "flip-flop". However, Comfurius et al[20] have postulated that the exposure of phosphatidylserine may be due to membrane perturbations, caused by the formation of diglycerides and phosphatidic acid by the phosphatidylinositol cycle. The phosphatidylinositol cycle is intimately linked to the process of platelet activation (shape change, aggregation and release), and therefore if this cycle is responsible for the exposure of phosphatidylserine, H179 would have the same effect on both.

The major effect of H179 is more likely to be on the activation of factor V as factor V is present at low concentrations and appears to have a low affinity for thrombin. The overall effect of H179 on the generation of platelet prothrombinase activity is due to its inhibition of both platelet activation and factor V activation, this combined effect contributing to the very low concentrations needed for the inhibition.

CONCLUSION

The isostere peptide inhibitors have very low toxicity (unpublished observations), good specificity and potency and high bioavailability. By further modification (eg) by a hirudin-like structure on the COOH side of the isostere, we consider that a highly effective inhibitor will be developed with potential as an antithrombotic pharmaceutical of choice.

REFERENCES

1. A.F. Spatola, Peptide backbone modifications: A structure activity analysis of peptides containing amide bond surrogates. Conformational constraints, in "Chemistry and Biochemistry of Amino Acids, Peptides and Proteins", B, Weinstein ed, Marcel Dekker, New York, 7:267 (1983).
2. Y. Sasaki, W.A. Murphy, M.L. Heiman, V.A. Lance, and D.H. Coy, Solid phase synthesis and biological properties of (CH_2NH) pseudopeptide analogues of a highly potent somatostatin octapeptide, J. Med. Chem. 30:1162 (1987).
3. P. Dauber-Osguthorpe, M.M. Campbell, and D. Osguthorpe, Conformational nalysis of peptide surrogates, Int. J. Pept. Prot. Res. 38:357 (1991).
4. S.J. Hocart, M.V. Nekola, and D.H. Coy, Effect of the CH_2NH and CH_2NAc peptide bond isosteres on the antagonistic and histamine releasing activities of a leutenizing hormone releasing hormone analogue, J. Med. Chem. 1:1820 (1988).
5. T. Ibuka, H. Habashita, A. Otaka, and N. Fujii, A highly stereoselective synthesis of (E)-alkene dipeptide isosteres via organo-cyanocopper-lewis acid mediated reaction, J. Org. Chem. 56:4370 (1991).
6. T. Allmendinger, P. Furet, and E. HungerBüller, Fluoroolefin dipeptide isosteres-1. The synthesis of Gly (CF=CH)Gly and racemic Phe CC(CF=CH)Gly, Tet. Lett. 31:7297 (1990).
7. G.S. Garrett, T. Emge, S.C. Lee, E.M. Fischer, K. Dyehouse, and J.M. McIver, General synthesis of polyfunctionalized fluoromethyleneketone retroamides as potential inhibitors of thrombin, J. Org. Chem. 56:4823 (1991).

8. J.M. Altenburger, and D. Schirlin, Synthesis of the monofluoro ketone peptide isostere,Tet. Lett. 32:7255 (1991).
9. R.E. TenBrink, A method for the preparation of stereochemically defined (CH_2O) pseudopeptides, J. Org. Chem. 52:418 (1987).
10. R. Herranz, M.L. Suárez-Gea, S. Vinuesa, M.T. García-Lopez, and A. Martínez, Synthesis of [CH(CN)NH] pseudopeptides. A new peptide bond surrogate, Tett. Lett. 32:7579 (1991).
11. S.M. Szelke, Ketomethylene pseudopeptide analogue of substance P: Synthesis and biological activity, Eur. Pat. Appl. EPO11828.OA1 (1984).
12. J.S. McMurray, and D.F. Dykes, A single and convenient method for the preparation of ketomethylene peptide analogues, J. Org. Chem. 50:1112 (1985).
13. A. Ewenson, R. Laufer, M. Chorev, Z. Selinger, and C. Gilon, Ketomethylene pseudopeptide analogues of substance P: synthesis and biological activity, J. Med. Chem. 29:295 (1986).
14. P. Bourdon, J. Tablonski, B.J. Chao, and J.M. Maraganore, FEBS Lett. 294:163 (1991).
15. G.A. Marguerie, and E.F. Plow, The fibrinogen-dependent pathway of platelet aggregation, Ann. N.Y. Acad. Sci. 408:556 (1983).
16. M.F. Scully, The use of an automated analyzer in the evaluation of antithrombin III and heparin, Semin. Thromb. Haemostas. 9:309 (1985).
17. V. Ellis, M.F. Scully, and V.V. Kakkar, Inhibition of prothrombinase complex by plasma proteinase inhibitors, Biochemistry 23:5882 (1984).
18. D. Collen, O. Mastuo, J.M. Stassen, C. Kettner, and E. Shaw, *In vivo* studies of a synthetic inhibitor of thrombin, J. Lab. Clin. Med. 99:76 (1982).
19. E.M. Bevers, J. Rosing, and R.F.A. Zwaal, Development of procoagulant binding sites on the platelet surface, in Mechanisms of Stimulus Response Coupling in Platelets, eds. J. Westwick, M.F. Scully, D.E. McIntyre, ad V.V. Kakkar, pp.359-372, Plenum Press (1985).
20. P. Comfurius, E.M. Bevers, and R.F.A. Zwaal, Stimulation of prothrombinase activity of platelets and erythrocytes by sub lytic treatment with phospholipid C from Clostridium welchii, Biochem. Biophys. Res. Commun. 117:803 (1983).

SYNTHETIC THROMBIN INHIBITORS AS ANTICOAGULANTS

PHARMACOLOGICAL ASPECTS

F. Markwardt and J. Hauptmann

Institute of Pharmacology and Toxicology
Medical Academy Erfurt
Federal Republic of Germany

INTRODUCTION

Thromboembolic disorders are a major cause of morbidity and mortality in industrialized countries. For prophylaxis of thrombosis at present the so-called indirect anticoagulants of the dicoumarol type, the biopolymer heparin and several antiplatelet drugs are used. The active principle of the medicinal leech, hirudin, is available now as a recombinant product and is being tested clinically. Each of the mentioned anticoagulants has its values and limitations, based either on the chemical nature or on the mode of action. Antithrombotic regimes that do not interfere with the coagulation system are proposed; their value, however, has not yet been fully documented.

An ideal anticoagulant should, therefore, fulfil requirements such as rapid onset of action, selectivity toward the target, no side-effects, easy monitoring and oral administration (which would be an advantage over heparin and hirudin).

Enzyme inhibition is a major mechanism of drug action. The target enzyme-directed development of enzyme inhibitors as potential drugs has gained increasing importance[1-47]. The biochemical basis for the development of inhibitors of blood coagulation enzymes has been reviewed with special emphasis on the catalytic mechanism of serine proteinases[5-10].

Recent rapid development of and increasing interest in inhibitors of enzymes of the blood proteolytic systems, stimulated by the progress in knowledge on the functions of thrombin, has prompted us to an updated review of the pharmacology of synthetic inhibitors of thrombin that is intended to cover especially the literature of the last decade.

The target enzyme for these synthetic, low molecular weight molecules is thrombin, the enzyme occupying a unique position in the coagulation cascade insofar as it not only proteolytically attacks soluble substrates but also acts on cellular receptors. Selective and potent inhibitors of thrombin are assumed to become clinical useful anticoagulant and antithrombotic drugs.

The Design of Synthetic Inhibitors of Thrombin
Edited by G. Claeson, *et al.*, Plenum Press, New York, 1993

Design and Biochemistry of Synthetic Thrombin Inhibitors

Thrombin with its trypsin-like substrate specificity cleaves in its natural substrates peptide bonds of arginine, which is fixed via its guanidinoalkyl side chain in the so called specificity pocket of the enzyme. Secondary binding sites in the vicinity of this pocket are able to interact with the leaving group of the substrate and with further groups in a way that the cleavable bond comes close to the active centre serine hydroxyl.

Blockade of the active site of the enzyme should eliminate its enzymatic activity which is the basis of most of its biological effects.

The catalytic mechanism of the serine proteinases and the substrate specificity allow the design of several types of inhibitors: From the biochemical point of view the synthetic thrombin inhibitors can be classified according to the kinetics of inhibition and the type of binding or according to the structural origin of the molecules. Table 1 presents a synopsis of thrombin inhibitors according to the mode of inhibition and kinetic characteristics.

Among the inhibitors listed are relatively simple substrate-derived competitive inhibitors and a number of compounds summarized as mechanism-based inhibitors. In this group the translation state inhibitors (tight binding inhibitors) are a link to the first group. The active site-directed affinity labels with a reactive function in the molecule form another interesting subgroup; the mechanism-based inhibitors (in a stricter sense) or inverse substrates and the enzyme-activated inhibitors or suicide substrates, in which the reactive group is unmasked by the enzymatic reaction, also belong to the latter group.

Increasing knowledge on the structure and function of the active site has considerably promoted the search for potent inhibitors. Vice versa, the use of different inhibitors as probes has enabled a mapping of the active site of thrombin and of binding sites adjacent to it.

The extracellular proteinase thrombin is present in blood in form of its inactive proenzyme and is activated after physiological or pathological stimuli. However, the possible substrates of thrombin are permanently present in blood or are exposed by such stimuli. Therefore, a thrombin inhibitor must be present in blood at concentrations

Table 1 Synposis of synthetic thrombin inhibitors.

Type	Chemistry	Characteristics
Substrate-derived competitive inhibitors	peptides arginine derivatives benzamidine derivatives lysine derivatives	noncovalent, reversible
Mechanism-based inhibitors Transition state inhibitors	peptide aldehydes peptide methyl ketones peptide boronic acids	noncovalent, reversible
Affinity labels	halomethyl ketones diazomethyl ketones	covalent, irreversible
Inverse substrates	peptide carbaminic acids "inverse" carbonic acids/amides after deacylation	covalent, reversible
Enzyme-activated inhibitors	heterocyclic esters/ amides	covalent, irreversible

Table 2 Chemical names and synonyms/abbreviations of thrombin inhibitors.

Arginine derivatives

OM 189, Nα-dansyl-L-arginine-4-methylpiperidide
OM 205, DAPA, Nα-dansyl-L-arginine-4-ethylpiperidide
OM 407, Nα-6,7-dimethoxynaphthyl-2-sulfonyl-L-arginyl-N-methoxy-ethylglycine
OM 805, MD-805, MCI-9038, Novastan°, argatroban, argipidine, 4-methyl-1-(N^2-(-methyl-1,2,3,4-tetrahydro-8-quinolinyl)-sulfonyl)-L-arginyl-2-piperidine carbonic acid

GYKI 14 166, D-Phe-Pro-Arg-H, D-phenylalanyl-L-prolyl-L-argininal
GYKI 14 451, BOC-D-Phe-Pro-Arg-H, butyloxycarbonyl-D-phenylalanyl-L-prolyl-L-argininal
GYKI 14 766, D-MePhe-Pro-Arg-H, N-methyl-D-phenylalanyl-L-prolyl-L-argininal

D-Phe-Pro-ArgCN, D-phenylalanyl-L-prolyl-L-arginine nitrile
D-Phe-Pro-ArgCH_2Cl, PPACK, FPRCH_2Cl, D-phenylalanyl-L-prolyl-L-arginine chloromethyl ketone
Ac-(D)-Phe-Pro-boroArg-OH, acetyl-D-phenylalanyl-L-prolyl-L-arginyl boronic acid

Benzamidine derivatives

APPA, 4-amidinophenylpyruvic acid
TAPAP, Nα-tosyl-3-amidinophenylalanine piperidide, Tos-(mAm)Phe-Pip
NAPAP, Thromstop°, Nα-tosylglycyl-4-amidinophenylalanine piperidide, βNas-Gly-(pAm)Phe)-Pip

Compounds of other structure

FOY°, gabexate mesilate, guanidinocaproic acid-4-carboxyethylphenyl ester
FUT 195, nafamostat mesilate, 4-amidino-2-benzoylphenyl-4-guanidinobenzoate
ACITIC, 7-amino-4-chloro-3-(3-isothioureidopropoxy)isocoumarin

(In the text, abbreviations referring to structural features of the compounds are preferentially used).

adequate for inhibition and must have an adequate affinity to the enzyme. In this context the question arises whether irreversible or reversible inhibitors would be of advantage. Irreversible inhibitors equipped with a reactive group blocking by covalent binding, unfortunately not only react with the enzyme, but also with other constituents of blood and tissue by virtue of this group. Mechanism-based inhibitors will, from the biochemical point of view, be of advantage for inhibition of an enzyme activated in the way thrombin is. But, up to now, no highly potent inhibitors of this type are available. Therefore, reversible, high-affinity inhibitors with high selectivity are preferred, even when dissociation of the enzyme-inhibitor complex and restoration of enzymatic activity may occur.

A key structural feature in the small, substrate-derived inhibitors is arginine or a group mimicking this amino acid, which is governing the primary affinity of thrombin substrates.

The sites in binding of an inhibitor to the enzyme are the same as for substrate binding: The specificity pocket to which the protonated side chain of arginine is attracted, the enzyme catalytic mechanism located at the entrance of the pocket, and the areas which accommodate the leaving group of the substrate and accept the group protecting the α-nitrogen.

Table 1 shows the chemical names and synonyms/abbreviations of selected compounds either typical of the various groups of inhibitors or frequently referred to in the literature.

One line of development started with small peptides imitating the structure of fibrinopeptides or sequences around the cleavable bond in the A-chain of fibrinogen. Arising from these studies, the sequence D-Phe-Pro-Arg has been found very effective in the design of inhibitors of various types. The group of Bajusz have been working on thrombin inhibitors of the D-Phe-Pro-Arg-sequence, especially on argininals. Much biochemical and pharmacological data have been collected by them and presented. D-Phe-Pro-Arg-H was shown to be an effective inhibitor of thrombin *in vitro* and *in vivo*[11-14]. Similarly, the BOC-derivative is a potent thrombin inhibitor. Upon storage of D-Phe-Pro-Arg-H in solution a loss of activity occurs, owing to cyclisation[15]. The N-methyl derivative D-MePhe-Pro-Arg-H) does not undergo such a change and was shown to be equally effective[15,16].

In the late seventies Kettner and Shaw reported on the highly selective affinity label for thrombin, D-Phe-Pro-ArgCH_2Cl (PPACK)[17]. The second order rate constant for inactivation of thrombin is several orders of magnitude lower than that for inhibition of other trypsin-like proteinases[18]. PPACK alkylates the active site histidine. The compound is effective as thrombin inhibitor in plasma and *in vivo*, as was first demonstrated by our group and by others later on[19-21]. However, the presence of a reactive group also brings about reactions with blood plasma components, leading to relatively rapid loss of activity *in vitro* and *in vivo*. Nevertheless, the inhibitor has been shown in several biochemical studies to irreversibly inactivate thrombin and also in pharmacological investigations using animal models.

The D-Phe-Pro-Arg-sequence was also equipped with a C-terminal nitrile moiety yielding a thrombin inhibitor of notable potency[22].

Recently, the sequence has been substituted with boronic acid. A compound of the structure Ac-(D)-Phe-Pro-boroArg-OH was shown to be a highly potent thrombin inhibitor *in vitro* and *in vivo*[23]. The most potent of these slow-binding inhibitors have K_i values in the picomolar range.

Benzoylarginylfluoroalkanes are reversible inhibitors of relatively low potency[24]. Starting from the synthetic thrombin substrate Nα-tosyl-arginine methyl ester (TAME), Okamoto and his group varied extensively the carbonyl and Nα-substituents and designed competitive inhibitors showing considerable potency and selectivity[25-28]. The most potent of a series of Nα-substituted arginine amides, so called 'tripode' inhibitors, is the compound OM 805 (MCI-9038) which bears in the Nα-position methyl tetrahydroquinoline sulfonyl, and in the amide portion methyl piperidine carboxylic acid. There are marked differences in inhibitory activity between the four stereoisomers indicating the importance of the proper binding of the amide portion of the inhibitor[27]. The K_i-value is 20 nm. The compound is highly effective as an antithrombotic in animal models. Recently, the first reports on its clinical use appeared. The compound, also known as argipidine and argatroban, was marketed as Novastan in Japan in 1990.

Our group focused on derivatives of benzamidine, allowing multiple structural variations of this core structure[9,10,29-31]. Benzamidine is able to mimic the protonated basic side chain of arginine and thus it is fixed in the specificity pocket with hydrophobic bonds contributing to binding. Benzamidine itself, however, is only a weak inhibitor with a K_i value of about 200 μM.

More potent inhibitors are benzamidine derivatives with a ß-carbonyl function in the side chain[10]. Compared to isosteric compounds the carbonyl function enhances the affinity for trypsin and thrombin. Electron-accepting substituents enhancing the positive charge of the carbonyl carbon increase the inhibitory strength of the molecule. Consequently, the positively charged carbonyl function was believed to interact with a nucleophilic group of the enzyme, possibly with the active centre serine alkoxide. Formation of a tetrahedral arrangement known from the enzyme-substrate-interaction was assumed.

Already in the late sixties a derivative of benzamidine, 4-amidinophenyl pyruvic acid (APPA) had been characterized as a relatively potent reversible inhibitor of serine proteinases. It became the first synthetic inhibitor of this type to be extensively studied with pharmacological methods[33].

Next, we were interested to develop inhibitors interacting with the other secondary binding site, too. Derivatives of amidinophenyl-alkyl-carboxylic acids were synthesized whose side chain is branched via a sulfonamide group. So a new class of compounds, the carboxyl substituted α-amidinophenyl-aminoalkyl-carboxylic acids were developed[33]. These compounds are isosteric to arginine derivatives; however, the basic guanidinoalkyl side chain is replaced by a benzamidine moiety.

For further designing of selective and potent inhibitors we used 4-amidino-phenylalanine as a key building block. The most striking increase in affinity of the piperidides of amidino phenylalanine was attained after variation of the N α-substituent. Replacement of the tosyl residue for a naphthylsulfonyl residue had only minimal enhancing effect on the antithrombin activity. Therefore, instead of further variation of the aromatic ring structure an amino acid was interpositioned. From the Nα arylsulfonylated compounds we came to structures with a more extended Nα-residue, namely to pseudopeptides possessing structures similar to peptidyl substrates and inhibitors[31].

The question arises which structural elements of these derivatives account for the selective inhibitory activity of certain compounds and which interactions occur. As mentioned before, degree and selectivity of inhibition depend on the type of the amide component, the Nα-substituent and the kind of the benzamidine containing amino acid. Obviously, in the binding of arylsulfonylglycyl derivatives of piperidides of 4-amidino-phenyl-alanine, all important substrate binding sites of thrombin are involved, such as the specificity pocket, the catalytic centre and the secondary binding sites.

Inhibitors of this type contain components selected to fulfil primary and secondary subsite requirements of the target enzyme. Nα-2-(naphthyl-sulfonylglycyl)-4-amidinophenyl-alanine piperidide (NAPAP) is the most potent competitive inhibitor reported so far with a K_i-value of 6.6 nM[35].

Prototypes of thrombin inhibitors of the benzamidinine-type to be studied experimentally in more detail were 4-amidino-phenyl pyruvic acid (APPA), Nα-tosyl-3-

amidino-phenylalanine piperidide (TAPAP) and Nα-naphthylsulfonyl-glycyl-4-amidino-phenylalanine piperidide (NAPAP)[29,33,35]. Figure 1 illustrates the structures of the most potent compounds of the above series of arginine- and benzamidine-derived inhibitors.

A French group synthesized a methyl-derivative of NAPAP and demonstrated its thrombin inhibitory potency[36]. In contrast to the Nα-arylsulfonyl-aminoglycyl-amidino-phenylalanine amides the corresponding arginine amides showed only very low thrombin

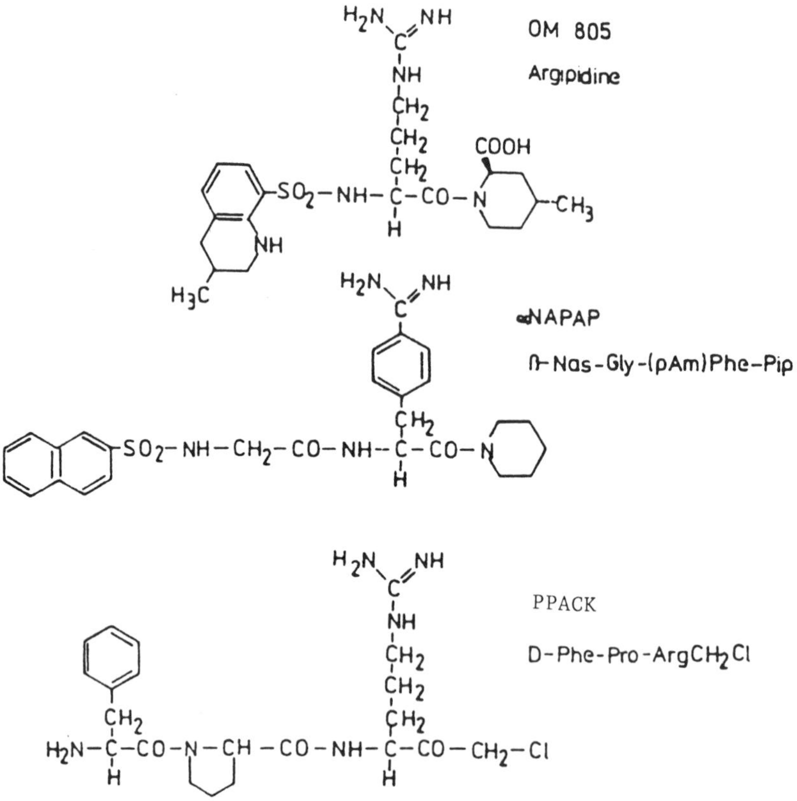

OM 805

Argipidine

αNAPAP

β-Nas-Gly-(pAm)Phe-Pip

PPACK

D-Phe-Pro-ArgCH_2Cl

Figure 1 Chemical structure of OM805, NAPAP and PPACK.

inhibitory potency[37]. Derivatives of guanidinophenylalanine, analogous to the amidinophenylalanine derivatives, were shown to be competitive inhibitors of similar or somewhat less potency[38].

Among aromatic bis-benzamidines relatively potent inhibitors of thrombin were found; their pharmacological properties, however, did not allow their *in vivo* use as anticoagulants.

The aromatic tri- and tetra-amidines reported by an Italian group are obviously not selective inhibitors of thrombin[39].

Quite another example for recent development in this field is the guanidino and isothioureide isocoumarin derivatives, proposed by Powers and co-workers[40]. Some of the compounds which represent true 'mechanism-based inhibitors or suicide substrates are readily hydrolyzed in buffer and blood plasma with this hydrolysis limiting the duration of the inhibitory action *in vitro* and *in vivo*. Chloromethylcoumarins also represent mechanism-based inhibitors[41].

It is questionable, however, whether these mechanism-based inhibitors are indeed suited as anticoagulants *in vivo* when one calculates on the basis of the kinetic constants and the concentrations necessary for anticoagulant effects *in vitro*. The competitive inhibitors of either the benzamidine- or the arginine-derivative type are 2-3 orders of magnitude more potent *in vitro*. A compound with a relatively broad inhibitory spectrum is the guanidinobenzoic ester gabexate mesilate (FOY). It was first characterized as a trypsin inhibitor and is not a selective thrombin inhibitor, nevertheless, the compound is proposed as an anticoagulant and antithrombotic agent[42-45].

Amidinophenylesters may also inhibit thrombin and prolong clotting times[46]. FUT 175, chemically amidinonaphthyl-guanidinobenzoate, a rather non-selective proteinase inhibitor, is also proposed for anticoagulant and antithrombotic use[47-49].

The above referred development shows that several groups are going beyond the synthetic and biochemical work in this field and are evaluating the potential of various types of thrombin inhibitors as anticoagulant agents *in vitro* and *in vivo*.

Characterization and Evaluation of Synthetic Thrombin Inhibitors

The various aspects of the pharmacological assessment of selective synthetic inhibitors of thrombin to be developed were included into a programme (or guidelines) by our group (Table 3). The several steps of characterization of a new potential inhibitor structure allow an overall evaluation of its potential usefulness. A considerable number of benzamidine derivatives were tested according to this programme.

Table 3 **Pharmacological screen for development and evaluation of synthetic thrombin inhibitors.**

	Subject	Procedure	Information
In vitro	isolated enzyme(s)	biochemical methods (inhibition of enzymatic activity towards substrates)	kinetics/type of inhibition affinity, specificity structure/activity relationship
Ex vivo	whole blood plasma platelets	coagulation assays (inhibition of enzymatic activity towards natural substrates in natural medium)	anticoagulant effect potency in vitro dose-effect relationship
In vivo	whole animal	standard pharmaco-toxicological methods experimental thrombosis	toxicity, side-effects pharmacodynamics pharmacokinetics

The first aspect is investigation on the stability of the inhibitor under study and its anticoagulant potency *in vitro*.

In vivo, the toxicity (lethal dose, acute non-specific pharmacodynamic effects), the pharmacokinetics and, last but not least, the antithrombotic action are of interest. Depending on the individual compound under study, special investigations, for instance on the metabolic fate, have to be carried out additionally.

According to literature data the following compounds have been studied in more detail *in vitro* and *in vivo*.

The arginine amide derivatives OM 189, OM 407, OM 805[25-27,50-53], the benzamidine derivatives APPA, TAPAP, NAPAP[6,29,33,54,55], the guanidinobenzoic ester, FOY[42,44,45] and the tripeptide derivatives D-Phe-Pro-Arg-H[11,13,21,56], D-MePhe-Pro-Arg-H[15,16], D-Phe-Pro-ArgCN[22,57,58], and D-Phe-Pro-Arg-CH2Cl[19-21].

Anticoagulant Action

The anticoagulant effect of a given thrombin inhibitor is the first biological effect to be tested after the biochemical study on inhibition of the enzymatic activity is finished. Here, the mechanism of inhibition and the inhibition constants determine the anticoagulant potency of the compound.

In order to be anticoagulantly effective, a thrombin inhibitor has to be stable in blood *in vitro* and *in vivo*. Chemical reactivity or easily hydrolyzable bonds of a compound may lead to a loss of inhibitory activity by interaction with blood constituents. This was demonstrated for several of the above mentioned thrombin inhibitors, e.g. for D-Phe-Pro-ArgCH_2Cl[19], FOY[42], mechanism-based chloromethyl-coumarins[41] and isocoumarins[40,159] and the benzamidine derivative TAPAM[60]. A loss of inhibitory activity may also be brought about by binding to plasma proteins or lipids, demonstrated for the arginine amide derivative OM 205 in a comparative study on this aspect several years ago[61].

Most of the competitive inhibitors of the arginine- and benzamidine-derivative type are stable in blood plasma and whole blood.

The inhibitors of thrombin reported that interactions with the active site of the enzyme are able to prevent the action of the enzyme on any of its substrates. Low molecular weight inhibitor molecules may reach also sites in activation complexes of coagulation factors inaccessible to the high molecular weight inhibitors of serine proteinases in plasma.

The *in vitro* analysis of the action of synthetic inhibitors reveals that all thrombin-mediated reactions are blocked.

Inhibition of thrombin simply finds its expression in the delay of fibrinogen clotting in isolated systems or in the delay of coagulation of blood samples irrespective whether thrombin is added to the sample or is generated from prothronbin via the extrinsic or the intrinsic activation pathways. More detailed studies on the influence of synthetic thrombin inhibitors on the procoagulant action of thrombin are found in a number of papers[47,62-65].

Synthetic thrombin inhibitors will also inhibit the reaction of the enzyme with its natural plasmatic inhibitor antithrombin III since enzymatic activity is involved in this reaction[66].

The question as to the importance of inhibition of thrombin-mediated feed-back reactions during prothrombin conversion for the overall anticoagulant effect has not yet been fully answered.

The K_i-values of the competitive inhibitors are predictive of their *in vitro* anticoagulant potency. Moreover, there is a parallelity of potencies in isolated systems either with synthetic substrates or with fibrinogen as substrate.

Simple coagulation assays as they are used in the clinical laboratory are quite well suited as measures of the anticoagulant potency of synthetic thrombin inhibitors.

The most sensitive clotting assay is the thrombin time, carried out commonly with 0.3-2.0 NIH units of thrombin per sample. Second in sensitivity is the activated partial thromboplastin time, whereas the prothrombin time is least sensitive. Selective thrombin inhibitors show rather uniform ratios of the effective concentrations (e.g. for doubling of clotting times) in the three assays *in vitro*.

Since the anticoagulant action of selective thrombin inhibitors is based solely on their effect on the enzymatic activity of thrombin, blockade of thrombin not only prevents plasmatic coagulation processes, but also the activation of blood platelets[11,67-75]. This is also true for tumor cell-mediated aggregation of platelets[76].

Inhibition of thrombin-induced platelet reactions, however, requires higher inhibitor concentrations than are necessary for preventing the reaction of thrombin with its substrate fibrinogen. This is due to the high affinity of thrombin to its receptors on platelets. Species-specific effects on platelets were reported for PPACK which agglutinated rabbit platelets in plasma[77]. The platelet inhibiting capacity of synthetic thrombin inhibitors will considerably contribute to their antithrombotic effects.

Irreversible inhibitors covalently blocking the active site of thrombin may principally alter the biological characteristics of the enzyme, as was demonstrated with PPACK-thrombin being able to antagonize the action of thrombin on platelets[69]. Such an effect would result in an anticoagulant action of the active-site blocked thrombin. On the other hand, thrombin irreversibly blocked by PPACK was still bound by the endothelium as is the case with active thrombin and was released from this binding by heparin[78].

For the competitive inhibitors it is obvious that the anticoagulant potency is strongly correlated to the K_i value. This is also evident *in vivo*. Provided that adequate doses and routes of administration are given, *ex vivo* blood samples show the same extent of anticoagulation for higher concentrations of less potent inhibitors compared to lower concentrations in samples of more potent inhibitors.

The desired selectivity of synthetic inhibitors is expressed in the ratio of the K_i values of a given compound for thrombin to the K_i values for other related serine proteinases of the blood, such as plasmin, kallikrein, factor Xa, urokinase, t-PA. DPhe-ProArgCH_2Cl was shown to strongly inhibit t-PA and is, therefore, used in studies on fibrinolytic parameters to prevent *ex vivo* artifacts[79-82].

OM 805 was also shown to prolong the clotting time of plasma clotted by the Bothrops atrox venom, indicating a limited selectivity of inhibitory action or high similarity of the enzyme with thrombin[62]. It was also able to promote *in vitro* fibrinolysis, possibly by influencing the clot structure and sensitivity towards plasmic digestion[83].

The effectiveness of selective thrombin inhibitors as anticoagulants was studied in comparison to inhibitors of other coagulation enzymes, especially those of factor Xa[84,85]. In these studies selective thrombin inhibitors of a given affinity proved to be anticoagulantly and antithrombotically more active than factor Xa inhibitors of similar affinity and selectivity. From a theoretical point of view, factor Xa inhibitors could be highly efficient since they would interfere in the coagulation cascade at a kinetically important point. However, the affinity to factor Xa has to be very high in order to inhibit the very efficient prothrombinase complex.

Pharmacological Profile

In contrast to the anticoagulant potency of the thrombin inhibitors, their pharmacological characteristics have been the subject of few comparative and systematic studies only. This applies first of all to the relationships between the chemical structure and the pharmacodynamics and pharmaco-kinetics.

Toxicity

The benzamidine-derived inhibitors show a relatively uniform acute toxicity upon i.v. administration to small laboratory animals. Bis-benzamidines ranked first with regard to toxicity[29]. APPA was the best tolerated compound with an acute LD_{50}-value of 150 mg/kg in mice. TAPAP and NAPAP had LD_{50}-values of about 40-5O mg/kg i.v. A compound, similar to the principal structure of NAPAP, bearing a carbonyl group in the amide moiety had considerably lower toxicity[19]. The acute toxicity was not related to the inhibitory strength of the compounds, it seems rather to be linked to the highly basic amidino group, irrespective of its position in the molecule[86]. Compounds with two amidino groups were less well tolerated. Signs of toxicity were a drop in blood pressure and respiratory disturbances occurring during i.v. injection of higher doses.

As regards the acute toxicity of the arginine amide derivatives LD_{50}-values of about 20, 210, and 600 mg/kg for intravenous administration in mice were reported for the compounds OM 189, OM 805, and OM 407, respectively[87].

The tripeptide inhibitors of the D-Phe-Pro-Arg-sequence show LD_{50}-values of more than 50 mg/kg intravenously for the chloromethyl ketone in mice[20], and of about 40 mg/kg for the nitrile[57].

Results of long-term toxicity studies of synthetic thrombin inhibitors have not yet been published. In earlier studies we found no abnormalities in mice and rats dosed with 150 and 100 mg/kg APPA orally over 6 and 4 weeks, respectively (unpublished results). For comparison, a rater nonselective irreversible proteinase inhibitor, tosyl-lysyl-chloromethylketone, was shown to be relatively well tolerated over a longer period of time[88].

Pharmacodynamics and Side Effects

Biochemical and haemostaseological investigations of synthetic thrombin inhibitors do not give information on possible pharmacodynamic effects and side-effects. Such effects have to be considered under two aspects: Are all thrombin-induced cellular responses inhibited, and may the compounds have effects of their own on cells and tissues (organs)?

Reactions of the animals observed in studies on the toxicity of the compounds may give hints towards effects which could also occur with antithrombotically effective doses and thus would be classified as "side-effects".

Pharmacological effects exerted by the thrombin inhibitors *in vitro* and *in vivo* clearly include effects brought about by the inhibition of thrombin-mediated reactions such as the release of t-PA[89,93], the release of prostacyclin[91], vasoconstriction[92-94], smooth muscle cell growth[93]. Another effect was assumed to be also based on inhibition of thrombin, the propagation of experimental metastasis by a tripeptide inhibitor[96].

These pharmacodynamic effects obviously are largely independent of the individual inhibitor compound.

Pharmacological effects unrelated to inhibition of thrombin were demonstrated for two arginine amide derivatives, which *in vitro* inhibited the action of 5-hydroxytryptamine on various isolated arterial vessels[97,98].

A number of benzamidine derivatives was shown to produce hypotensive effects of various duration[33,54]. Also, for the tripeptide compounds, a blood pressure lowering effect was seen[56].

Pharmacokinetics

There is no comprehensive coverage of the pharmacokinetics of synthetic thrombin inhibitors in the literature. Especially, comparative studies are rare[52,56,99].

Most studies focused on the time course of blood/plasma levels only, other aspects were seldom dealt with in detail. One of the reasons for this may be seen in the problems of analytical methods to be developed for each individual substance. So, often biological assays, based on the antithrombin activity, were used (Table 4). Coagulation variables such as the thrombin time or partial thromboplastin time were often used to quantify inhibitor levels in blood. Moreover, thrombin inhibition assays with natural or synthetic thrombin substrates were used[100]. Our group devoted several studies to the pharmacokinetics of benzamidine-derived inhibitors, including studies with radioactively labelled compounds[99,101].

A chemical method was used in an early pharmacokinetic study on 4-amidinophenylpyruvic acid[102]. HPLC methods are suited for selected compounds and selected biological samples only[42,103]. The various steps in the pharmacokinetics will be viewed separately:

Absorption

The benzamidine derivative APPA was shown to be absorbed enterally to a remarkable degree[55,102]. For other compounds only, insufficient or low absorption rates were reported which would in all probability not produce plasma levels efficient in anticoagulation[56,104]. In several studies a certain degree of enteral absorption of D-Phe-Pro-Arg-H and related compounds was reached by direct injection into the duodenum. In mice, the biovailability after oral administration was estimated to about 15% only[56]. Under such experimental conditions, also for a number of benzamidine derivatives similar to TAPAP, enteral absorption of a certain amount was seen (unpublished observations).

However, for drugs with such a strong concentration-response relationship as in the case of the thrombin inhibitors, enteral absorption should be almost complete in order to be on the safe side with a given oral dose. Moreover, experimental and clinical evidence indicate great individual variations in enteral absorption of drugs which are absorbed to a limited percentage only.

Time-course of blood levels

The importance of the concentration of a synthetic thrombin inhibitor in streaming blood is self-evident. Therefore, pharmacokinetic calculations with regard to compartments etc. are of secondary importance. The blood levels show direct correlation to the anticoagulant and antithrombotic effects. So, the time course of these effects is first related to the concentration *versus* time curve of the inhibitor in blood, only in the second line to the dose producing this blood level.

The distribution of the arginine- and benzamidine-derived compounds *in vivo* is obviously influenced by the physicochemistry of these relatively strong organic bases. Several compounds were shown to have relatively large apparent volumes of distribution.

Among the benzamidine derivatives APPA takes a special position with regard to several pharmacological characteristics[33,55]. So, its terminal elimination half-life in rabbits was found to be about 20 hrs, which is considerably longer than the half-lives of other benzamidine derivatives.

An ester derivative of a compound of the amidinophenylalanine series, TAPAM, was hydrolyzed in rat plasma, whereas it was rather stable in human plasma[60].

In our laboratory, the direct comparison of OM 805 with the most potent inhibitor of the benzamidine series, NAPAP, revealed that both are equally potent *in vitro*; *in vivo* OM 805 has a somewhat longer half-life of about 25 min in rabbits[99].

The apparent half-life of the tripeptide derivatives D-Phe-Pro-Arg-H, BOC-D-Phe-Pro-Arg-H, D-MePhe-Pro-Arg-H and D-Phe-Pro-ArgCN is in the range of about 20-50

Table 4 Experimental investigations on the pharmacokinetics of synthetic thrombin inhibitors

Compound	Species	Analytical procedure	Reference
APPA	Rabbit	Photometry	102
APPA (^{14}C)	Rabbit	Radioactivity measurement	55
TAPAP (^{3}H)	Rabbit	Radioactivity measurement	101
NAPAP	Rabbit	Biological assay*	99
NAPAP	Rat	Biological assay	105
ßNas-Gly-(pAm)Phe-Pro	Rabbit	Biological assay	108
Methyl-NAPAP	Rabbit	Biological assay	36
OM 189	Dog,rabbit	Biological assay	52
OM 407	Dog,rabbit	Biological assay	52
OM 805	Rat	Biological assay	50
OM 805	Rabbit	Biological assay	99
OM 805	Dog	Biological assay	119
FOY I^{14}C)	Rat	Radioactivity measurement	128
PPACK	Rabbit	Biological assay	20
D-Phe-Pro-ArgCN	Rabbit	Biological assay	57

minutes in rabbits and cats[56,57]. PPACK was also shown to have a very short biological half-life, obviously owing to its reactivity, so that in experimental studies infusions are necessary[19,20]. FOY, being chemically an ester, is rapidly hydrolyzed in blood[42]. When it is administered directly into the circuit, the short half-life of F0Y will not represent a drawback to its possible use as anticoagulant in extracorporeal circulation[43].

The isocoumarin-derivative ACITC reported recently also had a very short half-life in rabbits, irrespective of its relative high *in vitro* stability[40].

Elimination

Thrombin inhibitors of the above type are renally excreted and can be detected, at least in part, in active form in urine[52,54,55]. The low renal clearance of APPA in rabbits may be brought about by certain similarity of APPA to neutral amino acids which can be reabsorbed from the renal tubules by a high capacity transport system[55]. Also the arginine derivatives, OM 189 and OM 407, are eliminated via the kidneys; however, to varying degrees[52]. NAPAP, on the contrary, is renally eliminated only to a low percentage in rabbits[54].

The pharmacokinetic behaviour of NAPAP is characterized by a relatively short half-life of about 9 min in rabbits. The reason for this is rapid hepatic uptake of the compound and subsequent biliary excretion. In rabbits with interrupted hepatic circulation the plasma half-life was significantly longer[47]. Extensic hepatic uptake may also be connected with metabolic degradation of the compound. *In vitro*, benzamidine was shown to be metabolized to N-hydroxyl benzamidine[106]. For NAPAP, however, no metabolites were found in the bile[103].

The structure of the most potent inhibitor of the benzamidine series, NAPAP, was varied in order to achieve compounds with improved pharmacokinetic properties. A derivative with a N-hydroxyl group (benzamidoxime) was less active than NAPAP and was transformed *in vivo* by reductive biotransformation to the corresponding compound with the free amidino group[108]. A further derivative, bearing proline in the amide portion of the molecule, αNas-Gly-(pAm)Phe-Pro, was better tolerated and showed a longer half-life than NAPAP in rabbits, obviously owing to the proline carboxyl group[108].

The amidinophenylalanine amide derivatives, the arginine amides of similar structural characteristics and the tripeptide derivatives obviously share a common feature as regards the excretory route; they are eliminated via the bile to a varying but significant degree[52,105]. The relatively high molar masses of the compounds and structural elements such as the basic group, aromatic ring systems and a number of polar groups may be responsible for hepatic clearance and biliary excretion. However, the strongly basic amidino or guanidino group seems not to be the sole precondition for biliary excretion, since structurally related compounds with less strongly basic groups were also excreted via this way to a considerable extent. Hepatic extraction (and subsequent biliary excretion) could also be the reason for a first pass effect contributing to the low bioavailability in the systemic circulation after oral administration.

The various synthetic thrombin inhibitors are individual chemical entities the structures of which will govern their fate in the organism. The structural characteristic common to the inhibitors of the arginine and benzamidine derivative-type, however, seems to have a special influence on the pharmacokinetic profile.

For the more extensively studied compounds no considerable species differences in the pharmacokinetics were found.

A short half-life in the circulation might be considered advantageous from the point of view of drug safety since this brings about a rapid disappearance of the anticoagulant effect, for instance in cases of possible bleeding tendency. On the one hand, it necessitates repeated administration of the compound in order to achieve and maintain a constant blood level. On the other hand, from the results of the pharmacokinetic and pharmaco-dynamic studies it is evident that one of the requirements for an "ideal" anticoagulant, possible oral administration, is up to now not yet fulfilled with compounds of this type being relatively strong bases. The intestinal absorption of strong organic bases, which are almost fully ionized at physiological pH, is limited. The "zwitterionic" p-amidinophenylpyruvic acid, however, is absorbed readily.

Antithrombotic Action

Although there are various factors inducing the formation of intravascular thrombosis, the underlying process is closely linked to thrombin-induced platelet reactions and fibrin formation. Therefore, thrombin inhibitors are expected to prevent the formation of thrombi, either experimentally induced or arising from pathological processes in clinical situations, or to diminish the progression.

The potential antithrombotic action of the synthetic thrombin inhibitors was studied; in most cases in comparison to heparin (and hirudin), in animal models that largely correspond to the various pathomechanisms of venous and arterial thrombosis and of disseminated intravascular coagulation in man.

The thrombosis models preferred by our group are the venous, stasis-induced coagulation thrombus according to Wessler and the arterial deposition thrombus according to Hladovec[30,109]. Moreover, thrombin or thromboplastin-induced disseminated intravascular coagulation with microthrombosis in several organs was considered a useful model, an extracorporeal arterio-venous shunt model was also chosen to evaluate the antithrombotic effectiveness of the inhibitors[7,30]. Other experimental thrombosis models in studies on the antithrombotic action of synthetic thrombin inhibitors used mechanical, chemical and thermal injury to blood vessels (Table 5). The DIC-models included as triggers thrombin, thromboplastin, endotoxin, lactic acid and others. Table 6 presents an overview on the respective experimental models and the inhibitors studied.

Several studies on antithrombotic effects of thrombin inhibitors attempted to correlate the blood level of the inhibitor coagulation variables and the extent of the antithrombotic effect.

For the *in vivo* action of synthetic thrombin inhibitors one has also to take into consideration the ultimate fate of the target enzyme. In animal experiments it was demonstrated that thrombin irreversibly inactivated by tosyl-lysyl-chloromethyl ketone has a longer half-life in the circulation than the active enzyme[110]. This form of the enzyme could possibly lead to cellular events described to be still produced by the active-site blocked enzyme. The elimination of thrombin in the presence of a reversible competitive inhibitor, however, was shown to be characterized by an initial rapid endothelial binding of the enzyme and subsequent complexing with antithrombin III, which would not occur for active-site irreversibly blocked thrombin[110,111].

In antithrombin III-deficient animals, OM 805 was shown to be still antithrombotically effective when heparin failed[112]. OM 805 was also shown to improve cerebral micro-circulation after cerebral ischemia in rats, possibly owing to inhibition of platelet-mediated reactions[113].

PPACK inhibited DIC induced by Echis carinatus venom[114,115], the formation of platelet-dependent thrombi, in "inside out" vessel grafts[72] and in other platelet-dependent models, which often proved to be heparin-resistant[116,117].

In animal experiments high potency of a synthetic inhibitor is desired since this allows the administration of comparatively low doses and thus the avoiding of undesired non-specific pharmacodynamic effects which are not linked to the specific inhibitory action as we demonstrated for the benzamidine-type inhibitors[33,86].

The wide range of doses used in preventing experimental thrombosis has to be

Table 5 Antithrombotic effects of synthetic thrombin inhibitors in animal models of localized thrombosis.

Compound	Dose	Model, species	Reference
APPA	10-40 mg/kg iv	stasis, jugularis, rat	139
TAPAP	1-10 mg/kg/iv	stasis, jugularis, rat	34
(4-)TAPAP	5 and 10 mg/kg iv		
ßNas-(pAm)Phe-Pip	1-10 mg/kg iv		
TAPAP	18 µM/kg iv	stasis, V.jugularis, rat	84
NAPAP	0.001-0.05 mg/kg/min	stasis, V.jugularis, rabbit	54
NAPAP	0.005-0.05	stasis, V.jugularis,	140
	0.005-0.20 mg/kg/min	electrical injury, A.carotis, rat	
NAPAP	0.2 mg/kg iv	mesenteric vessels, laser	141
OM 805	0.3-0.6 mg/kg/hr	injury, rat	
Methyl-NAPAP	0.06-3.0 mg/kg/hr	stasis, rabbit	36
OM 805	10-50 mg/kg sc	copper coil, V.cava, rat	112
OM 805	100 and 200 µg/kg/min	A.femoralis eversion graft, rabbit	72
OM 805	1 µg/kg/min	chemical injury, A.carotis, rat	50
OM 805	0.5-7.0 mg/kg iv	coronary artery stenosis plus	142
	mechanical injury, dog		
PPACK	1 mg/kg iv	stasis, V.jugularis, rat	19
PPACK	100 nM/kg/min	endovascular stent,	117
		arteriovenous shunt, baboon	
D-Phe-Pro-Arg-H	0.5-2.0 mg/kg iv	stasis, V.jugularis, rabbit	139
BOC-D-Phe-Pro-Arg-H	2-6 mg/kg iv	V.cava, ligature, rat	104
ACITIC	1 mg/kg/min	chemical injury, V.femoralis, rabbit	40

Table 6 Antithrombotic effect of synthetic thrombin inhibitors in disseminated intravascular coagulation/microthrombosis and extracorporeal circulation.

Compound	Dose	Model, species	Reference
APPA	10-40 mg/kg iv	arteriovenous shunt, rat	139
NAPAP	0.05-0.40 mg/kg/min	arteriovenous shunt, rat	140
TAPAP	0.17 μmol/kg/min	thromboplastin infusion, rat	84
NAPAP	1.0 μmol/kg/hr	thrombin infusion, rat	135
TAPAP	0.5 μmol/kg/hr*	thrombin infusion, rat	7
OM 189	0.8 "	"	
NAPAP	0.2 "	"	
D-Phe-Pro-Arg-H	0.5 "	"	
PPACK	0.013 "	"	
OM 407	3.75-75 μg/kg/min	thrombin infusion, rabbit	71
OM 407	12.5-50 mg/kg sc	generalized SSP#, rabbit	53
OM 805	1-31.6 μg/kg/min	lactic acid and thromboplastin infusion, rabbit	143
PPACK	1 mg iv plus 4 mg/45 min	rabbit	
PPACK	57 nmol/kg/min	Echis carinatus venom infusion, dog	115
PPACK	100 nmol/kg/min	haemodialysis, baboon	116
D-Phe-Pro-Arg-H	10 and 50 μg/kg/min	thrombin infusion, rabbit	12
FOY	120 μmol/kg/h	thrombin infusion, rat	44
FOY	1-100 mg/kg ip**	endotoxin infusion, rat	144
FOY	0.04-0.07 mg/ml/min (regional infusion)	extracorporeal circulation, dog, sheep	43
FUT 175	0.001-10 mg/kg/4h	endotoxin infusion, rat	49

* ED_{100}

**ip: intraperitoneal. For other abbreviations see Table 5.

\# Sanarelli-Schwartzman phenomenon.

considered under various aspects. The potency of the thrombin inhibitors used is different, their pharmacokinetics are different, and the experimental models show different dose-response relationships. The ratios of effective doses of inhibitors varying in the inhibitory potency, therefore, are not the same as the ratios of their *in vitro* potencies. In general, most sensitive to thrombin inhibition - in terms of effective doses of thrombin inhibitors - is the model of stasis-induced venous thrombosis. Next comes thrombin-induced disseminated microthrombosis (with a clear-cut dependence on the thrombin dosage). Higher inhibitor doses are required in arterial thrombosis and extracorporeal shunt thrombosis. These differences might possibly be explained by the finding that, *in vitro*, higher concentrations are necessary for inhibiting thrombin-platelet reaction than for the thrombin-fibrinogen reaction.

Selective synthetic thrombin inhibitors have been shown to be potent antithrombotic agents, qualitatively, irrespective of the kind of thrombotic challenge. One of the main findings from such experiments was that the competitive inhibitors ranked in their antithrombotic potency in the order of their inhibition constants. As already mentioned above, the pharmacokinetics of the inhibitors was an important factor for the effective doses. Antithrombotically equivalent blood levels of the inhibitors, however, showed in every case also the same degree of anticoagulative effect, expressed for instance as prolongation of the plasma thrombin time.

After the importance of thrombin for early reocclusion of vessels, subsequent to successful thrombolysis had become evident, synthetic thrombin inhibitors were used to prevent reocclusion or even to increase the thrombolytic efficacy of t-PA[118,119]. The thrombin inhibitors were either given concomitantly or consecutively to the thrombolytic agent.

Use of Synthetic Thrombin Inhibitors as Biochemical and Pharmacological Tools

Selective thrombin inhibitors are widely used in studies on the properties and effects of thrombin. Prime criteria for their *in vitro* use are selectivity and potency of a compound; here, further pharmacological characteristics are of limited importance. A limited number of applications is listed in the following:

Investigations on the effects of thrombin in various isolated and cellular systems make use of thrombin inhibitors. For such purposes, besides hirudin, the irreversible synthetic inhibitor PPACK is often used. One has to be careful, however, with possible "hormone-like" effects of enzymatically inactive forms of thrombin in which binding domains are still fully active. Investigations on complex activation steps of proenzymes during blood clotting are facilitated by selective thrombin inhibitors. Selective inhibitors of thrombin are able to eliminate unwanted thrombin activity in factor Xa and factor VIIIC assays using chromogenic substrates[120-123]. For such a purpose NAPAP is used as Thromstop in a commercial kit.

Similarly, the determination of prothrombin in plasma was proposed to be carried out in the presence of a thrombin inhibitor[135]. Moreover, thrombin inhibitors may also be used as the sole anticoagulant for stabilization and preservation of blood samples[65,125-127].

The use of thrombin inhibitors in studies on possible thrombin-induced long-term effects *in vivo* probably might allow new insight into physiological and pathological processes of interest.

Clinical use of Synthetic Thrombin Inhibitors

Clinical reports, published up to now, comprise in most cases only data on a small number of patients.

Two compounds, argipidine (OM 805) and the broad-spectrum inhibitor FOY, have

been used clinically. Table 7 lists the compounds and the indications published so far.
At present the following clinical indications are taken into account:

- Short term perioperative thrombosis prophylaxis
- Adjunctive administration in thrombolysis
- Prevention of rethrombosis after thrombolysis (acute myocardial infarction)
- Anticoagulation in haemodialysis and extracorporeal circulation

The clinical indications may also cover replacemment of heparin treatment, as was shown for OM 805 in cases of heparin-induced thrombocytopenia[128,129].
A new and promising field of application could be the presently discussed "adjunctive" treatment in the thrombolytic therapy of myocardial infarction[72,118,119,130-132].

The aim here is to inhibit thrombin which might be released from the fibrin network of a thrombus during thrombolysis.

Table 7 Clinical use of synthetic thrombin inhibitors.

Compound	Dose	Indication	Reference
APPA	1 mg/kg iv every 6h	chronic consumption coagulopathy	145
OM 805	0.7 ug/kg/min	cardiovascular surgery	146
OM 805	20 mg/d iv	vascular surgery, postoperatively	147
OM 805	0.1 mg/kg/h	DIC in meningoencephalitis	148
OM 805	30-60 mg/d	cerebral thrombosis	149
OM 805	0.3 mg/kg/h	haemodialysis	74
OM 805	0.49 mg/kg/h	heparin-induced thrombocytopenia during haemodialysis	129
OM 805	10 mg iv followed by controlled infusion	heparin-induced thrombocytopenia during haemodialysis	128
FOY	300 mg iv followed by 100 mg/h	DIC in amniotic fluid embolism	150
FOY	900-2000 mg/h regional infusion	haemodialysis	151
FOY	1-2 mg/kg/h	pending or manifest DIC	152
FOY	1 mg/kg/h	DIC	153, 154
FOY	40 mg/kg/h (plus heparin)	open heart surgery	155

Released enzymatically active thrombin is assumed to account for the early reocclusion of coronary vessel after primary successful thrombolysis[133]. However, the selectivity of the inhibitor is a "must"; so one has to take into account the inhibition of t-PA by NAPAP and especially by PPACK[79,80]. Hirudin with its high selectivity might, therefore, be the thrombin inhibitor of choice for such an indication. As regards possible haemorrhagic effects, the thrombolytic agent and the thrombin inhibitor have to be considered[72]. The experimental results seem to be rather promising. Results of a clinical application, however, have not yet been published. The interesting aspect of the antithrombin III-independent mode of action and the "sparing effect" of synthetic thrombin inhibitors has still to be evaluated clinically.

Present State and Perspectives

The leading aspect in the development of selective synthetic thrombin inhibitors has been almost entirely the inhibitory potency of the compounds, other pharmacological aspects, obviously, did not play a primary role. The problems arising with the pharmacodynamics and pharmacokinetics of *in vitro* highly active compounds have been manifold. Only few systematic approaches to a design of inhibitor molecules taking structure-pharmacokinetics relationships into account have been made. Nevertheless, a number of principal questions regarding the potential usefulness of selective synthetic thrombin inhibitors as anticoagulants and antithrombotics have been clarified experimentally with compounds which, for various reasons, are not expected to become clinically applicable drugs.

Clinical Dose Finding

From the kinetical point of view the question as to the extent of inhibition of thrombin would be very interesting. However, there is no simple answer, since the contribution of thrombin to either localized or disseminated thrombotic processes, occurring acutely or chronically, has to be taken into account. In this connection, the different doses of one and the same inhibitor necessary for inhibition of different forms of experimental thrombosis are an indication of the different amounts of thrombin formed. Dose finding has, therefore, not only to consider the *in vitro* potency of an inhibitor, but also its pharmacokinetics and the clinical situation.

Monitoring of the administration of this new type of anticoagulant agents can easily be done by following the thrombin time (or the partial thrombo-plastin time). One must, however, be aware that the prolongation of APTT by a factor of 2 when monitoring heparin therapy may not necessarily express the same as the corresponding prolongation brought about by a synthetic competitive inhibitor.

Advantages and Shortcomings

There are several reports on increased bleeding times after argatroban, PPACK, D-Phe-Pro-Arg-H, and methyl-NAPAP[36,72,116,118,134], whereas others did not see an increased bleeding tendency at antithrombotically effective doses in animal experiments[109]. The clinical data up to now do not give a sufficient answer as to the question of a possible risk of bleeding. Synthetic inhibitors of thrombin have several major advantages in comparison to heparin:

- they are not dependent on plasmatic cofactors
- they are, unlike heparin, not neutralized by endogenous factors
- they have no direct or immune-mediated platelet-activating activity
- as low molecular compounds they should not be antigenic
- they should principally be orally effective

Compared to the oral anticoagulants of the dicoumarol type, synthetic thrombin inhibitors are directly acting, i.e. their anticoagulant effect is exerted immediately after they have reached the circulation.

A major advantage over heparin is the antithrombin III-independent mode of action. Thus, antithrombin III is protected from consumption as it may occur during heparin therapy[135]. Moreover, low molecular weight synthetic thrombin inhihitors are able to block thrombin to a thrombus, where it is inaccessible to inhibition by antithrombin III[135]. Synthetic thrombin inhibitors may become valuable alternatives to heparin in clinical situations when anticoagulation is necessary and heparin cannot be recommended, as in the case of heparin-induced thrombocytopaenia in patients on haemodialysis.

The administration for prophylaxis of such forms of thrombosis which are heparin-resistant may represent a therapeutic benefit. The above-mentioned potential clinical indications for selective synthetic thrombin inhibitors are very similar to those proposed for hirudin. A substantial advantage over hirudin would be the oral administration of the low molecular weight synthetic inhibitors. Up to now, however, most of the experimental data show that most of the inhibitors have low bioavailability only. For these low molecular weight compounds the percutaneous route of systemic administration could become of interest; moreover, topical administration might be useful in certain situations.

Newer developments competing with synthetic thrombin inhibitors as regards their possible clinical use are the LMW heparins, which are claimed to be free of certain drawbacks of heparin and, of course, hirudin.

With the development of relatively small synthetic hirudin peptides a link to the naturally occurring thrombin inhibitor has emerged[136-137].

In the so-called hirulogs, synthetic peptides containing the C-terminal portion of hirudin coupled to the D-Phe-Pro-Arg-sequence, which are bivalent inhibitors blocking the active centre and the anion binding site, the development of selective synthetic inhibitors of thrombin may have another interesting perspective[138].

The results of large-scale clinical trials will finally determine the perspectives of selective synthetic thrombin inhibitors as a new class of antithrombotic drugs.

REFERENCES

1. T.A. Krenitsky, and G.B. Elion, Enzymes as tools and targets in drug research, in: Strategy in Drug Research, Ed.: J.A. Keverling Buisman, Elsevier, Amsterdam, 65 (1982).
2. T.M. Penning, Design of suicide substrates: an approach to the development of highly selective enzyme inhibitors as drugs, Trends Pharmacol. Sci. 4:212 (1963).
3. H.J. Smith, Perspectives in the design of small molecule enzyme inhibitors as useful drugs, J. Theor. Biol. 73:531 (1978).
4. C.T. Walsh, Suicide substrates, mechanism-based enzyme inactivators: recent developments, Ann. Rev. Biochem. 53:493 (1984).
5. J.D. Geratz, and R.R. Tidwell, Current concepts of action of synthetic thrombin inhibitors, Haemostasis 7:170 (1987).
6. F. Markwardt, Pharmacological control of blood coagulation by synthetic, low-molecular weight inhibitors of clotting enzymes. A new concept of anticoagulants, Trends Pharmacol. Sci. 1:153 (1980).
7. F. Markwardt, G. Nowak, and J. Hoffmann, Comparative study of thrombin inhibitors in experimental microthrombosis, Thrombos. Haemostas. 49:235 (1983).
8. E. Shaw, Synthetic irreversible inhibitors of coagulation enzymes, Folia Haematol. (Lpz.) 109:33 (1982).
9. J. Stürzebecher, Inhibitors of thrombin, in: The Thrombin. Vol. I ed: R. Machovich, CRC Press, Boca Baton, Fl., 131 (1984).
10. J. Stürzebecher, and F. Markwardt, Synthetische Inhibitoren des Thrombins und andere Gerinnungsenzyme - Struktur und Wirkung, Beitr. Wirkst. Forsch. 16:1 (1983).

11. D. Bagdy, E. Barabas, L. Sebestyen, M. Dioszegi, S. Fittler, S. Bajusz, and E. Szell, Correlation between the anticoagulant and antiplatelet effect of D-Phe-Pro-Arg-H (RGH-2958), Thrombos. Haemostas. 58:177 (Abstr. 649) (1987).
12. D. Bagdy, E. Barabas, E. Szell, and S. Bajusz, Uber die biochemische Pharmakologie einger Tripeptidaldehyde, Folia Haematol. (Lpz). 109:22 (1982).
13. E. Barabas, S. Bajusz, D. Bagdy, and E. Szell, Studies on the inhibition by synthetic tripeptides of blood clotting, Acta Biochim. Biophys. 11:207 (1976).
14. J.I. Witting, C. Pouliott, J.L. Catalfamo, J. Fareed, and J.W. Fenton II, Thrombin inhibition with dipeptidyl argininals. Thromb. Res. 50:461 (1986).
15. S. Bajusz, E. Szell, D. Bagdy, E. Barabas, G. Horvath, M. Dioszegi, S. Fittler, G. Szabo, A. Juhasz, E. Tomori, and G. Szilagy G, Highly active and selective anticoagulants: D-Phe-Pro-Arg-H, a free tripeptide aldehyde prone to spontaneous inactivation, and its stable N-methyl derivative, D-MePhe-Pro-Arg-H, J. Med. Chem. 33:1729 (1990).
16. S. Bajusz, The story of D-MePhe-Pro-Arg-H, the likely anticoagulant and antithrombotic of the future, Biokemie (Budapest) XIV/3:127 (1990).
17. C. Kettner, and E. Shaw, D-Phe-Pro-ArgCH_2Cl - A selective affinity label for thrombin. Thromb. Res. 14:969 (1979).
18. H.R. Lijnen, M. Uytterhoeven, and D. Collen, Inhibition of trypsin-like serine proteinases by tripeptide arginyl and lysyl chloromethylketones, Thromb. Res. 34:431 (1994).
19. Bagdy, E. Barabas, M. Dioszegi, S. Bajusz, and A. Feher, Comparative studies on the anticoagulant effects of D-Phe-Pro-Arg-H /1/ and D-Phe-Pro-ArgCH_2Cl /2/, Thrombos. Haemostas. 50:53 (Abstr. 0146) (1983).
20. D. Collen, O. Matsuo, J.M. Stassen, C. Kettner, and E. Shaw, *In vivo* studies of a synthetic inhibitor of thrombin, J. Lab. Clin. Med. 99:76 (1982).
21. J. Hauptmann, and F. Markwardt, Studies on the anticoagulant and anti-thrombotic action of an irreversible thrombin inhibitor, Thromb. Res. 20:347 (1980).
22. W. Stuber, H. Kosina, and N. Heimburger, Synthesis of a tripeptide with a C-terminal nitrile moiety and the inhibition of proteinases, Int. J. Peptide Protein Res. 31:63 (1988).
23. C. Kettner, L. Mersinger, and R. Knabb, The selective inhibition of thrombin by peptides of boroarginine, J. Biol. Chem. 265:18289 (1990).
24. T. Ueda, C.M. Kam, and J.C. Powers, The synthesis of arginyl-fluoroalkanes, their inhibition of trypsin and blood-coagulation serine proteinases and their anticoagulant activity, Biochem. J. 285:539 (1990).
25. R. Kikumoto, Y. Tamao, K. Ohkubo, T. Tezuka, S. Tonomura, S. Okamoto and H. Hijikata, Thrombin inhibitors. 3. Carboxyl-containing amide derivatives of Nα substituted L-arginine, J. Med. Chem. 23:1293 (1980).
26. R. Kikumoto, Y. Tamao, T. Tezuka, S. Tonomura, H. Hara, K. Ninomiya, A. Hijikata, and S. Okamoto, Selective inhibition of thrombin by (2R,4R)-4-methyl-1-(N^2-((3-methyl-1,2,3,4-tetrahy-8. Quinolinyl)sulfonyl)-L-arginyl)-2-piperidinecarboxylic acid, Biochemistry 23:85 (1984).
27. S. Okamoto, A. Hijikata, R. Kikumoto, S. Tonomura, N. Hara, K. Ninomiya, Maruyama, S. Sugano, and Y. Tamao, Potent inhibition of thrombin by the newly synthesized arginine derivative No.805: the importance of the stereostructure of its hydrophobic carboxamide portion, Biochem. Biophys. Res. Commun. 101:440 (1981).
28. S. Tonomura, R. Kikumoto, Y. Tamao, K. Ohkubo, S. Okamoto, A. Kinjo, A. Hijikata, A novel series of synthetic thrombin inhibitors. II. Relationships between structure of modified OM inhibitors and thrombin-inhibitory effect, Kobe J. Med. Sci. 26:1 (1980).

29. J. Hauptmann, and F. Markwardt F, Pharmakologie synthetischer Thrombin-Inhibitoren. Beitr. Wirkst. Forsch. 26:1 (1986).
30. F. Markwardt, and J. Hauptmann, Synthetische Thrombininhibitoren als Antithrombotika. Z. Klin. Med. 41:540 (1986).
31. F. Markwardt, and J. Stürzebecher, Inhibitors of trypsin and trypsin-like enzymes with a physiological role, in: M.H. Sandler, J. Smith (eds.) Design of Enzyme Inhibitors as Drugs, Oxford University Press, Oxford, 619 (1989).
32. F. Markwardt, G. Wagner, J. Stürzebecher, and P. Walsmann, Nα arylsulfonyl-w(4-amidinophenyl)-α-aminoalkyl-carboxlic amides - novel selective inhibitors of thrombin, Thromb. Res. 17:425 (1980).
33. J. Hauptmann, Pharmacology of benzamidine-type thrombin inhibitors, Folia Haematol. (Lpz.) 109:89 (1982).
34. J. Hauptmann, B. Kaiser, F. Markwardt, and G. Nowak, Anticoagulant and antithrombotic action of novel specific inhibitors of thrombin, Thrombos. Haemostas. 43:118 (1980).
35. J. Stürzebecher, F. Markwardt, B. Voigt, G. Wagner, and P. Walsmann, Cyclic amides of Nα-arylsulfonaminoacylated 4-amidinophenyl-alanine - tight binding inhibitors of thrombin, Thromb. Res. 29:635 (1983).
36. Y. Cadroy, C. Caranobe, A. Bernat, J.P. Maffrand, P. Sie, and B. Boneu, Antithrombotic and bleeding effects of a new synthetic thrombin inhibitor and of standard heparin in the rabbit, Thrombos. Haemostas. 58:764 (1987).
37. G. Etemad-Moghadam, D. Delebasse, J.P. Maffrand, and D. Frehel, Syntheses of Nα-(ß-naphthyl-sulfonyl-aminoglycyl) argininamides as potential selective synthetic thrombin inhibitors, Eur. J. Med. Chem. 23:577 (1988).
38. G. Claeson, S. Gustavsson, and C. Mattsson, New derivatives of p-guanidino-phenylalanine as potent reversible inhibitors of thrombin. Thrombos. Haemostas. 50:53 (Abstr. 0147) (1983).
39. R. Ferroni, E. Menegatti, and P. Orlandini, Aromatic tetra-amidines: anti-proteolytic and antiesterolytic activities towards serine proteinases involved in blood coagulation and clot lysis, Farm. Ed. Sci. 41:464 (1986).
40. S.W. Oweida, D.N. Ku, A.B. Lamsden, C.M. Kam, and J.C. Powers, *In vivo* determination of the anticoagulant effect of a substituted isocoumarin (ACITIC),Thromb. Res. 58:191 (1990).
41. A. Mor, J. Maillard, C. Favreau, and M. Reboud-Ravauz, Reaction of thrombin and proteinases of the fibrinolytic system with a mechanism-based inhibitor, 3,4-dihydro-3-benzyl-6-chloromethylcoumarin, Biochim. Biophys. Acta 1038:119 (1990).
42. M.K. Nishijima, J. Takezawa, N. Taenaka, Y. Shimada, and I. Yoshiya, Application of HPLC measurement of plasma concentration of gabexate esilate, Thromb. Res. 31:279 (1983).
43. B. Oedekovan, R. Bey, K. Mottaghy, and H. Schmid-Schönberg, Gabexate mesilate (FOY[R]) as an anticoagulant in extracorporeal circulation in dogs and sheep, Thrombos. Haemostas. 52:329 (1984).
44. H. Ohno, J. Kamnayashi, S.W. Chang, and G. Kosaki, FOY: (Ethyl p-(6-guanidinohexanoyloxy)benzoate) methanesulfonate as a serine proteinase inhibitor. II. *In vivo* effect on coagulo-fibrinolytic system in comparison with heparin or aprotonin, Thromb. Res. 24:445 (1981).
45. H. Ohno, G. Kosaki, J. Kambayashi, S. Imaoka, and F. Hirata, FOY:(Ethyl p-(6-guanidinohexanoyloxy)benzoate) methanesulfonate as a serine proteinase inhibitor. I. Inhibition of thrombin and factor Xa *in vitro*, Thromb. Res. 19:579 (1980).

46. S.V. Pizzo, A.D. Turner, N.A. Porter, and S.L. Gonias, Evaluation of p--amidinophenyl esters as potential antithrombotic agents, Thrombos. Haemostas. 56:387 (1986).
47. Y. Hitomi, N. Ikari, and S. Fujii, Inhibitory effect of a new synthetic protease inhibitor (FUT-175) on the coagulation system, Haemostasis 15:164 (1985).
48. Y. Koshiyama, A. Kobori, M. Oginara, Y. Yokomoto, K. Ohtari, K. Shimamaura, and M. Iwaki, The effects of FUT-175 (nafamostat mesilate) on blood coagulation and experimental disseminated intravascular coagulation (DIC), Folia Pharmacol, Japan 90:313 (1987).
49. T. Yoshikawa, M. Murakami, Y. Furukawa, H. Kato, S. Takemura, and M. Kondo, Effects of FUT-175, a new synthetic protease inhibitor, on endotoxin-induced disseminated intravascular coagulation in rats, Haemostasis 13:374 (1983).
50. H. Ikoma, K. Ohtsu, Y. Tamao, R. Kikumoto, and S. Okamoto, Effect of a potent thrombin inhibitor, MCI 9038, on novel experimental arterial thrombosis, Blood & Vessel 13:72 (1982).
51. S. Nagano, S. Okamoto, K. Ikezawa, K. Mimura, A. Matsuoka, A. Hijikata, and Y. Tamao, Fluorescence studies on the mode of action of two synthetic thrombin inhibitors, No.206 and No.805, Thrombos. Haemostas. 46:(1981) 45 (Abstr. 0128).
52. K. Oda, K. Ohtsu, R. Tamao, R. Kikumoto, A. Hijikata, K. Kinjo, and S. Okamoto, Comparison of plasma levels and excretory routes between No.189 and No.407, potent thrombin inhibitors, Kobe J. Med. Sci. 26:11 (1980).
53. K. Ohtsu, Y. Tamao, R. Kikumoto, K. Ikezawa, A. Hijikata, and S. Okamoto, Effects of a potent thrombin inhibitor No. 407, on DIC models, Kobe J. Med. Sci. 6:61 (1980).
54. B. Kaiser, J. Hauptmann, A. Weiß, and F. Markwardt, Pharmacological characterization of a new highly effective synthetic thrombin inhibitor. Biomed. Biochim. Acta 44:1201 (1985).
55. F. Markwardt, J. Hauptmann, M. Richter, and P. Richter, Tierexexperimentelle untersuchungen zur Pharmakokinetik von 4-Amidinopheny-l-brenztraubensäure (APPA), Pharmazie 34:178 (1979).
55. F. Markwardt, A. Hoffmann, and J. Stürzebecher J, Influence of thrombin inhibitors on the thrombin-induced activation of human blood platelets, Haemostasis 13:227 (1983).
56. C. Mattsson, E. Eriksson, and s. Nilsson, Anticoagulant and antithrombotic effects of some protease inhibitors, Folia Haematol. (Lpz.) 109:43 (1982).
57. B. Kaiser, J. Hauptmann, and F. Marktwardt, Studies on toxicity and pharmacokinetics of the synthetic thrombin inhibitor D-phenylalanyl-L-prolyl-L-arginine nitrile, Pharmazie 46:131 (1991).
58. B. Kaiser, M. Richter, J. Hauptmann, and F. Markwardt, Anticoagulant and antithrombotic action of the synthetic thrombin inhibitor D-phenylalanyl-L-prolyl-L-arginine nitrile, Pharmazie 46:128 (1991).
59. G.M. Kam, K. Fujikawa, and J.C. Powers, Mechanism-based isocoumarin hibitors for trypsin and blood coagulation serine proteases: New anticoagulants, Biochemistry 27:2547 (1988).
60. J. Hauptmann, Degradation of a benzamidine-type synthetic inhibitor of coagulation enzymes in plasma of various species, Thromb. Res. 61:279 (1991).
61. W.B. Lawson, V.B. Valenty, J.D. Wos, and A.P. Lobo, Studies on the inhibition of human thrombin: Effects of plasma and plasma constituents, Folia Haematol. (Lpz.) 109:52 (1992).

62. D. Green, C-H. Ts'ao, N. Reynolds, D. Kahn, H. Kohl, and I. Cohn, *In vitro* studies of a new synthetic thrombin inhibitor, Thromb. Res. 37:145 (1985).
63. J. Hauptmann, B. Kaiser, and F. Markwardt, Anticoagulant action of synthetic tight binding inhibitors of thrombin *in vitro* and *in vivo*, Thromb. Res. 39:771
64. M. Niwa, C. Niwa, R. Yamagashi, S. Kondo, K. Takahashi, and N. Sakuragawa, The comparative study on the anticoagulant activities of the synthetic thrombin inhibitor (MD-805) and heparin, Blood & Vessel 16:421 (1985).
65. H. Sato, and A. Nakajima, Kinetic study of the initial stage of the fibrinogen-fibrin conversion by thrombin. (III) Effects of competitive inhibitors, Thromb. Res. 37:327 (1985).
66. A. Hijikata-Okunomiya, S. Okamoto, and K. Wanaka, Effect of a synthetic thrombin-inhibitor MD 605 on the reaction between thrombin and plasma antithrombin-III, Thromb. Res. 59:967 (1990).
67. A.P. Bode, and D.T. Miller, The use of thrombin inhibitors and aprotinin in the preservation of platelets stored for transfusion, J. Lab. Clin. Med. 113:753 (1989).
68. E. Glusa, A. Hoffmann, and F. Markwardt, Influence of benzamidine derivatives on thrombin-induced platelet reactions, Folia Haematol. (Lpz.) 109:86 (1982).
69. N.J. Greco, T.E. Tenner, and N.N. Tendon, PPACK-thrombin inhibits thrombin-induced platelet aggregation and cytoplasmic acidification, but does not inhibit platelet shape change, Blood 75:1983 (1990).
70. S.R. Hanson, and L.A. Harker, Interruption of acute platelet-dependent thrombosis by the synthetic antithrombin D-pheny-L-alanyl-L-prolyl-L-arginyl chloromethyl ketone, Proc. Natl. Acad. Sci. USA 85:3184 (1986).
71. H. Hara, Y. Tamao, R. Kikumoto, Y. Funahara, A. Hijikata, and S. Okamoto, Effect of a potent thrombin inhibitor, No. 407, on platelet function *in vitro* and *in vivo,* Kobe J. Med. Sci. 26:47 (1980).
72. I-K. Jang, H.K. Gold, A.A. Ziskind, R.C. Leinbach, J.T. Fallon, and Collen, Prevention of platelet-rich arterial thrombosis by selective thrombin inhibition, Circulation 81:219 (1990).
73. F. Markwardt, A. Hoffmann, and J. Stürzebecher, Influence of thrombin inhibitors on the thrombin-induced activation of human blood platelets, Haemostasis 13:227 (1983).
74. T. Matsuo, K. Nakao, T. Yamada, and O. Matsuo, Effect of a net anticoagulant (MD 805) on platelet activation in the hemodialysis circuit, Thromb. Res. 41:33 (1986).
75. E. Tremoli, P. Maderna, S. Colli, E. Agradi, A. Petroni, and R. Paoletti, GYKI 14,451, a synthetic tripeptide inhibitor of thrombin: activity on platelet aggregation and arachidonic acid metabolism, Pharmacol. Res. Commun. 13:339 (1981).
76. E. Pearlstein, C. Ambrogio, G. Gasic, and S. Karpatkin, Inhibition of the platelet-aggregating activity of two human adenocarcinomas of the colon and an anaplastic murine tumor with a specific thrombin inhibitor, dansylarginine N-(3-ethyl-1,5-pentanediyl)-amide, Cancer Res. 41:4535 (1981).
77. M. A. Packham, N. L. Bryant, M. A. Guccione, Agglutination of rabbit platelets in plasma by the thrombin inhibitor D-phenylalanyl-L-prolyl-L-arginyl chloromethyl ketone, Thrombos. Haemostas. 63:282 (1990).
78. M.W.C. Hatton, and S.L. Moar, Comparative behavior of thrombin and an inactive derivative, FPR-thrombin, toward the rabbit vascular endothelium. Heparin liberates FPR-thrombin from the endothelium *in vivo*, Circulat. Res. 7:221 (1990).
79. N. Gilboa, and J.W. Fenton II, Inhibition of tissue plasminogen activator (TPA)

by protein or synthetic thrombin inhibitors, Ann. N. Y. Acad. Sci. 485:414 (1986).

80. N. Gilboa, G.B. Villanueva, and J.W. Fenton II, Inhibition of fibrinolytic enzymes by thrombin inhibitors, Enzyme 40:144 (1988).

81. M.A. Mohler, C.J. Refino, S.A. Chen, A.B. Chen, and A.J. Hotchkiss, D-Phe-Pro-Arg-chloromethylketone: Its potential use in inhibiting the formation of *in vitro* artifacts in blood collected during tissue-type plasminogen activator thrombolytic therapy, Thrombos. Haemostas. 56:160 (1986).

82. E. Seifried, and P. Tanswell, Comparison of specific antibody, D-Phe-Pro-Arg-CH_2Cl and aprotinin for prevention of *in vitro* effects of recombinant tissue plasminogen activator on haemostasis parameters. Thrombos. Haemostas. 58:921 (1987).

83. Y. Tamao, T. Yamamoto, R. Kikumoto, H. Hara, J. Itoh, T. Hirata, K. Mineo, and S. Okamoto, Effect of a selective thrombin inhibitor MCI-9093 on fibrinolysis *in vitro* and *in vivo*, Thrombos. Haemostas. 56:28 (1986).

84. J. Hauptmann, B. Kaiser, G. Nowak, J. Stürzebecher, and F. Markwardt F. Comparison of the anticoagulant and antithrombotic effects of synthetic thrombin and factor Xa inhibitors, Thrombos. Haemostas. 63:220 (1990).

85. J. Stürzebecher, U. Stürzebecher, H. Vieweg, G. Wagner, J. Hauptmann, and F. Markwardt, Synthetic inhibitors of factor Xa and thrombin. Comparison of their anticoagulant efficiency, Thromb. Res. 54:242 (1989).

86. B. Kaiser, J. Hauptmann, and F. Markwardt, Ontersuchungen zur Pharmakodynamik synthetischer Thrombininhibitoren vom Typ basisch substituierter N -arylsulfonylierter Phenylalaninamide, Pharmazie 42:119 (1987).

87. S. Okamoto, A. Hijikata, K. Ikezawa, E. Mori, R. Kikumoto, Y. Tamao, K. Ohkubo, T. Texuka, and S. Tonomura, Structure-activity relationship in a series of synthetic thrombin inhibitors: No.205, No.407 and No.700, Blood & Vessel 11:230 (1980).

88. C.L. Litterst, Acute and subchronic toxicity of the protease inhibitor Nα-tosyl-L-lysyl-chloromethylketone (TLCK) in mice, Drug and Chemical Toxicol. 3:227 (1980).

89. A. Hijikata-Okunomiya, H. Kitaguchi, and M. Hirata, Effect of MD-805 on plasminogen activator release by thrombin from isolated perfused dog leg, Thromb. Res. 45:699 (1987).

90. H-P. Klöcking, A. Hoffmann, and F. Markwardt, Influence of α-NAPAP on thrombin-induced release of plasminogen activator, Thromb. Res. 52:71 (1968).

91. M. Miki, K. Ogawa, M. Hirata, H. Kitaguchi, and Y. Funahara, Prostacyclin release from the coronary vascular wall by vasoactive substances, Thromb. Res. 35:665 (1984).

92. E. Glusa, and U. Wolfram, The contractile response of vascular smooth muscle to thrombin and its inhibition by thrombin inhibitors, Folia Haematol. (Lpz.) 115:94 (1988).

93. D.D. Ku, Unmasking of thrombin vasoconstriction in isolated perfused dog hearts after intracoronary infusion of air embolus, J. Pharm. Exp. Ther. 243:571 (1987).

94. K. Nakamura, Y. Hatano, and K. Mori, Thrombin-induced vasoconstriction in isolated cerebral arteries and the influence of a synthetic thrombin inhibitor, Thromb. Res. 40:715 (1985).

95. M. Melzig, C. Harme, E. Teuscher, B. Voigt, and G. Wagner, Influence of inhibitors of thrombin on porcine aortic smooth muscle cells in primary culture, Biomed. Biochim. Acta 45:1199 (1986).

96. L.E. Ostrowski, A. Ahsen, B.P. Suthar, P. Pagast, D.L. Bain, C. Wong, A. Patel, and R.M. Schultz, Selective inhibition of proteolytic enzymes in an *in vivo* mouse model for experimental metastasis, Cancer Res. 46:4121 (1986).
97. W.K.W. Ho, K. Nakao, and S. Shibata, The inhibitory action of two thrombin inhibitors (TI-189 and TI-233) on the contractile response to 5-hydroxy-tryptamine and prostaglandin endoperoxide analogue (U-44069) in isolated vascular strips, Brit. J. Pharmacol. 71:399 (1990).
98. H. Karaki, K. Murakami, N. Nakagawa, U. Orakawa, The inhibitory effect of N^2-dansyl-L-arginine-4-t-butylpiperidide amide (TI 233) on contraction of vascular and intestinal smooth muscle, Brit. J. Pharmacol. 80:519 (1983).
99. J. Hauptmann, and B. Kaiser, *In vitro* and *in vivo* comparison of arginine- and benzamidine-derived highly potent synthetic thrombin inhibitors, Pharmazie 46:57 (1991).
100. T. Matsuo, and K. Nakao, Plasma antithrombin activity of MD 805 determined by chromogenic substrate during hemodialysis, Blood and Vessel 18:378 (1987).
101. J. Hauptmann, F. Markwardt, and M. Richter, Tierexperimentelle Untersuchungen zur Pharmakokinetik von Nα-Tosyl-3-amidino-phenyl-alaninpiperidid (TAPAP), einem neuen Thrombininhibitor, Pharmazie 37:430 (1962).
102. F. Markwardt, M. Richter, and G. Vogel, Zur Pharmakokinetik von 4-Amidinophenyl-brenztraubensäure, einem neuen synthetischer, Thrombinhemmstoff. Zbl. Pharm. 113:787 (1974).
103. M. Paintz, M. Richter, and J. Hauptmann, HPLC-determination of the synthetic thrombin inhibitor Nα-(2-naphthylsulfonylglycyl)-4-amidino-phenylalanine piperidide in biological material,Pharmazie 42:346 (1987).
104. E. Tremoli, G. Morazzoni, P. Maderna, S. Colli, and R. Paoletti, Studies on the antithrombotic action of BOC-D-Phe-Pro-ArgH (GYKI14,451), Thromb. Res. 23:549 (1981).
105. J. Hauptmann, B. Kaiser, M. Paints, and F. Markwardt, Biliary excretion of synthetic benzamidine-type thrombin inhibitors in rats, Biomed. Biochim. Acta 46:445 (1987).
106. B. Clement, and M. Zimmermann, Characteristics of the microsomal N-hydroxylation of benzamidine to benzamidoxime, Xenobiotika 17:659 (1987).
107. J. Hauptmann, M. Paintz, B. Kaiser, and M. Richter, Reduction of a benzamidoxime derivative to the corresponding benzamidine *in vivo* and *in vitro*, Pharmazie 43:559 (1988).
108. J. Hauptmann, B. Kaiser, M. Paintz, and F. Markwardt, Pharmacological characterization of a new structural variant of 4-amidinophenylalanine amide-type synthetic thrombin inhibitor, Pharmazie 44:282 (1989).
109. B. Kaiser, and F. Markwardt, Antithrombotic and haemorrhagic effects of synthetic and naturally occurring thrombin inhibitors, Thromb. Res. 43:613 (1986).
107. J. Hauptmann, B. Kaiser, M. Paintz, and F. Markwardt, Pharmacological characterization of a new structural variant of 4-amidino-phenyl-alanine amide-type synthetic thrombin inhibitor, Pharmazie 44:282 (1989).
110. C.N. Vogel, H.S. Kingdon, and R.L. Lundblad, Correlation of *in vivo* and *in vitro* inhibition of thrombin by plasma inhibitors, J. Lab. Clin. Med. 93:661 (1979).
111. M.A. Shifmann, and S.V, *In vivo* metabolism of reversibly inhibited-thrombin, Biochem. Pharmacol. 32:138 (1983).
112. T. Kumada, and Y. Abiko, Comparative study on heparin and a synthetic thrombin inhibitor No. 805 (MD-805) in experimental antithrombin III-deficient animals, Thromb. Res. 24:285 (1981).

113. T. Yamamoto, T. Hirata, M. Inagaki, R. Kikumoto, Y. Tamao and S. Okamoto, Effect of MCI-9038, a selective thrombin inhibitor on cerebral microcirculation after cerebral ischemia in rats, Thrombos. Haemostas. 58:108 (Abstr. 362) (1967).
114. R.C. Schaeffer, C. Briston, S-M. Chilton, and R.W. Carlson, Disseminated intravascular coagulation following Echis carinatus venom in dogs: Effects of a synthetic thrombin inhibitor, J. Lab. Clin. Med. 107:488 (1986).
115. R.C. Schaeffer, S-M. Chilton, T.J. Hadden, and R.W. Carlson, Pulmonary fibrin microembolism with Echis carinatus venom in dogs: effects of a synthetic thrombin inihibitor, J. Appl. Physiol. 57:1824 (1984).
116. A.B. Kelly, S.R. Hanson, L.W. Henderson, and L.A. Harker, Prevention of heparin-resistant thrombotic occlusion of hollow-fiber hemodialyzers by synthetic antithrombin, J. Lab. Clin. Med. 114:411 (1989).
117. W.C. Krupski, A. Bass, A.B. Kelly, U.M. Marzec, S.R. Hanson, and L.A. Harker, Heparin-resistant thrombus formation by endovascular stents in baboons. Interruption by a synthetic antithrombin, Circulation 82:570 (1990).
118. I.J.K. Jang, H.K. Gold, R.C. Leinbach, J.T. Fallon, and D. Collen, *In vivo* thrombin inhibition enhances and sustains arterial recanalization with recombinant tissue-type plasminogen activator, Circulat. Res. 67:1552 (1990).
119. M.J. Mellott, T.M. Conolly, S.J. York, and L.R. Bush, Prevention of reocclusion by MCI-9038, a thrombin inhibitor, following t-PA induced thrombolysis in a canine model of femoral arterial thrombosis, Thromb. Haemostas. 64:526 (1990).
120. K. Ikesawa, T.Yamashita, S. Okamoto, T. Ohara, K. Takemoto, A. Matsuoka, Improvement of the F VIII:C chromogenic method using the specific synthetic thrombin inhibitors, I-2581 and MD-805, Blood & Vessel 16:418 (1985).
121. S. Okamoto, S. Ikezawa, A. Nagano, A. Matsuoka, Y. Hijikata, and Y. Tamao, Selectivity increase of chromogenic assay of factor Xa by use of highly selective synthetic thrombin inhibitor having extremely potent stereostructure (No.605), Thrombos. Haemostas. 46:313 (Abstr. 0976) (1981).
122. J. Stürzebecher, and C. Klessen, Einsatz von synthetischen Inhibitoren bei gerinnungsphysiologischen Tests mit Peptidsubstraten, Folia Haematol. (Lpz). 109:157 (1982).
123. L. Svendsen, M. Brogli, G. Lindeberg, and K. Stocker K, Differentiation of thrombin- and factor Xa-related amidolytic activity in plasma by means of a synthetic thrombin inhibitor, Thromb. Res. 34:457 (1984).
124. A. Hijikata-Okunomiya, A new method for the determination of prothrombin in human plasma, Thromb. Res. 57:705 (1990).
125. M. Hiraishi, Z. Yamasaki, and K. Ichikawa, Plasma collection using nafamostat mesilate and dipyridamole as an anticoagulant, Int. J. Artif. Organs 11:212 (1988).
126. M.L. Rand, J. Neiman, D.M. Jakowec, and M.A. Packham, Effects of alcohol withdrawal from alcoholics - a study using platelet-rich plasma from blood anticoagulated with D-phenylalanyl-L-prolyl-L-arginyl chloromethyl ketone ($FPRCH_2Cl$), Thrombos. Haemostas. 63:178 (1990).
127. P. Stein, and J. Drawert, Anwendung eines 4-Amidinophenylbrenztrau-bensäure-Heparin-Stabilisators zur zitratfreien Stabilisierung von Blut, Dtsch. Ges. Wesen. 38:748 (1983).
128. T. Matsuo, Y. Chikahira, and Y. Yamada, Effect of synthetic thrombin inhibitor (MD 805) as an alternative drug on heparin-induced thrombocytopenia during hemodialysis, Thromb. Res. 52:165 (1988).
129. T. Matsuo, T. Yamada, T. Yamanashi, R. Ryo R, Anticoagulant therapy with MD 805 of a hemodialysis patient with heparin-induced thrombocytopenia, Thromb. Res. 58:663 (1990).

130. P.R. Eisenberg, Role of new anticoagulants as adjunctive therapy during thrombolysis, Amer. J. Cardiol. 67:A19 (1991).
131. H.K. Gold, I.K. Jang, R.C. Leinbach, J.E. McLary, J.T. Fallon, D. Collen, Acceleration of reperfusion by combination of rt-PA and a selective thrombin inhibitor, argatroban, Fibrinolysis Suppl. 3:15 (Abstr. 40) (1990).
132. T. Yasuda, H.K. Gold, H. Yaoita, R.C. Leinbach, J.L. Guerrero, I-K. Jang, R. Holt, J.T. Fallon, and D. Collen, Comparative effects of aspirin - synthetic thrombin inhibitor and a monoclonal antiplatelet glycoprotein IIb/IIIa antibody on coronary artery reperfusion, reocclusion and bleeding with recombinant tissue-type plasminogen activator in a canine preparation, J. Amer. Coll. Cardiol. 16:714 (1990).
133. I. Weitz, M. Hudoba, D. Massel, J. Maraganore, and J. Hirsh J, Clot-bound thrombin is protected from inhibition by heparin-antithrombin III but is susceptible to inactivation by antithrombin III-independent inhibitors, J. Clin. Invest. 86:385 (1990).
134. A.B. Kelly, Y. Cadroy, O.M. Marzec, C.M. Hanson, and L.A. Harker, Comparison of antithrombotic and antihemostatic effects produced by antithrombins in primate models of arterial thrombosis, Thrombos. Haemostas. 62:42 (Abstr. 93) (1989).
135. J. Hauptmann, E. Brüggener, and F. Markwardt, Effect of heparin, hirudin, and a synthetic thrombin inhibitor on antithrombin III in thrombin-induced disseminated intravascular coagulation in rats, Haemostasis 17:321 (1987).
136. C.G. Binnie, B.W. Erickson, and J. Hermans, Inhibition of thrombin by synthetic hirudin peptides, FEBS Lett. 270:85 (1990).
137. J.A. Jakubowski, and J.M. Maraganore, Inhibition of coagulation and thrombin induced platelet activities by a synthetic dodecapeptide modeled on the carboxy-terminus of hirudin, Blood 75:399 (1990).
138. X.J. Yang, M.A. Blajchman, S. Craven, L.M. Smith, N. Anvari, and F.A. Ofosu, Activation of factor V during intrinsic and extrinsic coagulation. Inhibition by heparin, hirudin and D-Phe-Pro-Arg-CH_2Cl, Biochem. J. 272:399 (1990).
138. J.M. Maraganore, P. Bourdon, J. Jablonski, K.L. Ramachandran, and J.W. Fenton II, Design and characterization of hirulogs: A novel class of bivalent peptide inhibitors of thrombin, Biochemistry 29:7095 (1990).
139. J. Hauptmann, A. Barth, F-P. Schönberger, and F. Markwardt, Comparative study on the antithrombotic effects of a synthetic thrombin inhibitor and of heparin in animal models, Biomed. Biochim. Acta 42:959 (1983).
140. B. Kaiser, and F. Markwardt, Experimental studies on the antithrombotic action of a highly effective synthetic thrombin inhibitor, Thrombos. Haemostas. 55:194 (1986).
141. K. Krupinski, K.N. Breddin, F. Markwardt, and W. Haarmann, Antithrombotic effects of three thrombin inhibitors in a rat model of laser-induced thrombosis, Haemostasis 19:74 (1989).
142. J.F. Eidt, P. Allison, S. Noble,J. Ashton, P. Golino, J. McNatt, L.M. Buja, and J.T. Willerson, Thrombin is an important mediator of platelet aggregation in stenosed canine coronary arteries with endothelial injury, J. Clin. Invest. 84:18 (1989).
143. H. Hara, Y. Tamao, R. Kikumoto, and S. Okamoto S, Effect of a synthetic thrombin inhibitor MCI-9038 on experimental models of disseminated intravascular coagulation in rabbits, Thrombos. Haemostas. 57:165 (1987).
144. T. Yoshikawa, Y. Furukawa, M. Murakami, S. Takemura, and M. Kondo, Protective effect of Gabexate mesilate against experimental disseminated intravascular coagulation in rats, Haemostasis 13:262 (1983).

145. G. Vogel, R. Huyke, and I. Heerklotz, Erste klinische Erfahrungen mit dem kleinmolekularen synthetischen Thrombininhibitor 4- amidino-phenyl-brenztraubensäure (APPA), Folia Haematol. (Lpz.). 104:785 (1977).
146. K. Kumon, K. Tanaka, N. Nakajima, Y. Naito, and T. Fujita, Anticoagulation with a synthetic thrombin inhibitor after cardiovascular surgery and for treatment of disseminated intravascular coagulation, Critical Care Med. 12:1039 (1884).
147. T. Ohshiro, J. Kambayashi, and G. Kosaki, Antithrombotic therapy of patient with peripheral arterial reconstruction - clinical study on MD-805, Blood & Vessel 14:216 (1983).
148. K. Yamada, A clinical trial of MD-605, a synthetic thrombin inhibitor. Bibliotheca Haematol. (Basel) 49:343 (1983).
149. S. Kobayashi, M. Kitani, S. Yamaguchi, T. Suzuki, K. Okada, and T. Tsunematsu, Effects of an antithrombotic agent (MD-805) or progressing cerebral thrombosis, Thromb. Res. 3:305 (1969).
150. N. Taenaka, Y. Shimada, M. Kawai, I. Yoshiya, and G. Kosaki, Survival from DIC following amniotic fluid embolism - successful treatment with a serine proteinase inhibitor - FOY, Anaesthesia 36:389 (1981).
151. N. Taenaka, Y. Shimada, T. Hirata, M. Nishijima, and I. Yoshiya, New approach to regional anticoagulation in hemodialysis using gabexate mesilate (FOY), Crit. Care Med. 10:773 (1982).
152. G. Kosaki, J. Kambayashi, and S. Imaoka, Application of a synthetic serine protease inhibitor in the treatment of DIC, Bibliotheca Haematol. (Basel) 49:317 (1983).
153. G. Palareti, M. Maccaferri, M. Poggi, F. Petrini, S. Coccheri, F. Haverkate, F. Montanari, and A.S. Corticelli, Effects of gabexate mesilate (FOY), a new synthetic serine protease inhibitor on blood coagulation in patients with DIC, Thrombos. Haemostas. 58:420 (Abstr. 1544) (1987).
154. S. Umeki, M. Adachi, and M. Watanabe, Gabexate as a therapy for disseminated intravascular coagulation, Arch. Intern. Med. 148:1409 (1968).
155. K. Tanaka, M. Takao, J. Yada, H. Yuasa, M. Kusagawa, and K. Deguchi, Application of gabexate mesilate (FOY) for open heart surgery. Blood & Vessel 19:612 (1988).

NEW PEPTIDE BORONIC ACID INHIBITORS OF THROMBIN

Said Elgendy, John Deadman and Goran Claeson

Thrombosis Research Institute
Manresa Road
London SW3 6LR

Peptides containing α-amino boronic acids bind to the active site of serine protease, forming highly effective inhibitors. The trigonal boron in these compounds contain a vacant 2p orbital that easily reacts with nucleotides such as the Ser hydroxyl or the His imidazole at the catalytic site of the enzyme to give a tetrahedral transition-state boron adduct as shown in Figure 1, similar to that observed with aldehydes which is stabilised in an analogous manner to the transition state of the substrate, by interactions with secondary binding sites (e.g. oxyanion binding pocket)

Thrombin, the final enzyme in the blood coagulation cascade, has arginine (or less commonly lysine)-directed specificity. This selectivity of thrombin for Arg in P1 position has been used for the design of a large number of synthetic substrates and inhibitors; D-Phe-Pro-Arg, imitating amino acid sequences of fibrinogen has been considered the best sequence, has very good affinity for the active site of thrombin[1]. Recently, by replacing Arg by the boronic acid analogue of Arg, tight-binding inhibitors of thrombin have been obtained[2]. The Ki value of these inhibitors, e.g. Ac-D-Phe-Pro-boroArg, are in the picomolar range.

We have found that if the guanidine group of boro Arg in these compounds is substituted for a methoxy group, then the inhibitors obtained are, surprisingly, still quite good (Ki in the nanomolar range)[3]. This result is very interesting as the side chain of the inhibitors is neutral, in strong contrast to the positively charged side chain of Arg which is bound by electrostatic attraction to the carboxylic group of Asp at the specificity pocket of thrombin. We therefore decided to investigate the role of the side chain of the C-terminal amino boronic acid in this type of thrombin inhibitor.

In the present study, we describe the synthesis of novel amino boronic acids (boroAa) and their use for the construction of the new inhibitors, D-Phe-Pro-boroAa[4]. We have made the side chain of the amino acid boroAa neutral and in a variety of sizes in order to optimize its influence on the properties of the inhibitors. In order to introduce these different side chains to the boronic acid in the P1 position we used two different routes, both having an α-chloro boronic ester as intermediate (see Scheme 1). The boronic acid group of the starting material is protected by pinanediol or pinacol. Matteson's facile route to α-amino boronic acids[6], route (b), uses (dichloromethyl) lithium and a boronic

The Design of Synthetic Inhibitors of Thrombin
Edited by G. Claeson, *et al.*, Plenum Press, New York, 1993

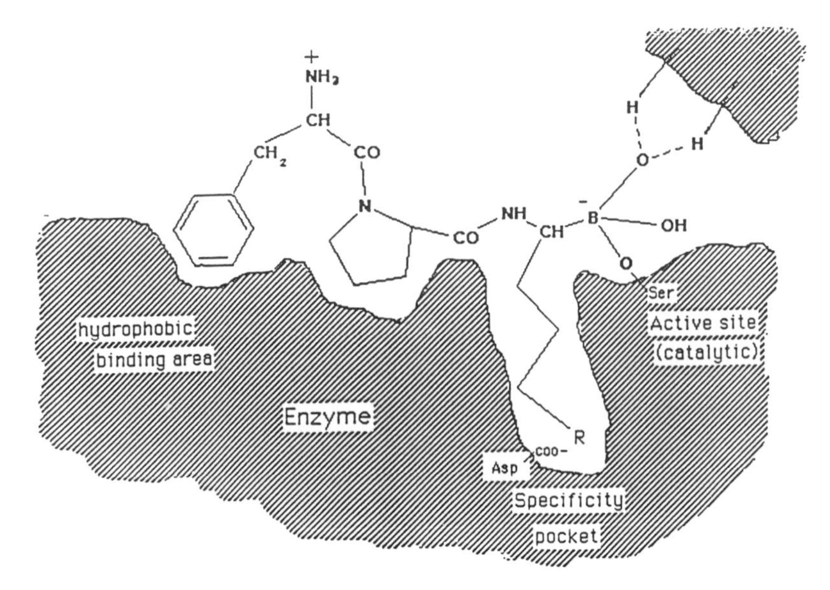

Figure 1 Inhibitor binding to active site of thrombin.

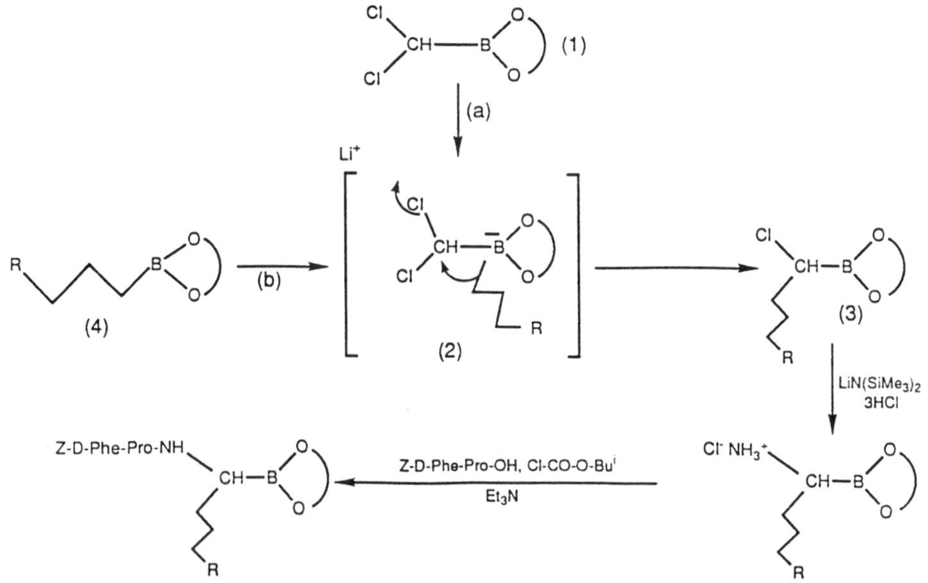

Cl
CH—B
Cl
(1)
(a)
Li⁺
Cl
CH—B
Cl
R
(2)
R
B
(4)
(b)
Cl
CH—B
(3)
R
$LiN(SiMe_3)_2$
3HCl
$Cl^- NH_3^+$
CH—B
R
Z-D-Phe-Pro-OH, Cl-CO-O-Bu^i
Et_3N
Z-D-Phe-Pro-NH
CH—B
R

Scheme 1

acid (4) to form the borate complex (2), which rearranges to give α-chloro boronic ester (3) in good yield. This route, however, requires hydroboration of an alkene to form (4) and necessarily gives only residues with a substituted C_2 alkyl chain. We therefore developed a route (a) from (dichloromethyl) borate (1) allowing the introduction of a range of substituents at the α-carbon *via* Grignard or alkyllithium reagents, and the α-chloro boronic esters (3) were isolated in excellent yields. A similar route has earlier been used by Rathke et al.[6,10]. However, their intermediates, α-chloro boronic esters, were not isolated but used for *in situ* oxidation to aldehydes. The dipeptide was then coupled to the α-aminoboronic ester hydrochloride *via* the mixed anhydryde method.

As shown in Table 1, Entry 2, the methoxy propyl side chain gave Ki=7 nM. By introducing the acetal with two oxygen atoms, the inhibition constant increased to 25 nM. Alkyl side chains such as *n*-pentyl and *n*-octyl gave 19 and 22 nM respectively. Branched and aromatic side chains seem to give the best results as shown in Table 1, Entry 6 and 7.

Table 1 Ki value and thrombin time for peptide amino boronic esters.

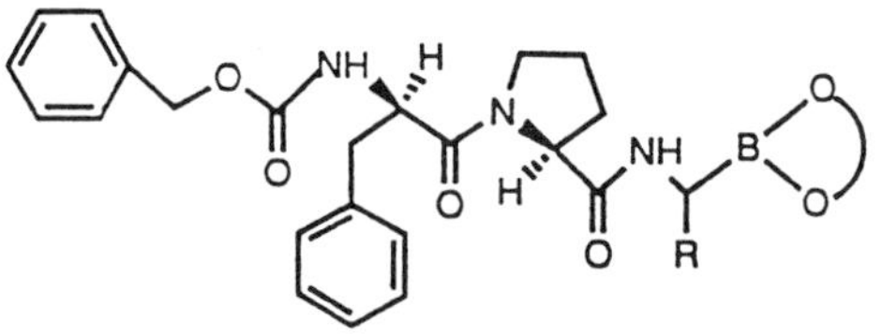

Entry	R	Ki(μM)	TT (μM)
1	$NH_2^+=C(NH_2)\text{-}NH\text{-}(CH_2)_3\text{-}$	0.001	0.066
2	$MeO(CH_2)_3\text{-}$	0.001	0.144
3	$CH_3(CH_2)_4\text{-}$	0.019	0.54
4	$CH_3(CH_2)_7\text{-}$	0.022	3.9
5	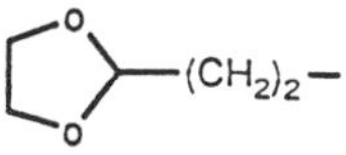 (1,3-dioxolan-2-yl)$(CH_2)_2$–	0.025	3.4
6	$CH_3CH_2C(CH_3)_2\text{-}$	0.007	0.56
7	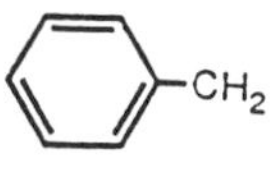 (phenyl)CH_2	0.012	0.33
8	$Br(CH_2)_3\text{-}$	0.007	ND

Table 2 Selectivity of the thrombin inhibitors[9].

Entry	Inhibitors	Ki(nM) Thrombin	Pla	Xa	Kal	IC_{50}(nM) UK	Chy
1	Z-D-Phe-Pro-boro**Irg**-OPin	1.4	8.4	14.1	13	75	1500
2	Z-D-Phe-Pro-boro**Mpg**-OPin	1	3200	1700	7300	3600	34300
3	Z-D-Phe-Pro-boro**Pgl**-OPin	19	ND	ND	ND	ND	165
4	Z-D-Phe-Pro-boro**Mbg**-OPin	7	11500	3900	12900	9100	454
5	Z-D-Phe-Pro-boro**Phe**-OPin	13	37300	7900	16700	6700	ND

Irg=isothiouronium analogue of Arg
Mpg=3-methoxypropylglycine
OPin=pinandiol ester
Mbg=2-methyl 2-butylglycine
Pgl=n-pentylglycine

Table 3 Ki value for peptide amino boronic esters.

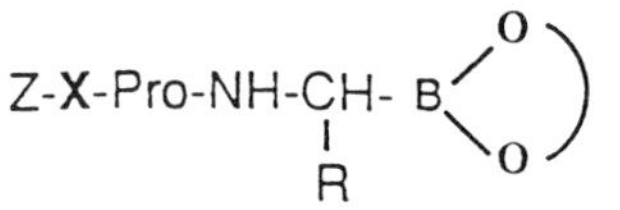

Z-**X**-Pro-NH-CH(R)-B(O)(O) (cyclic boronic ester)

Entry	R	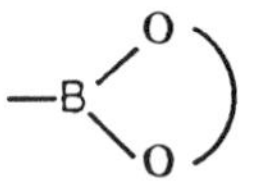 —B(O)(O)	Ki(μM) X= D-Phe	Ki(μM) X= D-Dpa[a]
1	$CH_3(CH_2)_4$-	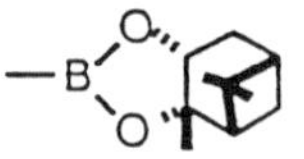 pinanediol ester	0.019	0.026
2	$CH_3CH_2C(CH_3)_2$-	,,	0.007	0.029
3	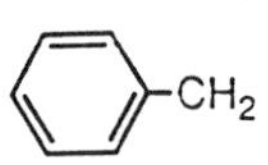 C_6H_5-CH_2	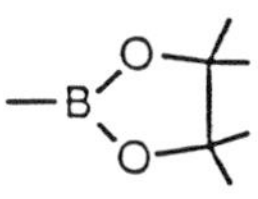 pinacol ester	0.012	0.049[b]
4	$MeO(CH_2)_3$-	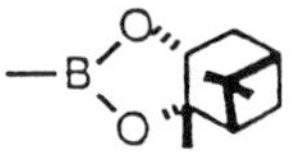 pinanediol ester	0.001	0.004
5	$Br(CH_2)_3$-	,,	0.007	ND
6	$^{+}NH_2$=C(NH_2)-S-$(CH_2)_3$-	,,	0.001	0.002

a) Dpa = β,β- diphenylalanine. b) L-Dpa form was used.

The reactivities of these peptides containing α-aminoboronic esters with neutral side chains were studied with other serine proteinases such as plasmin, factor Xa, urokinase, chymotrypsin and kallikrein to evaluate the selectivity in the inhibition of thrombin. The results show that all the inhibitors with a neutral side chain bind to thrombin at least 3 orders of magnitude tighter than their binding to other serine proteases. In general, as shown in Table 2, the selectivity for thrombin of these new inhibitors is clearly higher than that of the corresponding boroIrg inhibitor[2], the isothiouronium analogue of boroArg.

It is also interesting that by replacing D-Phe in the P3 position by D-Dpa gives a worse Ki value as shown in Table 3. This is in contrast to our results with peptidyl phosphonate diphenyl esters[7] and tripeptide aldehydes and ketones[4] where Dpa in the P3 position improves the binding and gives a better inhibition constant.

Reduction of blood pressure is a side-effect observed in many of the previous thrombin inhibitors containing Arg or Arg analogues[5]. This side-effect, which in some compounds can be disturbingly serious, is believed to depend on the positively-charged guanidino side chain of Arg.

Preliminary *in vivo* data (results not shown) indicate that the use of our neutral C-terminal side chain inhibitors reduces this toxicity. A large dose (4mg/kg) of the inhibitors in Table 1 were injected i.v. to rabbits, and the blood pressure was monitored before and after injection. The inhibitor with the positively-charged side chain gave serious drops in the blood pressure and the animals died. The other inhibitors also lowered the blood pressure but to a much lesser degree. To summarise, the new peptidometric thrombin inhibitors containing a P1 aminoboronic acid with neutral side chain show surprisingly good thrombin inhibition as well as selectivity for thrombin and are without serious side effect on blood pressure.

REFERENCES

1. G. Claeson, and L. Aurell, Small synthetic peptides with affinity for proteases in coagulation and fibrinolysis: An overview, Ann. N.Y. Acad. Sci. 370:798 (1981).
2. C. Kettner, L. Mersinger, and R. Knabb. The selective inhibition of thrombin by peptides of boroarginine, J. Biol. Chem, 265:18289 (1990).
3. G. Claeson, M. Philipp, and R. Metternich, Inhibitors of trypsin-like enzymes, US Patent Applic. 07-406 663 (1989).
4. G. Claeson, L. Cheng, N. Chino, J. Deadman, and S. Elgendy, Inhibitors and substrates of thrombin, UK Patent Applic. 9024129.0 (1990).
5. B. Kaiser, J. Hauptmann, and F. Markwardt, Studies of the pharmacodynamics of synthetic thrombin inhibitors of the basically substituted N-α-aryl sulfonylated phenylalanine amide type, Die Pharma. 42:119 (1987).
6. D.S. Matteson. α-Halo boronic esters: Intermediates for stereodirected synthesis, Chem. Rev, 89:1535 (1989).
7. L. Cheng, C. Goodwin, M.S. Scully,, V.V. Kakkar, and G. Claeson, Substrate-related phosphonopeptides: a new class of thrombin inhibitors, Tet. Lett, 32, 7333 (1991).
8. D.S. Matteson, and D. Majmudar, α-Chloro boronic esters from homologation of boronic esters, J. Am. Chem. Soc, 102:7588 (1980).
9. The inhibition constant (Ki) and the inhibitor concentrations required to inhibit 50% of the enzyme activity (IC_{50}) were measured without preincubation. The free enzyme activity was measured spectrophotometrically (405nm) by following the release of *p*-nitroanaline from the appropriate chromogenic substrate (Kabi

AB) for each enzyme. All experiments were carried out in 0.1 M phosphate, 0.2 M NaCl, 0.5% PEG, 0.02% NaN_3, pH 7.5

10. M.W. Rathke, E. Chao, and G. Wu, The preparation and reaction of esters of dichloromethane boronic acid, J. Organomet. Chem, 122:145 (1976).

SUBSTRATE-RELATED PHOSPHONOPEPTIDES AS THROMBIN INHIBITORS

L. Cheng*, C. A. Goodwin, M. F. Scully, J. Deadman and G. Claeson

Thrombosis Research Institute
Manresa Road
Chelsea, London SW3 6LR
*Present address: The Dyson Perrins Laboratory
University of Oxford, Oxford OX1 3QY, UK

INTRODUCTION

Phosphorus-containing inhibitors, such as diisopropyl phosphorofluoride (DFP), are classical inhibitors of serine proteases. These compounds react irreversibly with the active site serine hydroxy group, forming a covalent adduct. Nonpeptidyl phosphorylating agents do not discriminate between various serine proteases due to their lack of structural similarity with natural peptide substrates. Substrate-related peptidyl α-aminophosphonic acids have great potential for producing selective serine protease inhibitors[1-4] since they form stable transition-state analogues with the active site serine residue, similar to the tetrahedral intermediates formed during normal substrate hydrolysis.

Design of Substrate-Related Phosphonopeptide Inhibitors

D-Phe-Pro-Arg, a "fibrinogen-like" sequence, has been widely used in the design of substrates and inhibitors of thrombin due to its high affinity for the enzyme[5]. Recently, we found that the replacement of the N-terminal Phe in this sequence by a more lipophilic amino acid, Dpa (ß,ß-diphenyl-alanine), gave improved thrombin inhibitors having a C-terminal aldehyde or ketomethylene group[6]. It is proposed that the more lipophilic amino acid in P_3 position[7] (Figure 1) produced better affinity to the hydrophobic binding area of thrombin.

The rationale behind this present work is the idea that the replacement of the Arg carboxyl residue in the above sequence by the phosphonic acid moiety could form a very good transition-state analogue due to the presence of the tetrahedral phosphorus atom[8]. The following factors should be considered in the design of new therapeutically useful inhibitors: reactivity, chemical stability and specificity[9]. An ideal serine protease phosphorus inhibitor represents a delicate balance between several competing factors which

The Design of Synthetic Inhibitors of Thrombin
Edited by G. Claeson, *et al.*, Plenum Press, New York, 1993

S_3	S_2	S_1
D-Aa	Pro	Aa^P $(OPh)_2$
P_2	P_1	P_1

Figure 1 The nomenclature of Schechter and Berger[7] is used to designate the individual amino acid residues of a peptide inhibitor. $Aa^P(OPh)_2$ means diphenyl α-amino-phosphonate with neutral side chain.

determine its specificity and activity. The balance between reactivity and chemical stability is determined by the electrophilicity of the phosphorus atom. Very electro-negative-leaving groups on the phosphorus atom will increase activity toward serine proteases, but also decrease stability toward hydrolysis, which obviously is an obstacle toward development of practically useful inhibitors. Decreasing the electro-negativity of the leaving group substantially increases the chemical stability of the inhibitor, but at the cost of lower inactivation rates. Greater selectivity could be achieved when there are structural and stereochemical similarities between a natural serine protease substrate and the potential organophosphorus inhibitor, and could provide the ability to discriminate between different serine proteases.

The phenoxy group has been chosen as a leaving group on phosphorus for our inhibitors. If the choice of the leaving group is appropriate, these inhibitors will retain enough activity to inactivate their target enzyme effectively. Diphenyl esters of peptide phosphonates have been investigated[2,3] as inhibitors of a number of serine proteases and they demonstrate high selectivity and potency, but as yet no thrombin inhibitors of this type have been published.

Thrombin has distinct functions in haemostasis and it acts as a serine protease with an arginine- or lysine-directed specificity. We found that in some inhibitors with the D-Phe-Pro-Arg sequence, the positively charged side chains of Arg can be replaced by a neutral side chain without a serious loss of inhibitory effect and that this replacement produces enhanced selectivity for thrombin. Based on these results we have designed the new thrombin inhibitors shown in Figure 1.

Synthesis of Inhibitors

$Aa^P(OPh)_2$ (diphenyl α-aminophosphonates): Mpg^P (1a, α-amino-δ-methoxybutyl-phosphonyl), Pgl^P(1b, α-amino-n-hexyl-phosphonyl), Dpg^P (1c, α-amino-γ,γ- dimethyl-butylphosphonyl), Apg^P (1d, α-amino-p-aminobenzyl-phosphonyl), Epg^P (1e, α-amino-ß-ethylbutyl-phosphonyl) were prepared by a three-component reaction from triphenyl phosphite, the appropriate aldehyde and benzyl carbamate[10], then deprotected by hydrogenation on Pd/C (Figure 2). The phosphorus-containing tripeptides (3) were obtained through the coupling of dipeptides (2) with compound 1 by the mixed anhydride method[11] (Figure 3) and isolated by flash chromatography on a silica gel column (2% methanol in chloroform as eluent). Compounds 4 were obtained by removal of the Z-protecting group through hydrogenation on Pd/C and separation on a silica gel column (conditions as above). Compounds 5 and 6 were obtained by the cleavage of the diphenyl groups on PtO_2/H_2 from 3 and 4 respectively[12].

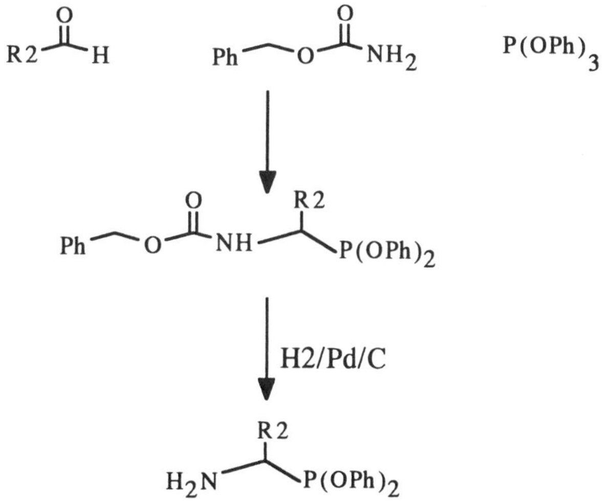

(1)

Figure 2 Synthetic route to diphenyl-α-amino phosphonate.

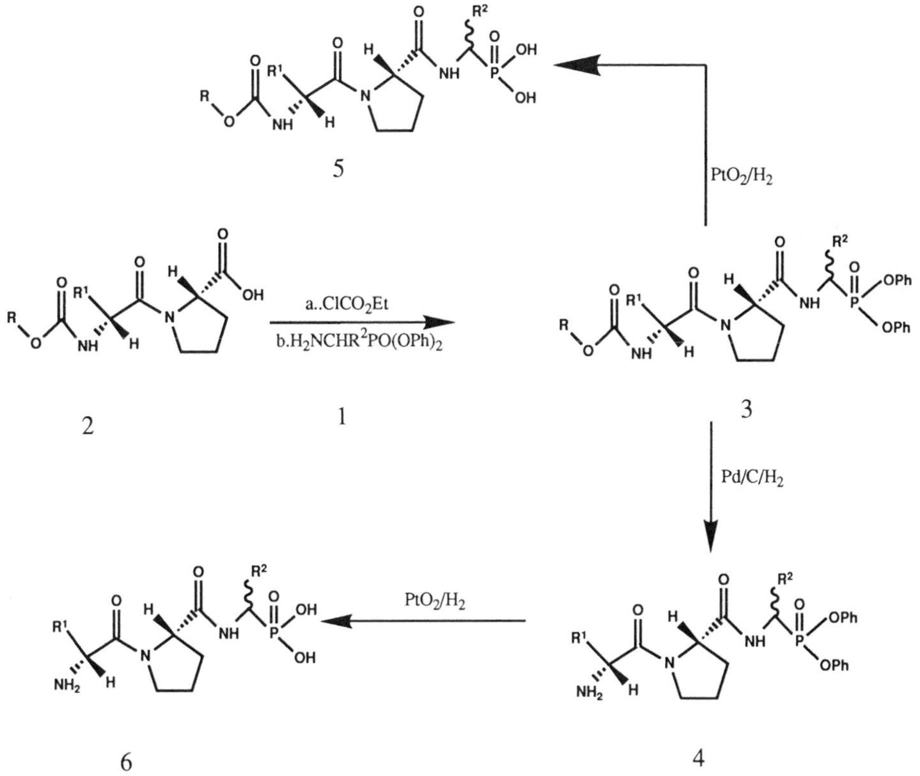

Figure 3 Synthetic scheme for the preparation of phosphorus-containing tripeptides.

Biological data

Table 1 shows the inhibitory activity of thrombin and Table 2 shows the selectivity of 4c and 4d for thrombin inhibition.

Table 1 Compounds and their inhibitory activity of thrombin.

	R	R^1	R^2	IC_{50} (μM) A	IC_{50} (μM) B	Ki (μM) A	Ki (μM) B
3a	$PhCH_2$	$PhCH_2$	n-pentyl	150	0.019	>31.0	-
3b	$PhCH_2$	Ph_2CH	n-pentyl			10.0	-
4a	-	$PhCH_2$	n-pentyl			1.7	-
4b	-	Ph_2CH	n-pentyl	0.19	0.00094	0.48	-
4c	-	$PhCH_2$	3-methoxypropyl	1.70	0.012	3.10	0.018
4d	-	Ph_2CH	3-methoxypropyl	1.50	0.0024		0.0048
4e	-	$PhCH_2$	2,2-dimethylpropyl	0.083	-	-	-
4f	-	$PhCH_2$	p-aminophenyl		3.96	-	-
4g	-	$PhCH_2$	1-ethylpropyl		0.12	-	-
5*	t-Bu	Ph_2CH	n-pentyl	5.30	0.084	59.0	-
6a	-	$PhCH_2$	n-pentyl	140	4.40	>98.0	-
6b	-	Ph_2CH	n-pentyl	5.2	0.082	2.3	-
6c*	-	Ph_2CH	n-pentyl	150	0.75	35.0	-

*The compound shows (HPLC) 1:1 ratio of diasteromers.
A: Preincubation time = 0; B: Preincubation time = 1h.

The inhibition constant (Ki) and the inhibitor concentration required to inhibit 50% of the enzyme activity (IC_{50}) were measured with preincubation (B) and without preincubation (A) of the enzyme and inhibitor at 37°C for 1h. The remaining enzyme activity was measured spectrophotometrically (405nm) by following the release of p-nitroaniline from the appropriate chromogenic substrate for each enzyme. All experiments were carried out in 0.1M phosphate buffer, 0.2M NaCl, 0.5% PEG, 0.02% NaN_3, pH7.5.

Table 2 Selectivity of 4c and 4d for thrombin inhibition expressed as relative IC_{50} Enzyme)/Thrombin.

Enzyme	Thrombin	Plasmin	Uro-kinase	Factor Xa	Kalli-krein	Chymotrypsin
4c	1	3,500	NE	NE	NE	22
4d	1	1,400	NE	NE	NE	620

NE=No effect was observed as a concentration of 20 μM.

Table 3 Effect on blood pressure after i.v. injection of 4mg/kg of H-D-Dpa-Pro-$Mpg^P (OPh)_2$.

Mean B.P. before (mmHg)	Mean B.P. after (mmHg)	% Change	Duration of response
135	142	+ 5	200 sec

*Dissolved in 50% DMSO-50% saline.

DISCUSSION

These new compounds were found to be potent inhibitors of thrombin (Table 1). They are more effective as phenyl esters than as free phosphonic acids and they function better without an N-protecting group. It is also clear that Dpa is preferable to Phe in the P_3 position which is in agreement with our previous observation that Dpa seems to improve binding to the apolar binding site of thrombin in the case of tripeptide aldehydes and ketones. Thrombin is known to cleave exclusively Arg and Lys bonds, and therefore it is interesting that good thrombin inhibition can be obtained with a P_1 amino acid having a neutral side chain. Compounds 4b and 4d were found to have the highest inhibitory activity which has been brought about by a normal five-atom side chain. The new compounds show slow binding behaviour. More than 100-fold higher potency was observed (measured as IC_{50} and Ki) after 1h incubation of enzyme and inhibitor before addition of substrate. The inhibitors presumably form covalent enzyme-inhibitor complexes (Figure 4). The diphenyl esters of peptidyl phosphonates are chemically very stable: at neutral conditions (pH 5.8) they are quite inert against hydrolysis and nucleophilic substitution at phosphorus.

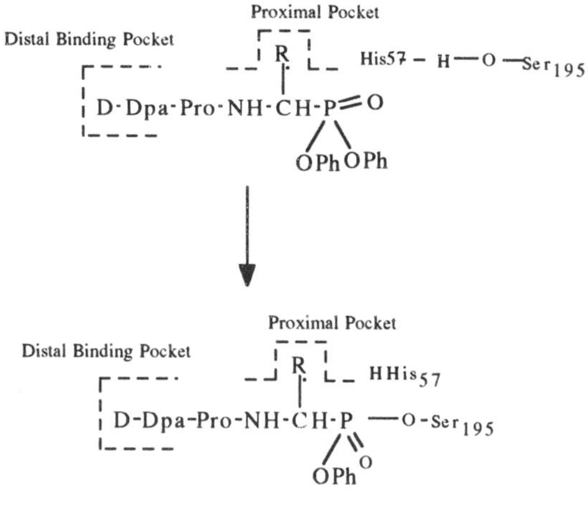

Proximal Pocket
Distal Binding Pocket
R
His57 – H—O—Ser_{195}
D-Dpa-Pro-NH-CH-P=O
OPh OPh

Proximal Pocket
Distal Binding Pocket
R
$HHis_{57}$
D-Dpa-Pro-NH-CH-P —O-Ser_{195}
OPh O

Figure 4 Putative interaction of thrombin with compound 4.

For this reason, the enzyme's catalytic site serine must be used to activate the phosphorus atom for a substitution reaction to occur. The phosphorylated enzyme should be a very good transition-state analogue because of the tetrahedral phosphorus which resembles the tetrahedral intermediate formed during hydrolysis of natural peptide substrate. When the complex between 4c and thrombin was highly diluted with buffer, 50% of thrombin activity was slowly regained (4 days) but the remaining thrombin seemed to be irreversibly bound.

This suggests that there might be two mechanisms involved in the interaction between enzyme and inhibitor, one giving a more stable enzyme-inhibitor complex. Table 2 shows the selectivity of the inhibitors. Both 4c and 4d are very poor inhibitors of plasmin, urokinase, kallikrein and factor Xa. The preliminary *in vivo* test of one of the inhibitors shows no significant blood pressure-lowering effect (Table 3). Thus, these new thrombin inhibitors show properties which make them interesting as potential antithrombotic agents.

REFERENCES

1. J. Oleksyszyn, and J.C. Powers, Irreversible inhibition of serine proteases by peptide derivatives of (α-aminoalkyl) phosphonate diphenyl ester, Biochem. 30:485 (1991).
2. N.S. Sampson, and P.A. Bartlett, Peptidic phosphonylating agents as irreversible inhibitors of serine proteases as models of the tetrahedral intermediates, Biochem. 30:2255 (1991).
3. J. Fastrez, L. Jespers, D. Lison, M. Renard, and E. Sonveaux, Synthesis of new phosphonate inhibitors of serine proteases, Tetrahedron Lett. 30:6861 (1989).
4. P. Kafarski, and B. Lejczak, Biological activity of aminophosphonic acids, Synthesis of new phosphonate inhibitors of serine proteases, Phosphorous, Sulfur and Silicon, 63:193-215 (1991).
5. G. Claeson, and L. Aurell, Small synthetic peptides with affinity for proteases in coagulation and fibrinolysis: an overview, Ann. N.Y. Acad. Sci. 370:789 (1981).
6. G. Claeson, L. Cheng, N. Chino, M.F. Scully, V.V. Kakkar, J. Deadman, and S. Elgendy, Thrombin inhibitors, Patent Appl. GB9024129.0. (1991).
7. I. Schechter, and A. Berger, On the size of the active site in proteases I. Papain, Biochem. Biophys. Res. Commun. 27:157 (1967).
8. A.A. Kossiakoff, and S.A. Spencer, Direct determination of the protonation states of aspartic acid-102 and histidine-57 in the tetrahedral intermediate of the serine proteases: Neutron structure of trypsin, Biochem. 20:6462 (1981).
9. J. Oleksyszyn, and J.C. Powers, Irreversible inhibition of serine proteases by peptidyl derivatives of α-aminoalkylphosphonate diphenyl ester, Biochem. Biophys. Res. Commun. 161:143 (1989).
10. J. Oleksyszyn, L. Subotkowska, and P. Mastralerz, Diphenyl amino-alkane-phosphonates, Synthesis, 985: (1979).
11. B. Lejczak, and P. Kafarski, Transesterification of diphenyl phosphonates using the potassium fluoride crown ether/alcohol system; Part 2, The use of diphenyl 1-aminoalkane-phosphonates in phosphonopeptide synthesis, Synthesis 412 (1982).
12. L. Cheng, C.A. Goodwin, M.F. Scully, V.V.Kakkar, and G. Claeson, Substrate -related phosphonopeptides: A new class of thrombin inhibitor, TET. Lett. 32:7333 (1991).

THE SYNTHESIS AND ANTICOAGULANT ACTIVITY OF NOVEL PEPTIDYLFLUOROALKANES

B. Neises, C. Tarnus, R.J. Broersma, C. Bald, J.M. Rémy, and E.F. Heminger

Marion-Merrell-Dow Research Institute
16, rue d'Ankara, 67084 Strasbourg Cedex (France)
and 9550 N.Zionsville RD, Indianapolis
Indiana 46268-0470, USA

INTRODUCTION

Thrombin - a protease with trypsin like specificity - plays a central role in thrombosis and haemostasis. Its conversion of fibrinogen into insoluble fibrin is accompanied by activation of other blood coagulation factors such as V, VIII, XIII, and protein C[1].

Inhibitors of this enzyme are potential drugs for the control of blood coagulation[2]. In recent years various types of low molecular weight inhibitors have been developed[3-5] along with the evaluation of natural inhibitors hirudin[6], TICK-peptide[7].

Peptidylfluoroalkanes have been shown to be potent inhibitors of serine-dependent proteases[8-10]. Based on the sequence required for a high affinity to the thrombin binding sites[10-11], we designed and synthezised the tripeptide analogs D-Phe-Pro-Arg-CF_3 and D-Phe-Pro-Arg-CF_2CF_3 (MDL 73756 and 74063; structures Fig.1), as potent thrombin inhibitors.

Synthesis

Key steps in the synthesis of MDL 73756 and 74063 from ornithine are 1) a modified Dakin-West reaction to introduce the trifluoromethyl (or pentafluoroethyl) ketone moiety, and 2) the guanylation in the δ-amino residue of the intermediate ornithine tripeptide analog as depicted in scheme 1.

Anticoagulant Activity

The anticoagulant activity of the peptidylfluoroalkanes (MDL 74063 and 73756) was determined (see Table 1) by measuring their ability to prolong clotting of human

The Design of Synthetic Inhibitors of Thrombin
Edited by G. Claeson, *et al.*, Plenum Press, New York, 1993

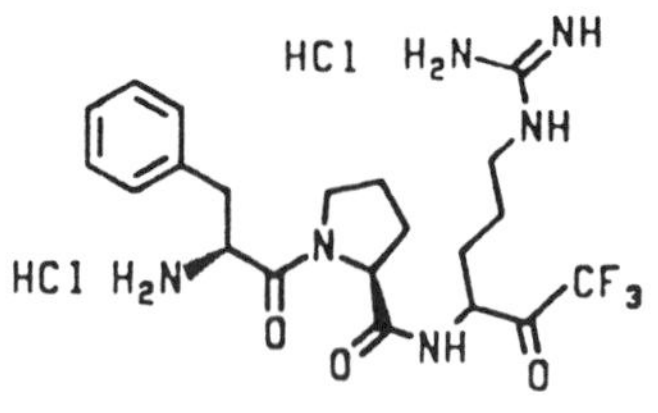

MDL 73756 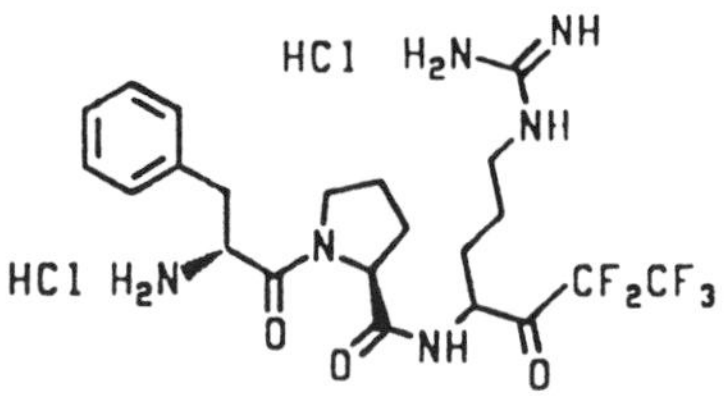 MDL 74063

Fig.1

plasma in the activated partial thromboplastin time (APTT) and thrombin time (TT) assays. Both inhibitors prolonged clotting in a dose-dependent manner and concentrations required to double the APTT were 2.5 μM for MDL 74063 and 18.5 μM for MDL 73756. MDL 74063 and MDL 73756 also increased thrombin times 2-fold at 1.7 and 37.1 μM, respectively. MDL 74063 and MDL 73756 selectively inhibited thrombin-induced aggregation of human platelets by half (IC_{50}) at 1.3 and 26.3 μM, respectively. Thus MDL 74063 was more effective than MDL 73756.

Table 1 Inhibitory activity and selectivity

Thrombin	Kon = $7.1x10^{3}$ M^{-1} s^{-1} Koff << 10^{-5} s^{-1}	Kon = $9.8x10^{3}$ M^{-1} s^{-1} Koff = $7.3x10^{-4}$ s^{-1}
	KI << 10^{-9} *M*	*KI* = $7.4x10^{-8}$ *M*
Plasmin	Kon = $1.3x10^{3}$ M^{-1} s^{-1} Koff = $1.1x10^{-4}$ s^{-1}	Kon = $6.5x10^{2}$ M^{-1} s^{-1} Koff = $1.4x10^{-3}$ s^{-1}
	KI = $8.5x10^{-4}$ *M*	*KI* = $2x10^{-6}$ *M*
Trypsin	Kon = $3.7x10^{3}$ M^{-1} s^{-1} Koff = $5x10^{-4}$ s^{-1}	Kon = $2.1x10^{4}$ M^{-1} s^{-1} Koff = $5x10^{-4}$ s^{-1}
	KI = $1.4x10^{-7}$ *M*	*KI* = $2.4x10^{-8}$ *M*

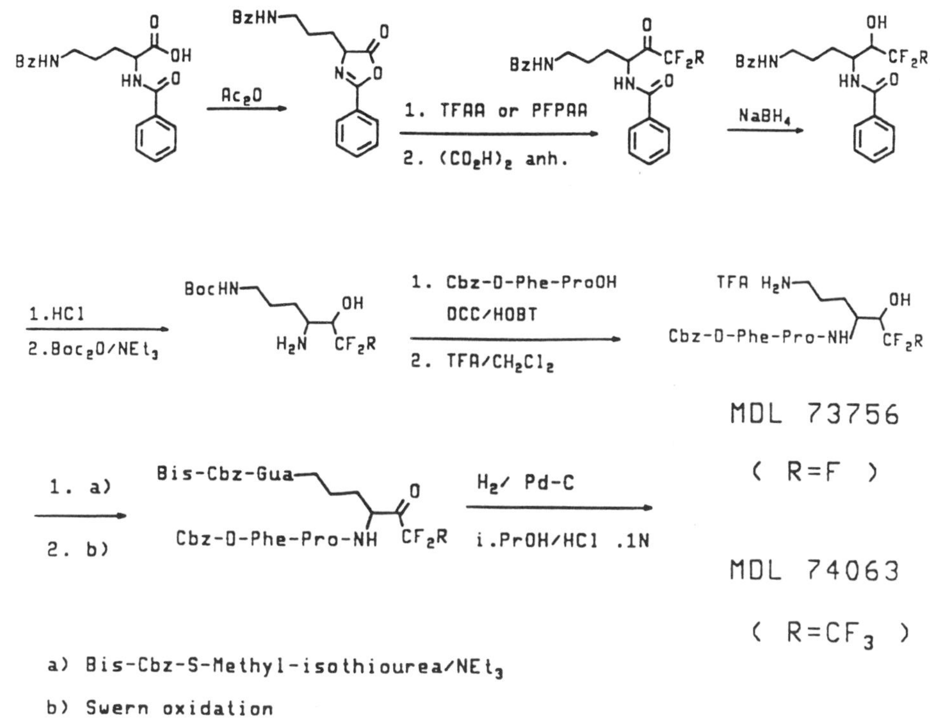

BzHN OH HN O — Ac_2O → BzHN N O O — 1. TFAA or PFPAA 2. $(CO_2H)_2$ anh. → BzHN CF_2R HN O — $NaBH_4$ → BzHN OH CF_2R HN O

1. HCl 2. Boc_2O/NEt_3 → BocHN OH H_2N CF_2R — 1. Cbz-D-Phe-ProOH DCC/HOBT 2. TFA/CH_2Cl_2 → TFA H_2N OH Cbz-D-Phe-Pro-NH CF_2R

1. a) 2. b) → Bis-Cbz-Gua O Cbz-D-Phe-Pro-NH CF_2R — H_2/ Pd-C i.PrOH/HCl .1N →

MDL 73756 (R=F)

MDL 74063 ($R=CF_3$)

a) Bis-Cbz-S-Methyl-isothiourea/NEt_3

b) Swern oxidation

Scheme 1

DISCUSSION

MDL 73756 and MDL 74063 are potent inhibitors of human thrombin. While the trifluoromethyl ketone displayed "irreversible"-like inactivation of the enzyme (with an interesting selectivity versus the two other enzymes tested) we observed only slow binding kinetics on thrombin with the pentafluoroethyl ketone. The kinetic data indicate a less stable EI complex of MDL 74063 than for 73756, a lower selectivity towards trypsin, but a high selectivity compared to plasmin. Thus we expected MDL 73756 to be the better candidate for the test of anticoagulant activity! Surprisingly, the results we obtained stand in a strong contrast to the expected. MDL 74063 appears to be a highly active inhibitor *in vitro* as determined by its effectiveness on clinical assays of coagulation using human plasma. This suggests that MDL 74063 may be useful as a treatment for thrombosis. This is particularly relevant in view of the pivotal role of thrombin in activation and aggregation of platelets and arterial thrombosis[12].

REFERENCES

1. J.W. Fenton II, Thrombin specificity, Ann. N.Y. Acad. Sci. 370:468 (1981).
2. F. Markwardt, Pharmacologic control of blood coagulation by synthetic, low molecular weight inhibitors of clotting enzymes, in: D.A. Walz, L.E. McCoy (Eds), Ann. N.Y. Acad. Sci. 370:757 (1981).

3. Review by R.B. Wallis, Inhibition of thrombin, a key step in thrombosis, in: Drugs of Today 25:597 (1989).
4. W. Stuber, H. Kosina, and H. Neimburger, Synthesis of a tripeptide with a C-terminal nitrile moiety and the inhibition of proteinase, Int. J. Pep. Prot. Res.31:63 (1988).
5. C. Kettner, L. Mersinger, and R. Knabb, The selective inhibition of thrombin by peptides of boroarginine, J. Biol. Chem. 265:18289 (1990).
6. F. Markwardt, in: Methods in Enzymology, S.P. Colowick, and N.O. Kaplan, Eds, 19:924 Academic Press, New York (1970).
7. A. Hoffmann, P. Walsmann, G. Riesener, M. Paintz, and F. Markwardt, Isolation and characterization of a thrombin inhibitor from the tick ixodes ricinus, Pharmazie, 46(3):209 (1991).
8. a) R. Abeles, and T.A. Alston, Enzyme inhibition by fluoro compounds, J. Biol. Chem. 265:16705 (1990), and references cited therein;
   b) Ch.P. Govardhan, and R.H. Abeles, Structure-activity studies of fluoroketone inhibitors of alpha-lytic protease and human leukocyte elastase, Arch. Biochem. Bioph. 280:137 (1990).
9. a) N.P. Peet, J.P. Burkhart, M.R. Angelastro, E.L. Giroux, S. Mehdi, P. Bey, M. Kolb, B. Neises, and D.G. Schirlin,Synthesis of peptidyl fluoromethyl ketones and peptidyl alpha-keto esters as inhibitors of porcine pancreatic elastase, human neutrophil elastase, and rat and human neutrophil cathepsin G, J. Med. Chem. 33:394 (1990); M. Kolb, J.P. Burkhart, M.J. Jung, F.E. Gerhart, E.L. Giroux, B. Neises, D.G. Schirlin, Patent application US 697987 (1985); EPO publication No. 0195212 (1986);
   b) P. Bey, N.P. Peet, M.R. Angelastro, and S. Mehdi, Preparation of amino acid and peptide pentafluoroethyl ketone derivatives as proteinase inhibitors, Patent application US 385624 (1991).
10. T. Ueda, C. Kam, and J.C. Powers, The synthesis of arginylfluoroalkanes, their inhibition of trypsin and blood-coagulation serine proteinases and their anticoagulant activity, Biochem. 265:539 (1990).
11. S. Bajusz, E. Szell, D. Bagdy, E. Barabas, G. Horvath, M. Dioszegi, Z. Fittler, G. Szabo, A. Juhasz, E. Tomori, and G. Szilagyi, Highly active and selective anticoagulants: D-Phe-Pro-Arg-H, a free tripeptide aldehyde prone to spontaneous inactivation, and its stable N-methyl derivative, D-MePhe-Pro-Arg-H, J. Med. Chem. 33:1729 (1990).
12. S.R. Hanson, and L.A. Harker, Interruption of acute platelet-dependent thrombosis by the synthetic antithrombin D-phenylalanyl-L-prolyl-L-arginyl chloromethyl ketone, Proc. Natl. Acad. Sci. USA 85:3184 (1988).

TRANSITION STATE ANALOGUE INHIBITORS OF THROMBIN: SYNTHESIS, ACTIVITY AND MOLECULAR MODELLING

W. Jetten, J.A.M. Peters, A. Visser, P.D.J. Grootenhuis
and H.C.J. Ottenheijm

Organon International B.V.
P.O. Box 20,5340 BH Oss
The Netherlands

Thrombin, an important enzyme in the blood coagulation system, cleaves four Arg-Gly peptide bonds in fibrinogen in order to form flbrin monomers, which form blood clots after polymerisation. In thrombosis the blood coagulation system is disturbed, resulting in an excess of blood clot formation. Thrombosis can be inhibited by inactivation of thrombin.

We aim at the development of peptide-derived thrombin inhibitors in order to modulate the blood coagulation processes. The scissile Arg-Gly bond is the focus of our attention. We want to validate the hypothesis that inhibition of thrombin can be accomplished by replacing the scissile Arg-Gly bond by a transition state analogue. Such a transition state analogue should mimic the transition state of the peptide "bond" during hydrolysis.

We selected the ketomethylene, hydroxyethylene and hydroxymethylene moieties as such peptide bond replacements that mimic the Arg-Gly bond in the transition state. In order to validate the concept, it was required to firstly find a short peptide sequence embedding the Arg-Gly bond. At the N terminus of Arg we built in H-D-Phe-Pro- as it had been shown by others that D-Phe-Pro-Arg is a substrate with high affinity for thrombin. Elongation at the C terminus of Gly deserves some more comments. In the many previously published studies on thrombin inhibitors the P' region has hardly received any attention. We expected that proper positioning of amino acid side chains on the C terminus of the Gly may contribute to the enhancement of the inhibitory activity. We built in several amino acid residues on the P2' position. Elongation of compound 1 with phenylalanine[5] resulted in an enhanced inhibitory activity.

Subsequently, in the optimal peptide sequence the transition state analogues were built in. The results of the *in vitro* assays done with the compounds prepared, show that inhibition can be increased indeed by mere substitution of the amide bond by a ketomethylene moiety (compound 2, see Table I). Introduction of the other moieties did not improve the inhibitory activity[3,4,5].

In conclusion, the ketomethylene amide bond replacement in combination with elongation with a phenylalnine at P2'[6] resulted in the most potent compound in this study. These results clearly indicate that it is important to elongate the inhbitors in the P' region in order to obtain enhanced inhibitory activities towards thrombin.

The Design of Synthetic Inhibitors of Thrombin
Edited by G. Claeson, *et al.*, Plenum Press, New York, 1993

Table 1 Results of the transition state analogues in the antithrombin assay.

H-D-Phe-Pro-Arg- A -Gly- Q

No.	A	Q	IC_{50}, mM a-IIa
1	C(O)-NH	OH	4.08
2	C(O)-CH_2	OH	0.84
3	S- CH(OH)-CH_2	OH	7.93
4	S- CH(OH)	OH	>10
5	C(O)-NH	Phe-OH	0.41
6	C(O)-CH_2	Phe-OH	0.037
7	S- CH(OH)	Phe-OH	>10

Recently, the X-ray structure of the human thrombin - H-D-Phe-Pro-Arg-chloro-methyl-ketone (PPACK) complex was published by Bode et al. (EMBO J. 1989,8, 3467). This structure served as a starting point for model-building the thrombin H-D-Phe-Pro-Arg-Gly-Phe[5] complex. In the figure one of the possible orientations of the pentapeptide that fit well in the active site cleft of thrombin is depicted. Apparently, the active site cleft of thrombin is well-suited to harbour specific peptides that have been extended at the C-terminal end of the scissile bond. This observation substantiates the results described above.

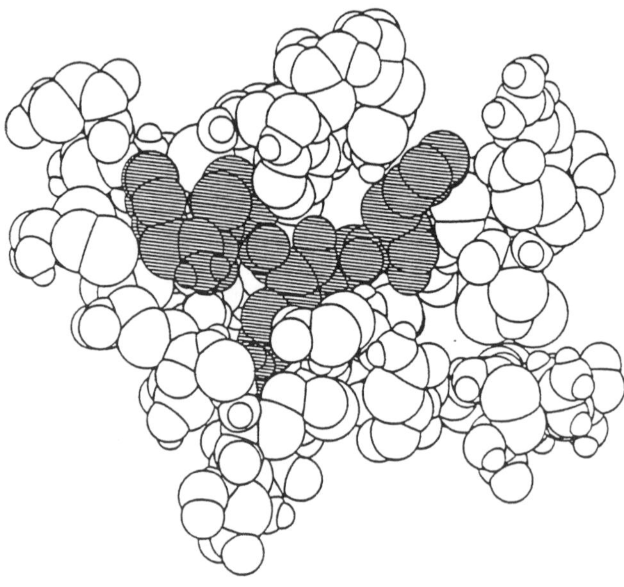

Figure 1 Space filling representation of the active site region of thrombin showing H-D-Phe-Pro-Arg-Gly-Phe-OH with shaded spheres.

HIRUDIN: THE FAMOUS ANTICOAGULANT AGENT

Fritz Markwardt

Institute of Pharmacology and Toxicology
Medical Academy Erfurt
Nordhauser Str. 74,0-5010 Erfurt, FRG

INTRODUCTION

Hirudin, the Anticoagulant from Medicinal Leeches

It has been known for a long time already that medicinal leeches (Hirudo medicinalis) contain a substance with anticoagulant properties. In early studies, aqueous extracts and dry preparations from leeches containing minute amounts of the active substance were used to prevent blood from clotting. Only after protein chemistry had developed and the biochemistry of blood coagulation had been elucidated, the anticoagulant agent named hirudin was isolated and its chemical nature and mode of action could be clarified.

The historical background of hirudin isolation as well as the individual steps of its biochemical and pharmacological characterization have already been described in several reviews[1-5].

Biochemistry

Hirudin is extracted from the homogenized heads of medicinal leeches and enriched by precipitation procedures followed by ion exchange chromatography and gel filtration[6-8]. Affinity chromatography on matrix bound thrombin can be used to obtain highly purified hirudin preparations[10-11]. The pure anticoagulant agent obtained is a carboxyhydrate-free single chain miniprotein containing 65 amino acids with a molecular weight of about 7 kDa. The amino-acid composition is characterized by a remarkably high content of acidic amino acids at the C-terminal, the absence of arginine, methionine, and tryptophan, a sulphated tyrosine residue as well as by three intramolecular disulfide bridges[3,7,12]. Considerable progress in the isolation, purification and analysis of hirudin has shown that the medicinal leech generates several variants of the miniprotein in trace amounts. The isoinhibitor forms share 85-90% amino acid sequence homology. Their main structural elements such as the N-terminal hydrophobic amino acids, the core region linked through three disulfide bridges and the unique clustering of acidic amino acid residues in the C-

The Design of Synthetic Inhibitors of Thrombin
Edited by G. Claeson, *et al.*, Plenum Press, New York, 1993

terminal half, remain constant[13-16]. The N-terminal amino acids Val-Val could be replaced by Ile-Thr (Figure 1). The analysis of the anticoagulant effect of hirudin showed that it is a selective inhibitor in that it is highly specific for α-thrombin in contrast to other closely related serine proteinases of the blood.

It forms tight stoichiometric complexes with thrombin whereby the active centre of the enzyme is blocked (apparent, assay-dependent inhibitor constants, Ki 10^{-12} to 10^{-14} M) (Figure 2). The formed complex is very stable in the physiological pH-range but can be dissociated by acidification or heating whereby thrombin is denaturated and hirudin released in its active form[3,7].

Already the first investigations on the thrombin-hirudin interaction had shown that the complex formation between hirudin and its target enzyme is unique among other serine proteinase inhibitors[3,17].

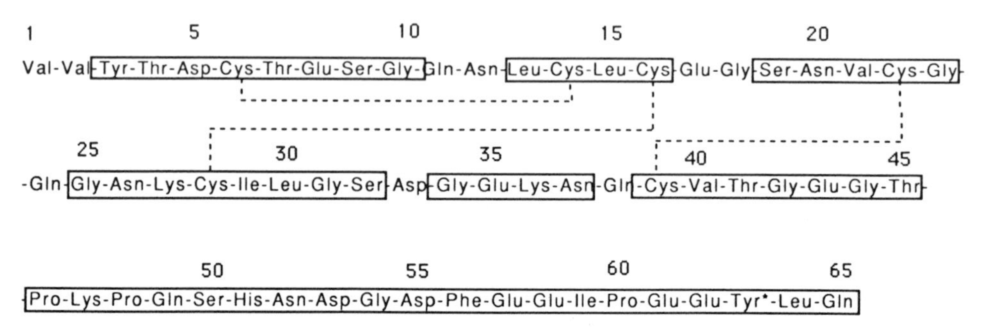

Figure 1 Primary structure of hirudin. Encircled residues are classified as conservative regions. Tyr* = sulfated tyrosine.

The secondary and tertiary structures of hirudin are basic for its antithrombin function which is lost during reduction of the disulfide bridges, splitting off the acidic C-terminal segment, or by proteolytic degradation. From binding studies with thrombin chemically modified at its active site serine it was concluded that the catalytic site is not necessarily required for hirudin-thrombin complexing. Indeed hirudin is able to bind the active site blocked thrombin[18-22].

The fast and specific reaction of hirudin with thrombin was utilized for quantitative determination of the inhibitor[3]. One mole hirudin complexes with one mole thrombin. This corresponds to the antithrombin activity of pure hirudin of which 1 mg inhibits about 5 mg of human thrombin. Therefore, the activity of hirudin is measured in antithrombin units (AT-U) where 1 AT-U is the amount of hirudin which neutralizes 1 IU of thrombin. Pure hirudin contains approximately 10000-15000 AT-U/mg protein, depending on the thrombin used for standardization.

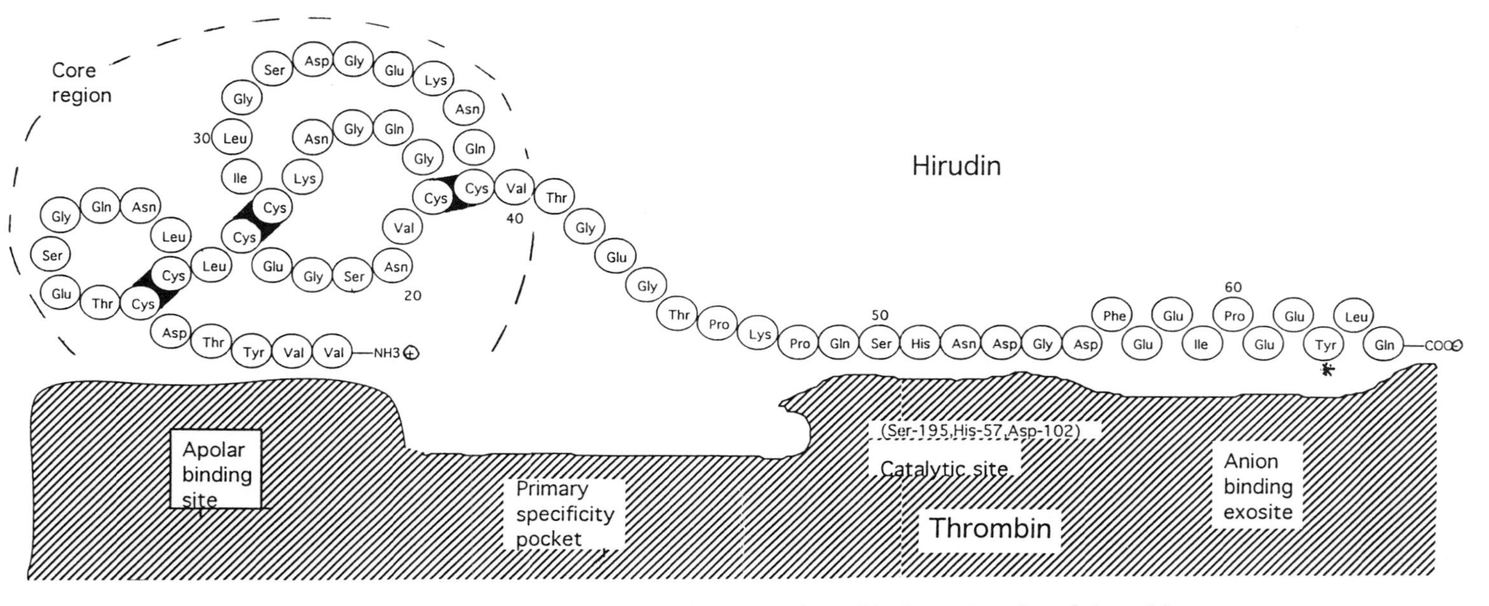

Figure 2 Scheme of the hirudin interaction with the active site of thrombin.

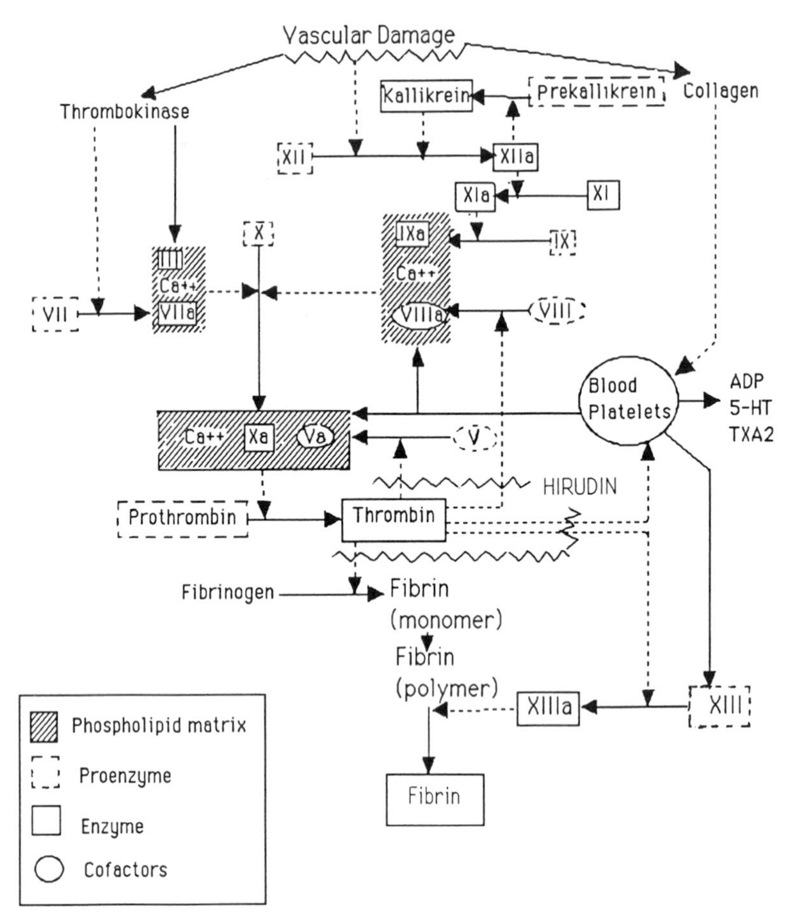

Figure 3 Blockade of the clotting system by hirudin.

The importance of the thrombin inhibitor hirudin as anticoagulant becomes especially evident when one takes into account that thrombin takes a central position in the coagulation system. The clotting enzyme catalyzes not only the formation of fibrin but also activates clotting factors and blood platelets. Furthermore, thrombin has direct effects on the vascular endothelium and mediates non-haemostatic cellular events. Therefore, the inhibition of thrombin by hirudin represents not only an effective interference in the coagulation process, but also modulates the multiple bioregulatory effects of the enzyme (Figure 3).

Depending on the hirudin concentration in blood, coagulation is retarded or completely inhibited. Correspondingly, the clotting variables change. From 1 ml of human blood, approximately 100-150 units of thrombin may be generated on activation with thromboplastin. The inhibition of this amount of thrombin therefore required 100-150 AT-U of hirudin. Therefore, an uncoagulable state in human plasma was reached at concentrations of more than 100 AT-U or 10 mg hirudin/ml blood.

Since the isolation and biochemical analysis, and after the specific reaction of hirudin with thrombin had been elicited, hirudin preparations have been employed for experimental research work in haemostaseology, selectively to interrupt thrombin action in diagnostic or preparation procedures and for blood collection; whereby hirudin was shown to provide a unique opportunity to unravel the complex physiological and pathological role of thrombin[23-28].

Pharmacology

Already at the very beginning of our pharmacological studies with the isolated naturally occurring thrombin inhibitor we showed that due to its pronounced and specific antithrombin effect, hirudin is an antithrombotic agent of high quality[29-32].

Only after isolation of the pure active substance and after elucidation of its mode of action did the in vivo use of hirudin become possible. Toxicology studies in experimental animals and in human volunteers had shown that, apart from its anticoagulant effect, hirudin is pharmacodynamically inert and is well tolerated *in vivo*. Haemorrhagic complications and signs of sensitization were not observed. The intravenous injection was tolerated without changes in heart and respiratory rates. The blood pressure remained constant. Hirudin did not cause any changes in platelet counts, fibrinogen level and plasma haemoglobin[30,33,34].

Pharmacokinetic data were obtained through methods for the determination of the activity. Pharmacokinetic data resulting from the evolution of its blood level and urinary excretion have shown that hirudin given intravenously to experimental animals is eliminated relatively rapidly. After the initial disappearance which was interpreted as a distribution phenomenon, first-order elimination kinetics followed. The elimination half-life was approximately 1 h. Hirudin was distributed into the extracellular space and a large percentage of the polypeptide was eliminated in active form through the kidneys by glomerular filtration[32,34-38]. Hirudin displays its antithrombotic action in dependence on the blood level. The efficacy of hirudin in preventing venous and arterial thrombosis and DIC was demonstrated in various animal models[17,30,34,37-42]. Clinical pharmacological studies corroborated the specific pharmacodynamic and pharmacokinetic properties of hirudin found in animal experiments[43,44]. The clinical use of hirudin remained limited to a few pilot studies since pure hirudin was not available in adequate amounts[45].

The Comeback of Hirudin - Initiated by Progress in Biotechnology

Beginning in 1986 and based on information about the structure of hirudin, several specialists in biotechnology studied cloning and expression of a cDNA coding for hirudin and succeeded in processing hirudin (for reviews see 46,47). Recombinant hirudin (rH) was now produced in different biological systems and expressed in bacteria (E. coli) or yeast cells (S. cerevisiae), through which it had become posssible to provide sufficient quantities of hirudin and hence the therapeutical use of the anticoagulant hirudin was no longer a subject of theoretic interest only. Here was the starting point for hirudin to return into the focus of general interest (for reviews see 5,48,49,50).

Biochemistry

Except for the absent sulphate residue on Tyr-63, the recombinant products showed the same configuration as natural hirudin occurring in leeches. Recombinant DNA-desulfato hirudins and native hirudin have both target and mechanism of action in common. The absence of the sulphate group leads only to an approximately tenfold increase in the Ki-value. Their anticoagulant activity in clotting assays was the same as known from previous studies with hirudin.

Classically, r-H has been determined as n-H by its thrombin inhibitory activity as expressed in antithrombin units (AT-U). A standardized thrombin is required for this procedure and the difficulties involved in thrombin standardization affect hirudin determination. Thus, assay conditions and techniques needed to be specified for determination purposes and that was the point where we developed a hirudin standard which can also be used as a crosscheck on the thrombin standard[51]. One mole of hirudin complexes with one mole of thrombin via a fast and specific reaction. Consequently, the titration of r-H with thrombin is possible. Therefore, the hirudin concentration may be expressed on molar or weight basis, too. Additional valuable information about the structure-function relationship of hirudin and the tight-binding mechanism was obtained from studies with the genetically engineered variants and fragments of hirudin (for review see 52).

The results obtained from amino acid variation by site-directed mutagenesis indicate that the N-terminus of hirudin contributes to the formation of the thrombin - hirudin complex[9,15]. Detailed studies were carried out to clarify the structure-activity relationships by investigating systematically the role of each single residue within the hirudin peptide sequence[55-57]. The minimal sequence with maximum anticoagulatory activity was identified to contain 12 residues which correspond to the aminoacids in positions 53-64 of desulfato hirudin. The tyrosine-sulfated peptide was ten times more potent in inhibiting coagulation of human plasma. It followed from this knowledge that smaller synthetic peptides based on the functional C-terminal fragment of hirudin could be considered for anti-coagulation[58,59]. Studies on the crystallographic structure of the hirudin-thrombin complex elucidated the hirudin-thrombin interaction at the molecular level and confirmed the established high affinity and specificity of the inhibitor. Moreover, they have shown which structure regions of the enzyme and the inhibitor are involved in the thrombin-hirudin interaction[60,61]. These investigations prompted the design of synthetic hirudin analogues as a novel class of peptides, called hirulogs[62]. They consist of an anion-binding exosite-recognition sequence such as that in the hirudin C-terminus and a linker of glycyl residues which forms a connecting bridge to an 'active site' specificity sequence (D-Phe-Pro-Arg-Pro) with a restricted Arg-Pro scissile peptide bond. Hirulog - having 4 glycyl residues as a linker - appeared to be highly specific for thrombin like its parent molecule hirudin.

As already found in studies on native hirudin, the anticoagulant effect of recombinant hirudins results from its blocking the clotting enzyme thrombin which takes a central position in the clotting system. In the hirudin-thrombin complex all proteolytic functions of the enzyme are blocked[21,63,64]. Thus, hirudin prevents not only fibrinogen clotting but also other thrombin-catalyzed haemostatic reactions. Therefore, by instantaneous inhibition of the initial small amount of thrombin generated after activation of the coagulation system, the positive feedback on prothrombin activation is prevented that otherwise would lead to accelerated generation of further thrombin[64,65].

In accordance with other studies on native hirudin, thrombin loses its effect on platelets after complexing with r-hirudin, i.e. the thrombin-induced platelet aggregation and release reaction are prevented in hirudinized blood[57,66-68].

It is of importance that hirudin has no effect on platelet aggregation and release reaction induced by other agents than thrombin, such as ADP, collagen, arachidonic acid and PAF. In hirudinized plasma, the response of platelets to aggregating substances differed from that in citrated plasma especially with regard to serotonin release and formation of thromboxane[69,69a,70]. After complexing with hirudin, thrombin loses its cellular nonhaemostatic effects, such as proliferation of fibroblasts, stimulation of endothelial cells and contraction of smooth muscle cells[58,71-75]. For thrombogenesis it is of importance that hirudin is able to inhibit the vasoconstrictory action of thrombin in de-endothelialized vessels[72].

Pharmacology

Based on our experience concerning native hirudin, and after pharmaco-toxicological profiling of recombinant hirudin, we were the first to start with preclinical and clinico-pharmacological investigations in 1988[76-81]. These studies and subsequent ones from other research workers showed that the pharmacological properties of recombinant hirudin are very similar to those of native hirudin. This applies to pharmaco-dynamic as well as to pharmacokinetic characteristics[48,82-89].

Pharmacokinetics

The efficacy of hirudin as an anticoagulant agent *in vivo* decisively depends on the maintenance and control of an adequate level in blood. Therefore, the knowledge of its blood level course as well as of the induced changes of clotting parameters are essential prerequisites for the therapeutic use of hirudins. The pharmacokinetic behaviour of recombinant hirudin has been studied in detail[38,76,77,90,91]. The analytic technique used for the estimation of hirudins in blood and urine is based on hirudin-thrombin complexing (for review see 92). Both clotting and chromogenic substrate methods are suited to measure thrombin activity[93,94]. Radiolabelled hirudin, immunoassays using antisera specific to hirudin, or monoclonal antihirudin antibodies were used[6,64,69,13,43,48,95-99].

The pharmacokinetic data were in good agreement with those of native hirudin and can be best described by an open two-compartment model with first-order kinetics. After intravenous bolus administration of hirudins in experimental animals, values of 10 to 15 min for the distribution and 60 min for the elimination half-lives were obtained. After subcutaneous administration peak, plasma hirudin levels were reached after 1 to 2 hours. The hirudins are distributed into the extra-cellular space and are predominantly excreted with the urine (Table 1). The renal clearance of hirudin approximates the creatinine clearance which suggests a glomerular filtration. Most of the injected dose of hirudin was recovered in the urine in unmodified form[100]. The percentage of the dose which was renally excreted differed in various species. In dogs the inhibitor was almost completely eliminated via glomerular filtration while the renal excretion of the administered amount of hirudin amounted to about 40 % in baboons, 20% in pigs and 10% in rats[96].

In nephrectomized dogs, about 80% of an intravenously administered r-hirudin were distributed into the extravascular compartments within 60 min. After this distribution phase the blood levels of hirudin remained nearly constant showing almost complete renal excretion of hirudin[101].

In order to prolong its duration of action and to control the pharmacokinetic

Table 1 Mean pharmacokinetic data of recombinant hirudin (r-H) in different animal species after a single intravenous bolus injection.

		Rat	Rabbit	Dog	Monkey
	r-H (mg/kg)	1.0	1.0	0.5	0.5
$t_{1/2}$ α	(h)	0.15	0.10	0.25	0.35
$t_{1/2}$ β	(h)	1.10	1.15	1.20	1.10
AUC	(μg/ml x h)	1.60	3.30	1.55	1.45
V_{dss}	(l/kg)	0.65	0.25	0.28	0.25
Cl_{tot}	(ml/min)	3.05	12.40	173.00	22.45

Table 2 Mean pharmacokinetic data of recombinant Hirudin (r-H) after single intravenous bolus injection of 0.5 mg r-H/kg in human volunteers.

$t_{1/2}$ α	(h)	0.2
$t_{1/2}$ β	(h)	1.8
AUC_{0-12}	(μg/ml x h)	3.2
V_{dss}	(l/kg)	0.3
Cl_{tot}	(ml/min)	190.0
Cl_{ren}	(ml/min)	130.0

behaviour, native and r-hirudin were covalently bound to biomacromolecules and prosthetic biomaterials[102]. Heterobifunctional crosslinking reagents were used to form covalent crosslinks between hirudins, dextranes and polyethyleneglycol producing active conjugates. For instance, dextran-bound hirudin was solely distributed in blood plasma and eliminated with a half-life of about 7 hours[103,104].

Haemorrhagic Effects

It is no surprise that hirudins do increase the bleeding time. But pronounced prolongation was only found at doses higher than required for the antithrombotic effect[37,76]. Normally, the rapid clearance of r-hirudin suffices to manage the haemostatic system so that additional neutralization of hirudin by a special antidote is not required. In cases where the elimination of hirudin is impaired, methods such as haemoperfusion, peritoneal dialysis, or haemodialysis with membranes with a sufficient large pore size can be used to remove large amounts of circulating hirudin[105].

A clinically useful antagonist of hirudin is not yet available[106,107]. The administration of prohaemostatic agents like prothrombin complex concentrates may provide sufficient antagonizing thrombin which readily form a complex with the circulating hirudin. Hirudin complexes with acylated thrombin derivatives and meizothrombin which have no clotting effect themselves, hirudin may be antagonized with chemically modified thrombin; for example, after i.v. injection of DIP-thrombin or other acyl thrombins the blood level of active hirudin was diminished[108]. Also, monoclonal or polyclonal antibodies against hirudin may be considered as antidotes for hirudin.

Antithrombotic Effects

Numerous studies with experimental animal models have been performed which corroborate the antithrombotic efficacy of hirudin (for reviews see 76,109). The antithrombotic effect of hirudin was corroborated by the prevention of the development of clotting thrombi experimentally induced in the jugular vein [41,87,110-112] and was also demonstrated in arterial thrombosis, especially in coronary thrombosis[113-116] induced by vessel wall lesions and in extracorporeal circulation[105]. The action of hirudin in disseminated intravascular coagulation (DIC) was repeatedly demonstrated in animal experiments[111,117-119]. The dose and plasma concentration of r-hirudin required for antithrombotic action depend on the kind and strength of the thrombogenic stimulus. It is most interesting that the dose of r-hirudin required to inhibit coagulation-dependent thrombosis (venous stasis and fibrin deposition in experimental DIC) is about ten times lower than that required for the inhibition of platelet aggregation-dependent thrombosis (arterial thrombosis and platelet deposition in experimental DIC).

The present state of development of recombinant hirudin analogues justifies pharmacological investigations focussing on the potential indications for this anticoagulant agent. Therefore, pharmacological studies on antithrombotic effects have to be performed using animal model systems in which the pathological mechanism largely corresponded to that of thromboembolic disorders in man, such as arterial thrombosis, venous thrombosis and DIC. Scrutinizing the possible indications (Table 3), further preclinical studies concentrated on the following three main fields:

Diffuse Microthrombosis

Hirudin may be useful for the prevention and treatment of venous thrombosis under clinical conditions with an increasing risk for development of deep vein thrombosis (DVT) and disseminated intravascular coagulation (DIC).

The action of r-hirudin in case of DIC was repeatedly demonstrated in animal experiments (for review see 118). Diffuse activation of the clotting system induced by infusion of active coagulation factors was prevented. Of special interest was the influence of hirudin on the clinically relevant endotoxin induced localized or generalized microthrombosis[43]. When hirudin was given, the endotoxin-induced consumption of clotting factors which is typical of Generalized Shwartzman Reaction (GSR) was less pronounced.

It is of importance that hirudin inhibits not only the consumption of clotting factors but also that of antithrombin III. As shown in special experiments during thrombin infusion in rats, consumption of plasma antithrombin III occurs only if heparin is administered whereas hirudin exerts an antithrombin-sparing effect without leading to reduced antithrombin III levels[63,120].

Table 3 Clinical aspects of recombinant hirudin therapy.

Pharmacodynamics

- Specific direct blockade of the clotting enzyme
- No effect on blood cells (platelets), plasma proteins, or enzymes
- Potent anticoagulant, no endogenous cofactors required
- No bleeding complications at antithrombotically effective dosages
- No side effects after acute or chronic administration
- Weak immunogen

Pharmacokinetics

- High bioavailability after s.c. administration
- Distribution in the extracellular space
- No deposition in organs
- Renal excretion in active form
- Biologic half-life of 1-2 h after i.v. injection

Indications

- Prophylaxis of postoperative venous thrombosis
- Prevention of reocclusion after intravascular (intracoronary) lysis, angioplasty, bypass operations
- Prevention of fibrin deposition in extracorporeal circulation, haemodialysis, and haemoperfusion
- Interruption of acute or chronic disseminated intravascular coagulation
- Effective in patients with antithrombin deficiencies. May be also used in the presence of thrombocytopenia

The number of fibrin deposits in the lungs, kidneys, liver and spleen was diminished under hirudin treatment. Both the increase in the right ventricular pressure and the progressive respiratory insufficiency were delayed. The animals survived the otherwise lethal endotoxin infusion (Figure 4).

Arterial Thrombosis and Reocclusion

Hirudins are effective in those models where vascular damage is caused by mechanical means, electric, or chemical lesions. For example a considerable damage to the endothelial cells was induced in the dorsal aorta in rats and in the rabbit femoral artery by experimental angioplasty. In these models, hirudin dose-dependently inhibited both platelet and fibrin deposition and decreased the amount of mural thrombi in deeply injured arterial segments[113,121,115,122].

One of the major limitations of the efficacy of fibrinolytic or mechanical dissolution of thrombi is rethrombosis. Occluding thrombi in arteries, particularly in coronary vessels, developed on the basis of previously existing arteriosclerotic stenosis which still continues after angioplasty or intravascular lysis and thus predisposes the vessels to rethrombosis. Rethrombosis produces coronary reocclusion and thus it precludes improvement of the left ventricular function and increases mortality[123]. During angioplasty, an active surface, due to the exposure of thrombogenic material, is produced and it appears to play an essential role in continued thrombosis. Hirudin was capable of preventing the reocclusion of the vessels after experimental angioplasty in the femoral arteries of rabbits which was confirmed by the reduced incidence of thrombotic occlusion[37,121-125]. The activation of the

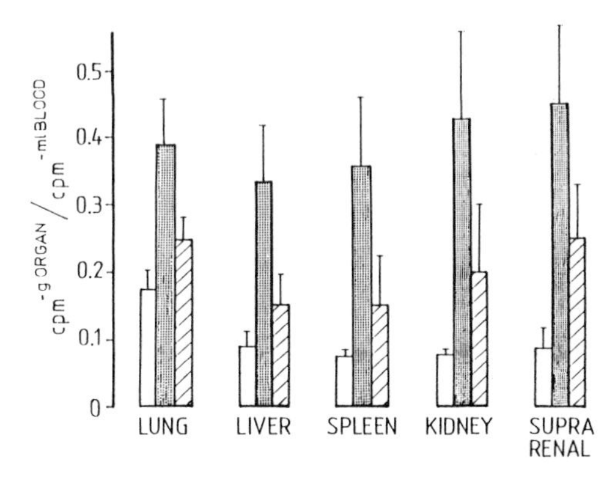

Figure 4 Influence of r-hirudin (50 μg/kg/h) on endotoxin-induced microthrombosis in weaned pigs pretreated with radiolabelled fibrinogen measured by the increase in radioactivity in the organs. Infusion of Seline (□), endotoxin (▩), endotoxin + hirudin (▨).

coagulation system during the thrombolytic therapy with tissue-type plasminogen activators or streptokinase is of major importance for the pathogenesis of thrombosis after thrombolysis.

The mechanisms contributing to rethrombosis after thrombolytic therapy have not been completely elucidated although increasing attention has been paid to the possible role of thrombus bound thrombin . Thrombin incorporated in a thrombus is enzymatically active. It has been shown that thrombolytic agents, removing successive fibrin layers, expose inaccessible molecules of active thrombin on the surface of the residual thrombus. Contrary to plasma-free thrombin, thrombin bound to fibrin is poorly accessible to heparin -antithrombin Ill complex[126]. Thus the bound thrombin remains active. Actually it has been shown that thrombolytic therapy is associated with thrombin generation and this also contributes to rethrombosis. Indeed, the activation of blood coagulation during thrombolytic therapy has already been reported. Therefore, hirudin which efficiently inhibits fibrin-bound thrombin proved to be particularly useful in preventing reocclusion after intravascular lysis[95,124,127]. This was illustrated by lysis of thrombi induced by vessel wall lesions in the coronary artery of pigs and the jugular artery of rabbits. For example, l recanalized by streptokinase, the vessel was reoccluded after a short time. Hirudin administration resulted in continuous patency of the vessel[48] (Table 4).

Extracorporeal Circulation

The efficiency of recombinant hirudin in preventing clot formation on artificial surfaces was confirmed by experiments in which materials with foreign surfaces of various degrees of thrombogenicity were placed in blood flowing through an extracorporeal circuit *in vivo*. For example, the development of occlusive thrombi in an arterio-venous shunt

Table 4 Prevention of primary thrombus formation and reocclusion of the femoral artery by r-hirudin in rabbits.

r-Hirudin dose (mg/kg sc)	n	Plasma concentration (μg/ml, x ± SD)	Incidence* of primary thrombosis	reocclusion of after angioplasty
-	6	-	6/6	6/6
0.5	4	0.8±0.06	4/4	4/4
1.0	5	0.13±0.06	2/5	4/5
2.0	5	0.25±0.06	0/5	3/5
4.0	4	0.60±0.13	0/4	0/4

*Expressed as number of animals with thrombotic occlusion over total number of animals treated.

model measured by means of occlusion time was prevented. But a significant prolongation of bleeding, measured by the amount of blood escaping from a standardized incision wound by means of conductometry was caused only by dosages higher than required for shunt patency[78].

Furthermore, r-hirudin was used for anticoagulation in experimental haemodialysis in nephrectomized dogs. After administration of hirudin, in contrast to untreated animals, no fibrin deposits were found in the dialyzer. Platelet count and fibrinogen level remained unchanged[101,128].

Clinical Pharmacological Studies

Toxicologic studies in experimental animals and tolerance studies in human volunteers had shown that, apart from its anticoagulant effect, pure recombinant hirudin is pharmaco-dynamically inert and is well tolerated *in vivo*[76,80]. Recombinant hirudin was given to healthy volunteers in single i.v. doses to study its anticoagulant effect, its pharmacokinetics and how it was tolerated.

Pharmacodynamic effects were achieved with 0.05 mg/kg and above, up to 1.0 mg/kg. In conclusion it may, therefore, be stated that all the hirudin doses were tolerated very well. No signs of bleeding tendencies were observed and safety measurements, such as clinical chemistry tests and haematological values remained unaffected. Due to the fact that hirudin shows no interaction with other blood constituents, immune or allergic reactions in experimental animals or in man were absent. That is why hirudin may be characterized as a weak immunogen and the risk of allergenicity appears to be low. No signs of sensitization were found. Hirudin-specific antibodies were not found in human

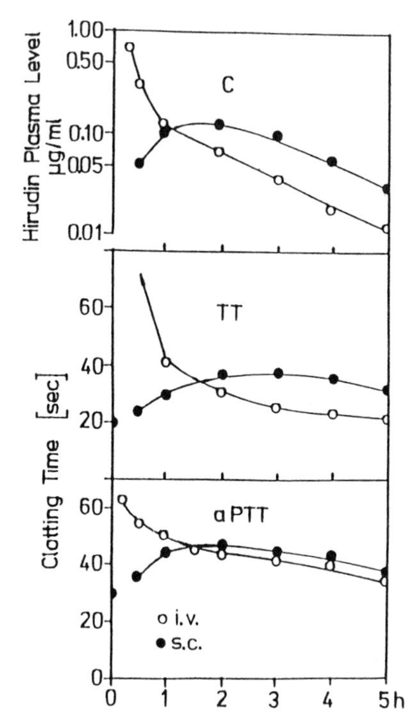

Figure 5 Mean plasma concentration of r-hirudin (C), thrombin time (TT), (4 IU/ml) and partial thromboplastin time (aPTT) after sc and iv injection of 0.1 mg r-hirudin/kg in human volunteers.

sera after treatment with recombinant hirudin, not even in volunteers who received two or three times the recombinant hirudin dose at intervals of three months. The same was true of volunteers who had a previous dose of native hirudin.

The results of the coagulation tests obtained at different times after administration of hirudin varied with the course of the hirudin level in plasma. An example is given in Figure 5. Administration of hirudin had no influence on fibrinogen level, platelet count or fibrinolytic activity; neither were any changes seen in euglobulin lysis time nor could fibrinogen degradation products be found. Bleeding time was not significantly prolonged, not even in the case of the highest recombinant hirudin concentration in plasma observed.

Important information on the pharmacokinetics of hirudins was obtained by estimation of the inhibitor in blood and urine after parenteral administration in human volunteers[48,80,85]. The pharmacokinetics of hirudins examined in phase studies after single intravenous or subcutaneous doses corresponded to those obtained in animal experiments. The pharmacokinetic parameters were calculated from the blood levels and urinary excretion by means of standard methods. The pharmacokinetic data analyzed from the semi-logarithmically plotted course of the hirudin plasma concentration determined after a single therapeutically relevant dose are listed up in Table 3. According to the data it is clear that hirudin is eliminated by a two-compartment body model but the existence of a third compartment cannot be excluded. A mean elimination half-life of 1-2 hours was calculated. Hirudin was distributed into the extracellular space and a large percentage of the polypeptide was eliminated in active form through the kidneys by glomerular filtration. The pharmacokinetic parameters C_{max}, $AUC_{o-1.5h}$, AUC_{o-5h} were dose-related - the mean values increasing with the dose in a linear fashion. In addition, since the mean $t^{½}\alpha$ and $t_{½}\beta$, total clearance, and volume of distribution remained reasonably constant as the dose increased, it may be assumed that hirudin obeys linear pharmacokinetics. This was confirmed by the fact that the total urinary excretion of hirudin also increased in an approximately linear fashion with an increasing dose. After subcutaneous administration of r-hirudin, a relatively high plasma level was found which still increased and reached a maximal concentration (C_{max}) after 2 to 3 hours (t_{max}). The AUC value corresponds to a resorption value of 75 % and demonstrates the bioavailability of r-hirudin after s.c. injection. Of special interest is the hirudin blood level after repeated s.c. injection which allowed to maintain a relatively constant r-hirudin level in blood over an extended period of time. The high recovery of active r-hirudin in urine identified renal excretion as the predominant route of clearance. This also became obvious in patients suffering from chronic renal malfunctions in whom a prolonged elimination phase was found. There was a significant correlation between half-lives of hirudin and creatinine clearance[129].

Recombinant hirudin was not accumulated after s.c. injection of therapeutically relevant dosages at 8 h intervals.

CONCLUSIONS

The clinical pharmacological studies concerning the pharmaco-kinetic and pharmacodynamic characteristics of recombinant hirudin demonstrated not only the antithrombin efficiency but also the outstanding tolerance of this specific thrombin inhibitor. Hence it follows that recombinant hirudin is a potent anticoagulant. In addition to the possible indications given in Table 3, r-hirudin may be recommended as the drug of choice in cardiovascular surgery and is going to be a promising anticoagulant where anticoagulant activity is needed for a short period of time. That may be particularly relevant in patients who have been sensitized to heparin. In addition to coronary artery bypass graft surgery, heart transplantation and use of artificial hearts may be clinically useful areas for r-hirudin. Because r-hirudin has only a minimal effect on platelets, it may not induce bleeding. Furthermore, neutralization of r-hirudin may not be required because of its short half-life and lesser bleeding potential compared to heparin. Particularly in patients deficient in antithrombin Ill, hirudin may be an interesting anticoagulant since heparin is ineffective in such patients. On the other hand, r-hirudin would be an effective anticoagulant since it does not require the cofactor AT III to exert its effects. When platelets are activated, platelet factor 4 is released which acts as an antiheparin agent.

Under the same conditions, hirudin will remain active since platelet factor 4 has no interfering effect.

Furthermore, r-hirudin may be used as an adjunct drug to enhance the antithrombotic properties of other anticoagulant or thrombolytic agents which may come in useful to prevent reocclusion during thrombolysis. R-hirudin may also be the drug of choice in other clinical situations , such as in angioplasty, haemodialysis and microsurgery irrigation.

Apart form that, hirudin may be useful in the coating of artificial devices to provide a non-thrombogenic surface. Catheters, tubings, membranes, extra-corporeal pumps, haemodialysis units, blood collection units and the like are all potential areas of biomedical application.

As a laboratory reagent, r-hirudin may have many potential applications. In the clinical laboratory, this anticoagulant for blood collection will be useful in haematology, coagulation, chemistry, immunology and cytology. For platelet studies, r-hirudin is advantageous in that it has little effect on platelet function. Sodium citrate and EDTA which chelate calcium, actually change the state of the blood sample because native calcium levels cannot be reestablished and calcium is required for proper coagulant and platelet function.

Most of the clinical developments are in the area of interventional cardiology. Hirudins have been proposed to prevent reocclusion during PTCA and thrombolysis. Continued investigations along several new research lines provided knowledge about the beneficial effects of the anticoagulant agent hirudin. Since none of the commonly known disadvantages and potentially life-threatening side effects of currently available antithrombotics is true for hirudin, now it only remains to be shown in large-scale clinical trials that the encouraging results obtained so far will finally contribute to decisive progress in antithrombotic therapy.

REFERENCES

1. S. Hoet, P. Close, J. Vermylen, and M. Verstraete, Hirudo medicinalis and hirudin. In: Recent Advances in Blood Coagulation, Poller L (ad.). Vol 3, Publ. Churchill Livingstone pp 223-44 (1990).
2. F. Markwardt, Hirudin, der blutgerinnungshemmende Wirkstoff des medizinischen Blutegels, Blut 4:161 (1958).
3. F. Markwardt, Hirudin as an inhibitor of thrombin, Methods Enzymol. 19:924 (1970).
4. F. Markwardt, Biochemistry and pharmacology of hirudin, in: Pirkle Ht. Markland Jr FS (eds), Hemostasis and animal venoms, New York, Basel, Dekker pp 255 (1988).
5. F. Markwardt, Past, present and future of hirudin, Haemostasis 21: (suppl 1)11 (1991).
6. D. Bagdy, E. Barabas, L. Graf, T.E. Peterson, and S. Magnusson, Hirudin. Methods Enzymol. 45:669 (1976).
7. F. Markwardt, Die Isolierung und chemische Charakterisierung des Hirudins, Hoppe-Seylers Z. Physiol. Chem. 308:147 (1957).
8. F. Markwardt, and P. Walsmann, Reindarstellung und Analyse des Thrombin inhibitors Hirudin, Hoppe-Seylers Z. Physiol. Chem. 348:1381 (1967).
9. P. Walsmann, Isolation and characterisation of hirudin from Hirudo medicinalis, Sem. Thromb. Haemost. 17:83 (1991).
10. P. Walsmann, Untersuchungen zur Affinitatschromatographie von Hirudin an tragerfixiertem Thrombin, Pharmazie, 36:860 (1981).

11. P. Walsmann, and F. Markwardt, On the isolation of the thrombin inhibitor hirudin, Thromb. Res. 40:563 (1985).
12. F. Markwardt, Untersuchungen Über Hirudin. Naturwissenschaften 42:537 (1955).
13. I.P. Baskova, O.U. Cherkesova, and V.V. Mosolov, Hirudin from leech heads and whole leeches and pseudo-hirudin from leech bodies, Thromb. Res. 30:459 (1983).
14. J. Dodt, H. Machleidt, U. Seemüller, R. Maschler, and H. Fritz, Isolation and characterization of hirudin isoinhibitors and sequence analysis of hirudin PA, Biol. Chem. Hoppe Seyler, 367:803 (1986).
15. J. Dodt, U. Seemüller, R. Maschler, and H. Fritz, The complete covalent structure of hirudin. Localization of the disulfide bonds, Biol. Chem. Hoppe Seyler 366:379 (1985).
16. M. Scharf, J. Engels, and D. Tripier, Primary structures of new "iso-hirudins", FEBS. Lett. 255:105 (1989).
17. P. Walsmann, and F. Markwardt, Biochemische und pharmakologische Aspekte des Thrombin inhibitors Hirudin, Pharmazie 36:653 (1981).
18. J.Y. Chang, The functional domain of hirudin, a thrombin-specific inhibitor, FEBS. Lett. 164:307 (1983).
19. S. Konno, J.W. Fenton II, and G.B. Villanueva, Analysis of the secondary structure of hirudin and the mechanism of its interaction with thrombin, Arch. Biochem. Biophys. 67:158 (1988).
20. F. Markwardt, Untersuchungen über den Mechanismus der blutgerinnung-shemmenden Wirkung des Hirudins, Naunyn Schmiedebergs Arch Pharmacol 229:389 (1956).
21. F. Markwardt, and P. Walsmannn, Die Reaktion zwischen Hirudin und Thrombin, Hoppe-Seylers Z. Physiol. Chem. 312:85 (1958).
22. S.R. Stone, and J. Hofsteenge, Kinetics of the inhibition of thrombin by hirudin, Biochemistry 25:4622 (1986).
23. F. Markwardt, Die Bestimmung des Thrombins durch Titration mit Hirudin, Arch. Pharmaz. 290:280 (1957).
24. F. Markwardt, Der Hirudintoleranztest, Klin. Wschr. 37:1142 (1958).
25. F. Markwardt, Hirudin - Biochemisch-pharmakologische Wirkung und Anwendung zur gerinnungsphysiologischen Diagnostik, In: Marx R, Thies HA (eds) controlle von Antithrombotika, Basel, Editiones Roche pp 105 (1980).
26. R. Schmutzler, and F. Markwardt, Der Hirudintest, eine Mikromethode zur Kontrolle des Prothrombinspiegels bei der Antikoagulantientherapie, Klin. Wschr. 40:796 (1962).
27. K. Stocker, Laboratory use of hirudin, Semin. Thromb. Hemost. 17:113 (1991).
28. P. Walsmann, Uber den Einsatz des spezifischen Thrombin Inhibitors Hirudin für diagnostische und biochemische Untersuchungen, Pharmazie 43:737 (1988).
29. F. Markwardt, Die antagonistische Wirkung des Hirudins gegen Thrombin *in vivo*, Naturwissenschaften 43:111 (1956).
30. F. Markwardt, Versuche zur pharmakologischen charakterisierung des Hirudins, Naunyn Schmiedebergs Arch. Pharmacol. 234:516 (1958).
31. F. Markwardt, Le genie genetique conduit au retour de l'hirudine comme anticoagulant, Sang Thrombose Vaisseaux 3:519 (1991).
32. F. Markwardt, Development of hirudin as an antithrombotic agent, Semin. Thromb. Hemost. 15:269 (1989).
33. H.P. Klöcking, Toxicology of hirudin, Semin. Thromb. Hemost. 17:126 (1991).
34. F. Markwardt, J. Hauptmann, C. Nowak, C. Kleßen, P. Walsmann, Pharmacological studies on the antithrombotic action of hirudin in experimental animals, Thromb. Haemost. 47:226 (1982).

35. A. Henschen, F. Markwardt, and P. Walsmann, Identification by HPLC analysis of the unaltered forms of hirudin and desulfated hirudin after kidney passage (Abstract), Thromb. Res. 7: (suppl) 37 (1987).
36. S. Kaiser, and F. Markwardt, Antithrombotic and haemorrhagic effects of the naturally occurring thrombin inhibitor hirudin, Folia Haematol. 115:41 (1988).
37. S. Kaiser, and F. Markwardt, Antithrombotic and haemorrhagic effects of synthetic and naturally occurring thrombin inhibitors, Thromb. Res. 613 (1986).
38. F. Markwardt, C. Nowak, U. Stürzebecher, and P. Walsmann, Studies on the pharmacokinetics of hirudin, Biomed. Biochim. Acta 46:237 (1987).
39. E. Bucha, G. Nowak, and F. Markwardt, Prevention of experimental coronary thrombosis by hirudin, Folia Haematol. 115:52 (1988).
40. A. Ishikawa, R. Nafter, U. Seemüller, J.M. Gokel, and H. Graeff, The effect of hirudin on endotoxin induced disseminated intravascular coagulation (DIC), Thromb. Res. 19:351 (1990).
41. K. Krupinski, H.K. Breddin, F. Markwardt, and W. Haarmann, Antithrombotic effects of three thrombin inhibitors in a rat model of laser-induced thrombosis, Haemostasis 19:74 (1989).
42. F. Markwardt, C. Nowak, and J. Hoffmann, Comparative studies on thrombin inhibitors in experimental microthrombosis, Thromb. Haemost. 49:235 (1983).
43. J. Bichler, St. Fichtl, H.I. Siebeck, and H. Fritz, Pharmacokinetics and pharmacodynamics of hirudin in man after single subcutaneous and intravenous bolus administration. Arzneim-Forsch 38:704 (1988).
44. F. Markwardt, C. Nowak, J. Stürzebecher, U. Grießbach, P. Walsmann and C. Vogel, Pharmacokinetics and anticoagulant effect of hirudin in man, Thromb. Haemost. 52:160 (1984).
45. C. Vogel, and F.Markwardt, Preliminary clinical reports on the antithrombotic action of hirudin (abstract), Thromb. Res. 7 (suppl) 42, (1987).
46. W.E. Marki, H. Großenbacher, M.C. Grütter, M.H. Liersch, S. Meyhack, and J. Heim, Recombinant hirudin: Genetic engineering and structure analysis. Semin. Thromb. Hemost. 17:88 (1991).
47. P. Walsmann, and B. Kaiser, Biochemical and pharmacological properties of recombinant hirudin, Drugs of Today 25:473 (1989).
48. J. Bichler, and N. Fritz, Hirudin, a new therapeutic tool? Ann Hematol 63:67 (1991).
49. J. Fareed, J.M. Walenga, L. Iyer, D. Hoppensteadt, and A. Pifarre, An objective perspective on recombinant hirudin: a new anticoagulant and antithrombotic agent, Blood Coag. Fibrinol. 2:135 (1991).
50. F. Markwardt, The comeback of hirudin as an antithrombotic agent, Semin. Thromb. Haemost. 17:75 (1991).
51. F. Markwardt, J. Stürzebecher, and P. Walsmann, The hirudin standard, Thromb. Res. 59:395 (1990).
52. J. Stürzebecher, and P. Walsmann, Structure-activity relationships of recombinant hirudins, Semin. Thromb. Hemost. 17:94 (1991).
53. I.J. Braun, S. Dennis, J. Hofsteenge, and S.R. Stone, Use of site-directed mutagenesis to investigate the basis for the specificity of hirudin, Biochemistry 27:6517 (1988).
54. E. Degryse, M. Acker, G. Defreyn, A. Bernat, J.P. Maffrand, C. Roitsch, and M. Courtney, Point mutation modifying the thrombin inhibition kinetics and antithrombotic activity *in vivo* of recombinant hirudin, Protein Eng. 2:459 (1989).

55. S. Dennis, A. Wallace, J. Hofsteenge, and S.R. Stone, Use of fragments of hirudin to investigate thrombin-hirudin interaction, Eur. J. Biochem. 188:61 (1990).
56. S.J.T. Mao, M.T. Yates, T.J. Owen, and J.L. Krstenansky, Interaction of hirudin with thrombin: identification of a minimal binding domain of hirudin that inhibits clotting activity, Biochemistry 27:8170 (1988).
57. F. Markwardt, J. Stürzebecher, and E. Glusa, Antithrombin effects of native and recombinant hirudins, Biomed. Biochim. Acta 49:399 (1990).
58. J.A. Jakubowski, and J.M. Maraganore, Inhibition of coagulation and thrombin-induced platelet activities by a synthetic dodecapeptide modeled on the carboxy-terminus of hirudin, Blood 75:399 (1990).
59. J.L. Krstenansky, and S.J.T. Mao, Antithrombin properties of C-terminus of hirudin using synthetic unsulfated N-acetyl-hirudin45-65, FEBS Letts. 211:10 (1987).
60. M.C. Crutter, J.P. Priestel, T. Rehuel, H. Groffbenbacher, W. Bode, J. Hofsteenge, and S.R. Stone, Crystal structure of the thrombin-hirudin complex: a novel mode of serine protease inhibition, Embo. J. 9:2361 (1990).
61. T.J. Rydel, K.C. Ravichandran, A. Tulinsky, W. Bode, R. Huber, C. Roitsch, and J.W. Fenton II, The structure of a complex of recombinant hirudin and human α-thrombin, Science 249:277 (1990).
62. J.M. Maraganore, P. Bourdon, J. Jablonski, K.L. Ramachandran, and J.W. Fenton II, Design and characterization of hirulogs: a novel class of biovalent peptide inhibitors of thrombin, Biochemistry 29:7095 (1990).
63. J. Hauptmann, E. Brüggener, and F. Markwardt, Effect of heparin, hirudin and a synthetic thrombin inhibitor on antithrombin III in thrombin-induced disseminated intravascular coagulation in rats, Hemostasis 17:321 (1987).
64. T. Lindhout, R. Blezer, and C. Hemker, The anticoagulant mechanism of action of recombinant hirudin (CCP39393) in plasma, Thromb. Haemost. 64:464 (1990).
65. J. Hofsteenge, H. Taguchi, and S.R. Stone, Effect of thrombomodulin on the kinetics of the interaction of thrombin and inhibitors, Biochem. J. 237:243 (1986).
66. E. Glusa, Hirudin and platelets, Semin. Thromb. Hemost. 17:122 (1991).
67. E. Glusa, and F. Markwardt, Platelet functions in recombinant hirudin-anticoagulated blood, Hemostasis 20:112 (1990).
68. A. Hoffmann, and F. Markwardt, Inhibition of the thrombin-platelet reaction by hirudin, Hemostasis 14:164 (1984).
69. M. Basic-Micic, K. Krupinski, and H.K. Breddin, r-Hirudin effects on various platelet functions, Angio. Arch. 18:11 (1989).
69a. F. Markwardt, A. Hoffmann, and J. Stürzebecher, Influence of thrombin inhibitors on the thrombin-induced activation of human blood platelets, Hemostasis 13:227 (1983).
70. E. Glusa, and F. Markwardt, Adrenalin-induced reactions of human platelets in hirudin plasma, Hemostasis 9:188 (1980).
71. R. Bizios, L. Lai, J.W. Fenton II, S.A. Sonder, and A.B. Malik, Thrombin-induced aggregation of lymphocytes; non-enzymic-induction by an hirudin blocked thrombin exosite, Thromb. Res. 38:424 (1985).
72. E. Glusa, and F. Markwardt, Studies on thrombin-induced endothelium-dependent vascular effects, Biomed. Biochim. Acta 47:623 (1988).
73. F. Markwardt Jr., T. Franke, E. Glusa, and S. Nilius, Pharmacological modification of mechanical and electrical responses of frog heart to thrombin, Naunyn-Schmiedebergs Arch. Pharmacol. 341:341 (1990).
74. M.A. Shuman, Thrombin-cellular interaction, Ann. N. Y. Acad. Sci. 485:228 (1986).

75. E. Van Obberghen-Schilling, R. Perez-Rodriguez, and J. Pouyssegur, Hirudin, a probe to analyze the growth promoting activity of thrombin in fibroblasts; re-evaluation of the temporal action of competence factors, Biochem. Biophys. Res. Commun. 105:79 (1982).
76. F. Markwardt, E. Fink, S. Kaiser, H.P. Klöcking, C. Nowak, M. Richter, and J. Stürzebecher, Pharmacological survey of recombinant hirudin, Pharmazie 43:202 (1988).
77. F. Markwardt, X.Q. Huan, J.M. Walenga, M. Münch, and D. Hoppensteadt, Pharmacokinetics of recombinant hirudin in primates (Abstract). Fed. Proc. 1:484 (1988).
78. F. Markwardt, S. Kaiser, and C. Nowak, Studies on antithrombotic effects of recombinant hirudin, Thromb. Res. 54:377 (1989).
79. F. Markwardt, C. Nowak, and J. Stürzebecher, Clinical pharmacology of recombinant hirudin, Haemostasis 21 (suppl 1) :133 (1991).
80. F. Markwardt, C. Nowak, J. Stürzebecher, and C. Vogel, Clinico-pharmacological studies with recombinant hirudin, Thromb. Res. 52:393 (1988).
81. C. Nowak, F. Markwardt, and E. Fink, Pharmacokinetic studies with recombinant hirudin in dogs, Folia. Haematol. 115:70 (1988).
82. J. Fareed, J.M. Walenga, D. Hoppensteadt, and A. Pifarre, Development perspectives for recombinant hirudin as an antithrombotic agent, Biol. Clin. Hematol. 11:143 (1989).
83. H.P. Klöcking, J. Güttner, E. Fink and F. Markwardt, Studies in toxicity of recombinant hirudin, Blut 60:129 (1990).
84. J.P. Maffrand, A. Bernat, D. Delebassce, M. Courtney, C. Roitsch, and C. Defreyn, Antithrombotic and haemorrhagic effects of rHV2-1ys47 hirudin compared with standard heparin in rabbits, Thromb. Haemost. 62:434 (1989).
85. S.H. Meyer, H.C. Luus, F.O. Müller, P.N. Badenhorst, and H.J. Röthig, The pharmacology of recombinant hirudin, a new anticoagulant, S. Afr. Med. J. 78:268 (1990).
86. M. Talbot, Biology of recombinant hirudin (DGP 39393): A new prospect in the treatment of thrombosis, Semin. Thromb. Hemost. 15:293 (1989).
87. M.D. Talbot, J. Ambler, K.D. Butler, V.S. Findley, K.A. Mitchel, R.F. Peters, M.F. Tweed, and R.B. Wallis, Recombinant desulphatohirudin (DGP 39393) anticoagulant and antithrombotic properties *in vivo*, Thromb. Haemost. 61:77 (1989).
88. J.M. Walenga, R. Pifarre, and J. Fareed, Recombinant hirudin as antithrombotic agent, Drug of the Future 5:267 (1990).
89. J.M. Walenga, R. Pifarre, D. Hoppensteadt, and J. Fareed, Development of recombinant hirudin as a therapeutic anticoagulant and antithrombotic agent: Some objective considerations, Semin. Thromb. Haemost. 15:316 (1989).
90. C. Nowak, Pharmacokinetics of hirudin, Semin. Thromb. Hemost. 17:145 (1991).
91. M. Richter, U. Cyranka, C. Nowak, and P. Walsmann, Pharmacokinetics of ^{125}I-hirudin in rats and dogs, Folia Haematol. 115:64 (1988).
92. J. Stürzebecher, Methods for determination of hirudin, Semin. Thromb. Hemost. 17:99 (1991).
93. U. Grießbach, J. Stürzebecher, and F. Markwardt, Assay of hirudin in plasma using a chromogenic thrombin substrate, Thromb. Res. 37:347 (1985).
94. M. Spannagl, J. Bichler, A. Birg, H. Lill, and H. Schramm, Chromogenic substrate assay for determination of hirudin in plasma, Blood Coag. Fibrinol. 2:121 (1991).

95. C. Agnelli, C. Pascucci, S. Cosmi, and C.C. Nenci, The comparative effects of recombinant hirudin (CCP39393) and standard heparin on thrombus growth in rabbits, Thromb. Haemost. 63:204 (1990).
96. J. Bichler, R. Gemmerli, and N. Fritz, Studies for revealing a possible sensitization to hirudin after repeated intravenous injections in baboons, Thromb. Res. 61:39 (1991).
97. J.M. Schlaeppi, S. Vekemans, H. Rink, and J.Y. Chang, Preparation of monoclonal antibodies to hirudin and hirudin peptides, Eur. J. Biochem. 188:463 (1990).
98. S. Spinner, F. Scheffauer, R. Maschler, and C.A. Stöffler, A hirudin catching ELISA for quantitating the anticoagulant in biological fluids, Thromb. Res. 51:617 (1988).
99. S. Spinner, C. Stoffler, and E. Fink, Quantitative enzyme-linked immunosorbent assay (ELISA) for hirudin, J. Immunol. Methods 87:79 (1986).
100. A. Henschen, F. Markwardt, and P. Walsmann, Evidence for the identity of hirudin isolated after kidney passage with the starting material. Folia Haematol. 115:59 (1988).
101. F. Markwardt, C. Nowak, and E. Bucha, Hirudin as anticoagulant in experimental haemodialysis, Haemostasis 21: (suppl 1) 49 (1991).
102. R. Ito, M.D. Phaneauf, and LoGerfo FW, Thrombin inhibition by covalently bound hirudin. Blood Coagulation Fibrinolysis, 2:77-81 (1991).
103. F. Markwardt, M. Richter, P. Walsmann, G. Riesener, and M. Paintz, Preparation of dextran-bound recombinant hirudin and its pharmacokinetic behaviour. Biomed. Biochim, Acta 49:1103 (1990).
104. M. Richter, P. Walsmann, and F. Markwardt, Plasma level of dextran-r-hirudin, Pharmazie 44:73 (1989).
105. E. Bucha, F. Markwardt, and G. Nowak, Hirudin in haemodialysis. Thromb. Res. 60:445 (1990).
106. J. Fareed, and J.M. Walenga, Do we need to neutralize hirudin's anticoagulant effects to minimize bleeding? Fed. Proc. 3:A 328 (1989).
107. J. Fareed, J.M. Walenga, D. Hoppensteadt, L. Iyer, and A. Pifarre, Neutralization of recombinant hirudin: Some practical considerations. Semin. Thromb. Hemost. 17:137 (1991).
108. E. Bruggener, P.I. Walsmann, and F. Markwardt, Neutralization of hirudin anticoagulant action by DIP-thrombin. Pharmazie 44:648 (1989).
109. S. Kaiser, Anticoagulant and antithrombotic actions of recombinant hirudin. Semin. Thromb. Hemost. 17:130 (1991).
110. C. Doutremepuich, E. Deharo, M. Guyot, M.C. Lalanne, J.M. Walenga, and J. Fareed, Antithrombotic activity of recombinant hirudin in the rat: a comparative study with heparin. Thromb. Res. 54:435 (1989).
111. M. Freund, J.P. Cazenave, M. Courtney, E. Debryse, C. Roitsch, A. Sernat, D. Delebassee, C. Defreyn, and J.P.Maffrand, Inhibition by recombinant hirudins of experimental venous thrombosis and disseminated intravascular coagulation induced by tissue factor in rats. Thromb. Haemost. 63:187 (1990).
112. E.P. Paques, and J. Römisch, Comparative study on the *in vitro* effectiveness of antithrombotic agents, Thromb. Res. 64:11 (1991).
113. J.H. Chesebro, M. Heras, J. Mruk, D. Grill, and V. Fuster, The critical role of thrombin in arterial thrombosis is demonstrated by hirudin. Angio. Arch. 18:9 (1989).
114. M. Just, D. Trobisch, and D. Tripier, r-DNA-hirudin infusion inhibits acute platelet thrombus formation in dog coronary arteries. Thromb. Haemost. 62:588 (1989).

115. A.B. Kelly, U.M. Marzec, and W. Krupski, et al, Hirudin interruption of heparin-resistant arterial thrombus formation in baboons. Blood 77:1006 (1991).
116. F. Markwardt, Prevention of coronary thrombosis by hirudin, in: Papp JG (ed) Cardiovascular pharmacology '87 Budapest, Budapest, Akademiai Kiado pp 449 (1987).
117. H. Hoffmann, M. Siebeck, M. Spannagl, M. Weis, R. Geiger, M. Jochum, and H. Fritz, Effect of recombinant hirudin, a specific inhibitor of thrombin, on endotoxin-induced intravascular coagulation and acute lung injury in pigs: Am. Rev. Respir. Dis. 142:782 (1990).
118. C. Nowak, and F. Markwardt, Hirudin in disseminated intravascular coagulation, Haemostasis 2(suppl 1):142 (1991).
119. M. Siebeck, H. Hoffmann, J. Weipert, M. Spannagl, M. Weis, and J. Bichler, Hirudin prevents intravascular coagulation in porcine endotoxin shock, in: Chirurgisches Forum '88 für Experimentelle und Klinische Forschung. Schriefers et al. (eds). Berlin, Heideslberg, pp 297-300; Springer (1988).
120. J. Hauptmann, and E. Brüggener, Influence of hirudin on the consumption of antithrombin III in experimental DIC. Folia Haematol. 115:83 (1988).
121. M. Heras, J.H. Chesebro, K.R. Baily, L. Badimon, and V. Fuster, Effects of thrombin inhibition on the development of the acute platelet-thrombus deposition during angioplasty in pigs. Heparin versus recombinant hirudin, a specific thrombin inhibitor. Circulation 79:657 (1989).
122. S. Kaiser, and F. Markwardt, Comparative studies of r-hirudin and heparin on arterial reocclusion after experimental thrombolysis with streptokinase and r-tPA, Arch. Pharmacol. (suppl) R97 (1992).
124. K. Rübsamen, and V. Eschenfelder, Reocclusion after thrombolysis: a problem solved by hirudin? Blood Coag. Fibrinol. 2:97 (1991).
125. I.J. Sarembock, S.D. Gertz, L.W. Gimple, E. Powers, and W.C. Roberts, Angiographic and pathologic study of the effect of recombinant desulphatohirudin on restenosis following balloon angioplasty in rabbits, Circulation 82:111 (1990).
126. J.I. Weitz, H. Hudoba, J.M. Maraganore, and J. Hirsh, Clot-bound thrombin is protected from inhibition by heparin-antithrombin III but is susceptible to inactivation by antithrombin Ill-independent inhibitors, J. Clin. Invest. 86:385 (1990).
127. M. Mirshahi, J. Soria, C. Soria, R. Faivre, H. Lu, M. Courtney, C. Roitsch, D. Tripier, and J.P. Caen, Evaluation of the inhibition by heparin and hirudin of coagulation activation during r'-tPA-induced thrombolysis, Blood 74:1025 (1989).
128. E. Bucha, F. Markwardt, and G. Novak, Hirudin in haemodialysis, Thromb. Res. 60:445 (1990).
129. C. Nowak, E. Sucha, T. Goock, H. Thieler, and F. Markwardt, Pharmacology of r-hirudin in renal impairment, Thromb. Res. 66:707 (1992).
130. C. Nowak, and F. Markwardt, Influence of hirudin on endotoxin-induced disseminated intravascular coagulation (DIC) in weaned pigs, Exper. Pathol. 18:438 (1980).
131. B. Kaiser, A. Simon, and F. Markwardt, Antithrombotic effects of recombinant hirudin in experimental angioplasty and intravascular thrombolysis, Thromb. Haemost. 63:44 (1990).

MECHANISMS FOR THE ANTICOAGULANT EFFECTS OF SYNTHETIC ANTITHROMBINS

F.A. Ofosu

Senior Scientist
Canadian Red Cross Society
Blood Transfusion Service, and Professor of Pathology
McMaster University, 1200 Main Street West
Hamilton, Ontario, Canada L8N 3Z5

SUMMARY

The important roles of thrombin in the development and propagation of thrombosis are well recognized. In addition to being the enzyme for clotting fibrinogen (the major protein component of blood clots), thrombin accelerates its own generation by activating factor V, factor VIII, factor XI and platelets. It accelerates the stabilization of clots by activating factor XIII to factor XIIIa, the enzyme which crosslinks fibrin. There are probably two major pathways for regulating the availability of thrombin *in vivo*: inactivation of thrombin (by antithrombin III/vessel wall heparan sulfate and perhaps by other endogenous antithrombins) and the inactivation of factor Va and factor VIIIa by activated protein C. Factor Va and factor VIIIa accelerate the production of thrombin. However, when thrombin becomes bound to fibrin (in clots or possibly on cell surfaces), the ability of antithrombin III/heparin to inactivate thrombin is then reduced significantly. Impairment by fibrin of thrombin inhibition by antithrombin III may account in part for the inability of unfractionated heparin to prevent post-operative deep vein thrombosis in up to 20% of patients who undergo major elective orthopaedic surgery, and may also explain the need for oral anticoagulants after unfractionated and low molecular weight heparins are used to initiate the treatment of established deep vein thrombi. The ineffectiveness of the antithrombin III/heparin pathway for inhibiting thrombin under some circumstances has been a contributory factor for the development, evaluation and identification of other inhibitors of thrombin which are more able than antithrombin III/heparin to inactivate thrombin when the enzyme is bound to fibrin. The focus of this review is to detail how these synthetic agents, by directly or indirectly inactivating thrombin, can also effectively inhibit prothrombin activation *in vitro* .

The Design of Synthetic Inhibitors of Thrombin
Edited by G. Claeson, *et al.*, Plenum Press, New York, 1993

INTRODUCTION

Unfractionated (UF) and low molecular weight (LMW) heparins achieve their *in vitro* anticoagulant and *in vivo* antithrombotic actions by catalyzing thrombin and factor Xa inhibition by antithrombin III[1-4]. As a direct consequence of these two catalytic actions, UF and LMW heparins can delay the onset of prothrombin activation *in vitro*[3,4]. Doses of UF and LMW heparins shown to prevent post-operative DVT also appear to inhibit the conversion of prothrombin to thrombin *in vivo*[5-7]. It is likely that UF and LMW heparins also inhibit the activation of prothrombin *in vivo* when they are used to inhibit the growth of established deep vein thrombosis. There are two significant drawbacks associated with the prophylactic and therapeutic uses of UF and LMW heparins. Up to 20% of patients who undergo elective or orthopaedic surgery and receive the recommended doses of UF and LMW heparins for prophylaxis of postoperative DVT develop this complication[8-14]. Secondly, therapeutic use of UF and LMW heparins must be accompanied by use of oral anticoagulants for 3 months or more if the frequency of recurrence of the DVT is to be minimized[15,16]. There is a third potential drawback associated with the prophylactic use of UF or LMW heparins after orthopedic surgery. There is evidence that patients who develop DVT after elective high risk orthopaedic surgery have higher endogenous prothrombinase activity *in vivo*, both pre- and post-operatively, than patients in whom prophylaxis is successful[6,7]. Thus, the prophylactic use of UF or LMW heparins in these patients does not result in the suppression of *in vivo* prothrombinase to the level required for prophylaxis. Significantly, the same catalytic concentrations of either UF or LMW heparins are found in the plasmas of DVT-positive and DVT-negative patients[7]. While increasing the currently recommended prophylactic dosages of UF or LMW heparins might reduce the post-operative DVT risk, the potential for higher post-operative bleeding rates with higher doses is a potential cause for concern.

The three drawbacks cited above may be related to observations that when thrombin binds fibrin in thrombi *in vivo*[17,18] or to fibrin *in vitro*[19-21], the enzyme becomes resistant to inhibition by antithrombin III and by antithrombin III/heparin. Based on the results of *in vitro*[20] and *in vivo*[18] studies, the limitations of UF and LMW heparins noted above can be overcome by the prophylactic and therapeutic use of some direct synthetic antithrombins. By a direct antithrombin is meant an agent which is able to directly bind thrombin at its catalytic centre or block access of substrates to this site and thereby abrogate enzymatic actions of thrombin. Unlike heparin and dermatan sulfate which catalyze the inhibition of thrombin by antithrombin III and heparin cofactor II, respectively, synthetic direct antithrombins do not require a cofactor. Examples of direct antithrombins are natural and recombinant hirudins[22-24], several synthetic truncated hirudins[25-27], several modified truncated hirudins[28-31], D-Phe-Pro-ArgCH_2Cl[32], Ac(D)-Phe-Pro-boro-(Arg)[33], and argatroban[34]. D-Phe-Pro-ArgCH_2Cl alkylates the reactive centre histidine of the enzyme to inactivate it[32] while D-PhePro-bor-(Arg-OH) and argatroban reversibly bind α-thrombin at its catalytic centre[33,34]. Argatroban, D-Phe-Pro-Arg-CH_2Cl, and Ac(D)-Phe-Pro-bor-(Arg-OH) are monovalent inhibitors of the enzyme[32-33]. Hirudins and modified truncated hirudins (known as hirulogs) bind α-thrombin at two distinct sites, and are thus divalent thrombin inhibitors[22-24,28-31]. The carboxyl termini of these divalent direct antithrombins bind α-thrombin at the fibrin(ogen) recognition exosite of the enzyme while their amino termini bind the enzyme either at its catalytic centre (hirulogs) or blocks access to that site (hirudins). Because these synthetic direct antithrombins do not require a cofactor to inactivate thrombin, they may be able to effectively inactivate thrombin under conditions which compromise the inhibition of thrombin by antithrombin III/heparin. In addition to antithrombin III, a second plasma serpin, heparin cofactor II, directly inactivates thrombin.

The inactivation of thrombin by heparin cofactor II in plasma is catalyzed by dermatan sulfate[35-37]. Unlike antithrombin III/heparin, inactivation of thrombin by heparin cofactor II/dermatan sulfate is apparently not impaired by fibrin(ogen) *in vivo*[17,18] or *in vitro*[21]. However, on a mole for mole basis, dermatan sulfate is 10 to 50-fold less potent than heparin in its ability to catalyze the inhibition of thrombin in plasma[36,37]. Recently, two bislactobionic acid amides, which catalyze thrombin inhibition by heparin cofactor II more effectively than dermatan sulfate, have been synthesized[38]. In this review, the capacity of one of the two bis-lactobionic acids, LW10082, to inhibit intrinsic and extrinsic prothrombin activation will be compared with the following three direct synthetic antithrombins: D-Phe-ProArgCH_2Cl; Ac-(D-Phe-Pro-bor-(Arg); and D-Phe-Pro-Arg-Pro-$(Gly)_4$ hirudin[54-65] (hirulog-1). As will become evident during the review, LW10082, by catalyzing the inhibition of thrombin by heparin cofactor II[39], or the three direct antithrombins, by directly inactivating thrombin, can effectively inhibit factor X and prothrombin activation. Thus all four are potent anticoagulants.

Table 1 Ki's of hirudin and synthetic antithrombins*.

ANTITHROMBIN	Ki
Hirudin	**0.2 pM**
Hirulog-1	**2 nM**
D-PHE-PRO-ARGCH_2C1	**2 nM**
D-PHE-PRO-BOR-(ARG-OH)	**< 4 pM**

Rate constants for the inhibition of the reactions of α-thrombin with its natural and synthetic inhibitors.
*** Adapted from references (30-34).**

Relative Effectiveness of D-Phe-Pro-Arg-CH_2Cl. Ac(d)-Phe-Pro-boro-(Arg), Hirulog-1 and LW0082/heparin Cofactor II as Inhibitors of α-thrombin

Two kinetic constants, the Ki and the second order rate constant (Ka), reasonably describe the relative concentrations of inhibitors needed to reduce the reactivites of enzymes with their substrates. The Ki (with units of moles/liter) is the concentration of inhibitor which will reduce the velocity of an enzyme-substrate reaction by 50%, while the Ka (with units of per mol/unit time) describes the rate of complex formation between an enzyme and its inhibitor. The Ki describes the affinity of the inhibitor for the enzyme and broadly stated, the lower the Ki, the higher the affinity of the inhibitor for the enzyme. Table 1 lists the reported Ki's of reactions of the direct thrombin inhibitors, namely hirudin, D-Phe-Pro-ArgCH_2Cl,

Hirulog-1 and Ac(D-Phe-Pro-boro-(Arg-OH) with α-thrombin. Of the three synthetic antithrombins, the last direct thrombin inhibitor binds most tightly to the catalytic centre of α-thrombin. The Ki describing the concentration of LW10082 to inhibit the release of thrombin-mediated fibrinopeptides release is unknown. However, in line with other polyanions which bind α-thrombin and thereby inhibit fibrinopeptide release directly[40], the concentration of LW10082 to directly inhibit α-thrombin-catalyzed cleavage of fibrinopeptide A from fibrinogen is likely to be approximately 1μM. Without exception,

Table 2 Association/inactivation constants for the reaction of α-thrombin with hirudin, three synthetic and two plasma-direct antithrombins[a].

ANTITHROMBINS		KASSOC/KINACT
Hirudin		$1.8 \times 10^8\ M^{-1}\ S^{-1}$
Hirulog-1	*	$1 \times 10^7\ M^{-1}\ S^{-1}$
D-PHE-PRO-ARGCH$_2$C1		$1.2 \times 10^7\ M^{-1}\ S^{-1}$
Ac(D)-PHE-PRO-BOR-(ARG-OH)		$8 \times 10^6\ M^{-1}\ S^{-1}$
ANTITHROMBIN III - HEPARIN	**	$4 \times 10^7\ M^{-1}\ S^{-1}$
HEPARIN COFACTOR II/- DERMATAN SULFATE OR HEPARIN COFACTOR II - LW10082	**	$9 \times 10^6\ M^{-1}\ S^{-1}$

The second order rate constants for complexation of thrombin with synthetic and two plasma inhibitors of thrombin (antithrombin III and heparin cofactor II).
*** - calculated**
**** - maximum achievable rates**
a - Adapted from references (30,31,33,34,36,43).

all the polyanions cited in Reference 40 also catalyze thrombin inhibition by antithrombin III and/or heparin cofactor II[40]. Thus each of the above direct thrombin inhibitors is at least orders of magnitude more effective than LW10082 with respect to the concentrations which will inhibit the reactivity of the enzyme with fibrinogen, one of the plasma macromolecular substrates of α-thrombin.

The second order rate constants describing the rate of formation of complexes of thrombin with each of the three direct inhibitors are listed in Table 2. It is important to note that D-Phe-Pro-ArgCH2Cl and Hirulog-1 directly form complexes with α-thrombin with a similar velocity. Hirulog-1 binds the enzyme at both its catalytic centre and the fibrin(ogen) recognition exosite whereas D-Phe-Pro-ArgCH$_2$Cl binds the enzyme solely at its catalytic centre[29,30,32]. The similarity of the rate constants describing the inhibition of α-thrombin by the two direct antithrombins suggests the possibility that the affinity of D-Phe-Pro-Arg-moieties for the catalytic centre of α-thrombin is a primary determinant for the inhibition of thrombin by the two synthetic direct antithrombins. However, the similarity of the two second order rate constants may also reflect the cleavage of the Arg-Pro bond of Hirulog-1 by the enzyme[41]. Cleavage of the Arg-Pro bond will convert Hirulog-1 to hirudin[54-65] which binds α-thrombin with a Ki a thousand-fold larger than Hirulog-1.

As noted in Table 2, the second order rate constants describing the inhibition of α-thrombin by antithrombin III/heparin and heparin cofactor II/LW10082 are similar to those achieved when D-Phe-Pro-ArgCH$_2$Cl and Hirulog-1 inactivate thrombin. It is thus not surprising that in spite of the differences in their structures and their modes of action, the synthetic antithrombins discussed in this review are nearly equipotent in their ability to inhibit prothrombinase formation in plasma (see below).

Inhibition of Prothrombin and Factor X Activation by Heparin

Before discussing how direct antithrombins and LW10082 inhibit prothrombin and factor X activation, it would be useful to briefly review the steps in coagulation most sensitive to inhibition by heparin, and to review how heparin influences prothrombin and factor X activation *in vitro*. As summarized in Figure 1, prothrombin activation in plasma is catalyzed by prothrombinase (the complex of factor Xa (enzyme) and factor Va (essential cofactor) bound to a coagulant surface)[44]. The intrinsic pathway activation of plasma prothrombin, initiated by adding micronized silica and CaCl$_2$ to plasma, proceeds to completion within 90 s of calcium addition[45-47]. Extrinsic prothrombin activation, initiated by adding tissue factor and CaCl$_2$ to plasma, also proceeds to completion within 60s of CaCl$_2$ addition[47]. There are two pathways for generating factor Xa in plasma: intrinsic tenase (factor IXa (enzyme) and factor VIIIa (essential cofactor) bound to a coagulant surface)[44] and extrinsic tenase (factor VIIa/tissue factor complex)[44].

The generation of prothrombinase and the two tenases are CaCl$_2$-dependent[45]. In the presence of coagulant phospholipid and Ca^{2+}, factor Xa alone can activate prothrombin. However, the rate of prothrombin activation by full prothrombinase is at least 1,000-fold faster than that of factor Xa in the presence of coagulant phospholipids and Ca^{2+} [44]. Similarly, intrinsic tenase activates factor X at a rate several orders of magnitude greater than the rate achievable with factor IXa, coagulant phospholipids and Ca^{2+} [44]. Thrombin plays a critical role in the generation of prothrombinase since it activates factor VIII, factor V, and factor XI in plasma by limited proteolysis[44]. For these reasons, exogenous thrombin accelerates prothrombin activation in plasma[45-47]. As noted above, factor VIIIa is an essential cofactor component of intrinsic tenase, and factor Va is an essential cofactor component of prothrombinase. Thus by catalyzing the inhibition of the initial thrombin generated by factor Xa alone (i.e. before saturating concentrations of factor Va become available *in situ*), approximately 0.1μM heparin can significantly delay the onset of intrinsic factor X and prothrombin activation[45-47]. A second coagulant enzyme for activating factor V in plasma is factor Xa, an enzyme as catalytically as efficient as thrombin for this reaction[48]. For this reason, exogenous factor Xa also accelerates prothrombin activation in plasma[45-47]. Additionally, when $>$1.0nM factor Xa becomes available *in situ* during intrinsic prothrombin activation, approximately 0.1 μM heparin can no longer delay the onset of intrinsic prothrombin activation[45-47]. The major action of heparin is then to catalyze the inhibition of the factor Xa and thrombin generated[45-47]. Intrinsic prothrombin activation is also readily accelerated by the addition of ≥ 1.0nM thrombin to plasma, and when the concentration of exogenous thrombin is ≥ 10.0nM, $\approx 0.1\mu$M heparin fails to inhibit intrinsic prothrombin activation. The main action of heparin in the plasma is then to catalyze the formation of factor Xa-antithrombin III and thrombin/antithrombin III *in situ*[45-47].

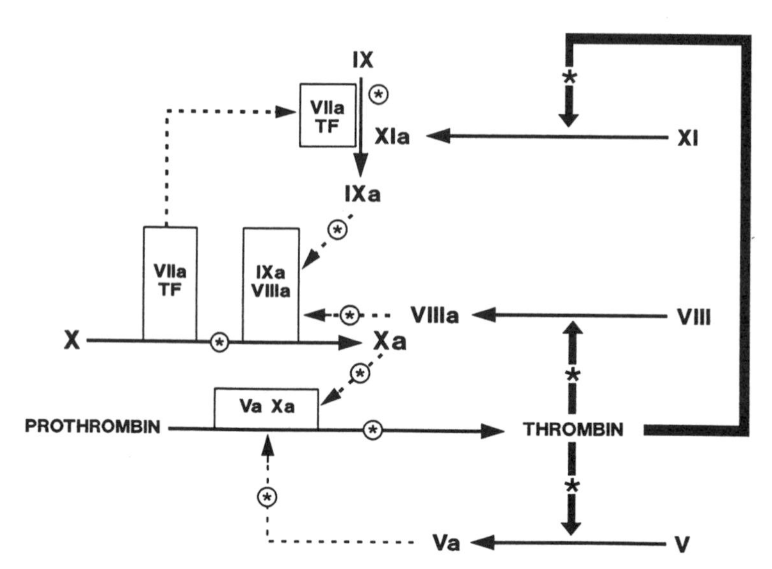

Figure 1 A summary of the coagulation scheme demonstrating the sites direct antithrombins and LW10082 act to inhibit thrombin-mediated activation of factor VIII, factor V and factor XI, and to ultimately prolong the time required to generate prothrombinase. The sites of action of direct action of the antithrombin are indicated by asterisks. Asterisks enclosed in open circles mark the subsequent reactions that are inhibited by the direct antithrombins.

Two important kinetic considerations explain why catalysis of thrombin inhibition by heparin provides a more efficient means for suppressing prothrombin activation than catalysis of factor Xa inhibition by heparin. First, heparin catalyzes the inhibition of thrombin, by antithrombin III more effectively than the inhibition of factor Xa[48]. Secondly, once factor Va becomes available *in situ* (after limited proteolysis of factor V by factor Xa or thrombin), the inhibition of factor Xa within prothrombinase by antithrombin III/heparin becomes even slower than the inhibition of free factor Xa[49-51]. For these reasons, the catalysis of endogenous thrombin inhibition is the most critical action of heparin which contributes to its ability to inhibit intrinsic factor X and prothrombin activation.

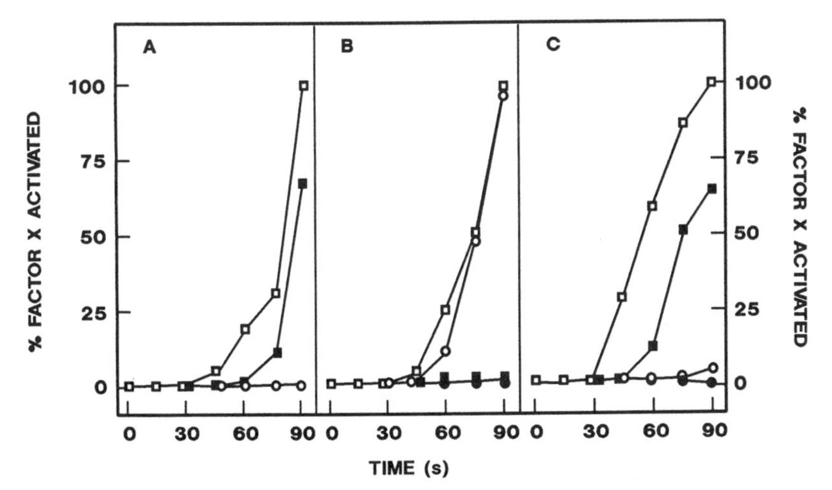

Figure 2 Inhibition intrinsic factor X activation by LW10082 (A), D-Phe-Pro-ArgCH$_2$Cl and Ac(D)-Phe-Pro-bor-(Arg-OH) (B) and D-Phe-Pro-Arg-Pro-(Gly)$_4$hirudin$^{54-65}$ (C). Symbols used: A: □, control plasmas; ■, 0.4 μM LW10082; ○, 4.0 μM LW10082. B: □, control plasma; ■, 1.0 μM H-D-Phe-Pro-ArgCH$_2$Cl; ○ and ● represent 0.1 and 1.0 μM Ac(D)-Phe-Pro-bor-Arg-OH respectively. C: □, control plasma; ■, ○ and ● represent 0.1, 1.0 and 10 μM D-Phe-Pro-Arg-Pro-(Gly)$_4$-hirudin$^{54-65}$ respectively.

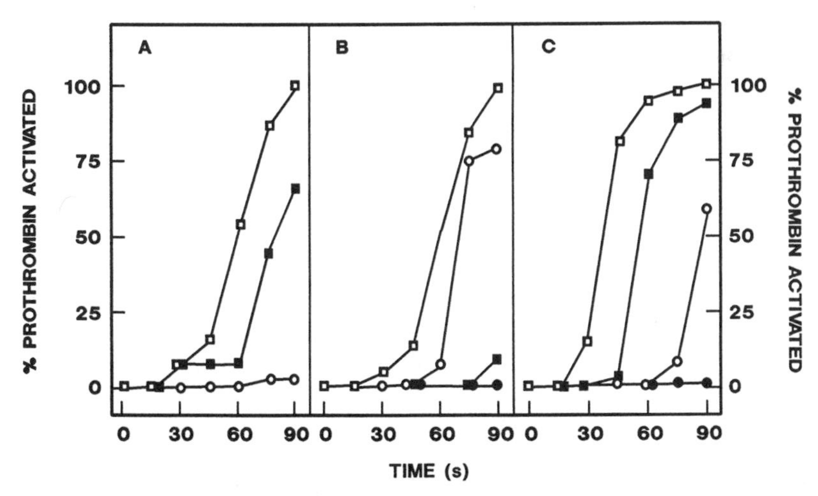

Figure 3 Inhibition of intrinsic prothrombin activation by LW10082 (A), D-Phe-Pro-ArgCH$_2$Cl or Ac(D)-Phe-Pro-bor(Arg-OH) (B) and D-Phe-Pro-Arg-Pro-(Gly)$_4$-hirudin $^{54-65}$ (C). See Figure 2 for an explanation of symbols used.

The inability of $0.1\mu M$ heparin to inhibit the onset and extent of intrinsic prothrombin activation in plasma containing $>1.0nM$ factor Xa explains in part why this concentration of heparin cannot inhibit extrinsic prothrombin activation[45-47]. During extrinsic coagulation, saturating concentrations of factor Xa readily become available independently of thrombin action on factor VIII as extrinsic tenase (factor VIIa/tissue factor complex) activates factor X directly during extrinsic coagulation[44]. In summary, therefore, the major anticoagulant actions of the concentration of heparin achievable both prophylactically and therapeutically ($<0.1\mu M$) is to delay the onset of the generation of prothrombinase. Once prothrombinase becomes available *in situ*, the major anticoagulant actions of heparin are the catalysis of factor Xa-antithrombin III and thrombin-antithrombin III formation. As will become evident presently, sub-micromolar concentrations of LW10082 and the direct synthetic antithrombins also inhibit prothrombin activation by delaying the onset of factor X and factor V activation (i.e. they delay the generation of prothrombinase *in situ*.

Inhibition of Factor X and Prothrombin Activation by Synthetic Antithrombins

We routinely use enzyme-linked immunosorbent assays to quantify factor X and prothrombin in plasma. The major advantage of this approach over chromogenic or clotting assays is that the enzyme-linked immunosorbent assays allows one to directly distinguish between inhibition of clotting factor activation and the increased inhibition of the enzymatic activities of activated clotting factors. Immunological measurement of prothrombin activation (consumption) or production of prothrombin fragment 1+2 provide the only direct means for assessing prothrombin activation in plasmas containing a direct antithrombin which can neutralize the thrombin generated. Following contact activation of and addition of $CaCl_2$ to plasma, or following addition of rabbit brain tissue factor and $CaCl_2$ to plasma, timed aliquots are added to a coagulation inhibitor containing EDTA, heparin and hirudin to inhibit further factor X and prothrombin activation. Hirudin inactivates any thrombin generated while the heparin catalyzes the inhibition of factor Xa by antithrombin III. Prothrombin activation is quantified as the change in the concentration of prothrombin, while the factor X activation is quantified as the concentration of total factor Xa-antithrombin III generated. Details of the enzyme-linked immunosorbent assays have been reported elsewhere[45-47].

The effects of LW10082, D-Phe-Pro-ArgCH_2Cl, Ac(D)-Phe-Pro-Bor-Arg, and Hirulog-1 on intrinsic activation of factor X activation are summarized in Figures 2A to C. At $0.1\mu M$, each of the 3 direct antithrombins can effectively suppress factor X activation, while $0.4\mu M$ is the minimum concentration of LW10082 to inhibit intrinsic factor X activation as effectively as $0.1\mu M$ of two of the direct antithrombins. As none of the four antithrombins significantly inactivates factor Xa in the presence of heparin, these results demonstrate that by directly inactivating thrombin (direct antithrombins) or by catalyzing the inhibition of thrombin (LW10082), intrinsic factor X activation can markedly be suppressed by these antithrombins. In line with their ability to inhibit intrinsic factor X activation, all our antithrombins effectively inhibit intrinsic prothrombin activation (Figures 3A through C).

Figure 4 summarizes the effects of the four antithrombins on extrinsic factor X activation. During the first 60s of incubation, none of the three direct antithrombins can inhibit extrinsic factor X activation as effectively as they can inhibit intrinsic factor X activation (Figure 2). Nonetheless, the significant inhibition of extrinsic pathway activation of factor X by the three direct antithrombins suggests that the enzymatic complex factor IXa-factor VIIIa (i.e. intrinsic tenase) contributes significantly to factor X activation when tissue factor is added to plasma to initiate extrinsic coagulation. Note that upon adding tissue factor to plasma, factor IX will be activated by factor VIIa/tissue factor or by factor

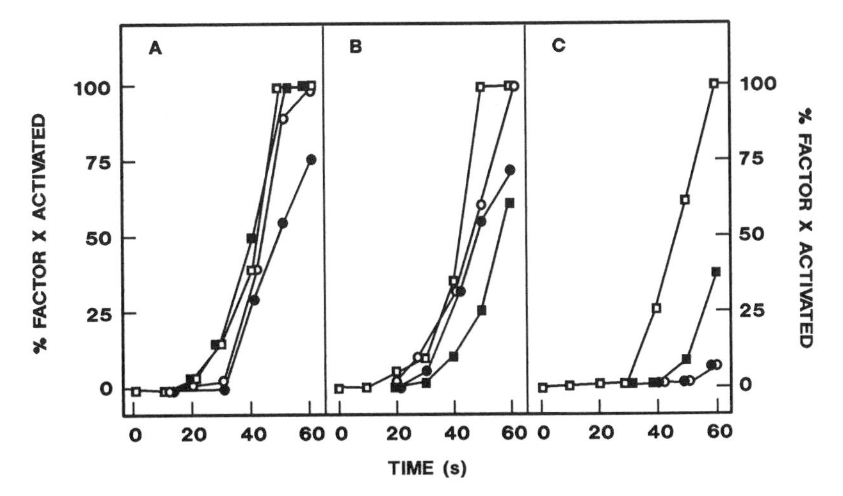

Figure 4 Inhibition of extrinsic activation of factor X by LW10082 (A), D-Phe-Pro-ArgCH$_2$Cl or Ac(D)-Phe-Pro-bor-(Arg-OH) (B) and D-Phe-Pro-Arg-Pro(Gly)$_4$hirudin$^{54\text{-}65}$ (C). Symbols: A: □, control plasma; ■, 0.4 μM LW10082; ○, 4.0; and ● 40 μM LW10082, respectively. B: ○, control plasma; ■, 1.0 μM D-Phe-Pro-ArgCH$_2$Cl; ○ and ●, 0.1 μM and 1.0 μM c(D)-Phe-Pro-bor-(Arg-OH), respectively. C: □, control; ○, and ● represent Phe-Pro-Arg-Pro-hirudin$^{54\text{-}65}$, respectively.

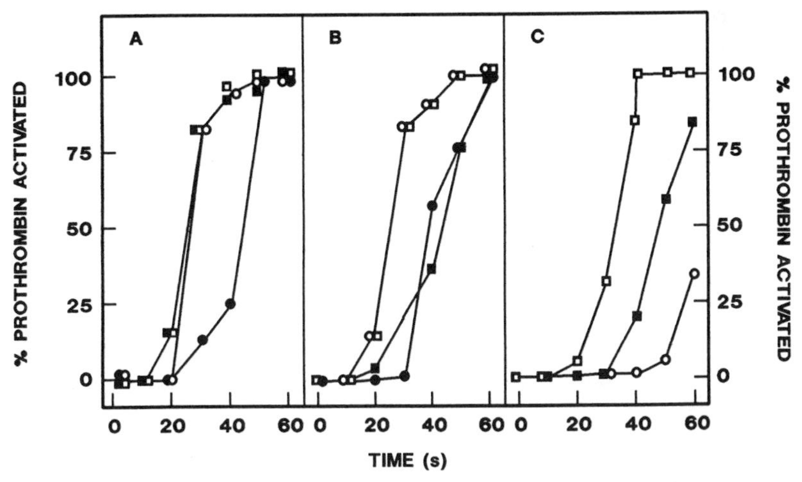

Figure 5 Inhibition of extrinsic prothrombin activation by LW10082 (A), D-Phe-Pro-ArgCH$_2$Cl or Ac(D)-Phe-Pro-bor-(Arg-OH) (B) and D-Phe-Pro-Arg-Pro-hirudin$^{54\text{-}65}$ (C). See Figure 4 for an explanation of symbols used.

XIa[44]. Factor XIa in turn is generated by the limited proteolysis of factor XI by thrombin[44]. Unlike the direct antithrombins, LW10082 cannot effectively inhibit extrinsic factor X activation. As summarized in Figure 5, none of the four anticoagulants can inhibit extrinsic prothrombin activation as effectively as they can inhibit intrinsic prothrombin activation. Failure of the lowest concentration of the 4 anticoagulants to effectively inhibit extrinsic activation of prothrombin reflects the ability of the factor Xa, generated endogenously by the factor VIIa-activation as effectively as they can inhibit intrinsic prothrombin activation. Failure of the lowest concentration of the 4 anticoagulants to effectively inhibit extrinsic activation of prothrombin reflects the ability of the factor Xa, generated endogenously by the factor VIIa-tissue factor complex (ie extrinsic tenase), to efficiently activate factor V[47,48]. The sites of coagulation sensitive to inhibition by LW10082 and the three direct antithrombins are shown schematically in Figure 1.

Consistent with their ability to inhibit factor X and prothrombin activation, the efficiency with which the anticoagulants inhibit intrinsic and extrinsic activation of factor V also parallel their ability to inhibit prothrombin and factor X activation[46,52]. With the exception of D-Phe-Pro-Arg CH_2Cl, none of the other antithrombins can inactivate factor Xa. The results summarized above clearly indicate that by directly inactivating thrombin, or by accelerating the inactivation of thrombin by heparin cofactor II, these synthetic antithrombins can markedly inhibit the activation of prothrombin *in vitro*. Coupled with the *in vivo* antithrombotic properties of the protein C pathway[54] and tissue factor pathway inhibitor[55], it is possible that these synthetic antithrombins will prove to be effective antithrombotic drugs in situations where heparin and low molecular weight heparins are not effective.

CONCLUSIONS

1. Synthetic and natural antithrombins delay the onset of prothrombin activation by inhibiting the three amplification reactions of coagulation which are usually catalysed by endogenous α-thrombin (Figure 1).

2. By prolonging the time required for the thrombin-catalyzed amplification reactions to occur, submicromolar concentrations of the three synthetic direct antithrombins can effectively delay the onset of intrinsic and extrinsic prothrombin activation.

3. By catalysing thrombin inhibition to the optimal rates achievable with heparin cofactor, LW10082 can also effectively delay the onset of intrinsic prothrombin activation.

4. LW10082, which is an indirect antithrombin like heparin, inhibits intrinsic prothrombin activation more effectively than extrinsic prothrombin activation for the following reasons.

5. Only thrombin can initiate the activation of factor VIII when factor VIII is bound to vWF[53]. In contrast, both factor Xa and thrombin can initiate factor V activation[48].

6.

The mandatory role of thrombin for activating factor VIII bound to von Willebrand factor probably accounts for the ability of LW10082 to effectively inhibit intrinsic prothrombinase formation, and the inability of LW10082 to effectively inhibit extrinsic prothrombinase formation.

ACKNOWLEDGEMENTS

The experimental work described in this review was funded in part by a grant-in-aid by the Ontario Heart and Stroke Foundation.

REFERENCES

1. D.A. Lane, I.R. MacGregor, R. Michalski, and V.V. Kakkar, Anticoagulant activities of four unfractionated and fractionated heparins, Thromb. Res. 12:257 (1978).
2. F.A. Ofosu, and T.W. Barrowcliffe, Mechanisms of action of low molecular weight heparin and heparinoids, Balliere's Clinical Haematology 3:505 (1990).
3. F.A. Fernandez, M.R. Buchanan, J. Hirsh, J.W. Fenton II, and F.A. Ofosu, Catalysis of thrombin inhibition provides an index for estimating antithrombotic potential of glycosaminoglycans in rabbits, Thromb. Haemostas. 57:286 (1987).
4. F.A. Ofosu, M.R. Buchanan, N. Anvari, L.M. Smith, and M.A. Blajchman, Plasma anticoagulant mechanism of heparin, heparan sulphate and dermatan sulphate, Ann. N.Y. Acad. Sci. 556:123 (1989).
5. J. Bogaty-Yver, and M. Samama, Thrombin-antithrombin III complexes for the detection of postoperative hypercoagulable state in surgical patients receiving heparin prophylaxis, Thromb. Haemostas. 61:538 (1989).
6. J.A. Hoeck, M.T. Normohamed, J.W. ten Cate, H.R. Buller, H.C. Knipscheer, H.I. Hamelynck, R.I. Marti, and Stark. Thrombin-antithrombin III complexes in the prediction of deep vein thrombosis following total hip replacement, Thromb. Haemostas. 62:1050 (1989).
7. F.A. Ofosu, Prophylactically effective doses of Enoxaparin and heparin inhibit prothrombin activation, in: Heparin and Related Polysaccharides, D.A. Lane, and U. Lindahl, ed. Plenum Press pp.231-236 (1992).
8. V.V. Kakkar, and W.J.G. Murray, Anticoagulant effect of two types of low molecular weight heparin administered subcutaneously, Br. J. Surg. 72:786 (1985).
9. P. Berquist, U. Hedner, E. Sjorin, and E. Holmer, Efficacy and safety of low molecular weight heparin (CY216) in preventing postoperative venous thromboembolism: a cooperative study, Thromb. Res. 32:381 (1983).
10. M. Holler, U. Schoch, P. Buchanan, F. Largiaden, A. von Felton, P.G. Frick, Low molecular weight heparin (KABI 2165) as thromboprophylaxis in elective visceral surgery. A randomized double blind study versus unfractionated heparin, Thromb. Haemostas. 56:243 (1986).
11. A.G.G. Turpie, M.N. Levine, J. Hirsh, C.J. Carter, R.M. Jay, P.J. Powers, M. Andrew, R.P. Hull, and M. Gent, A randomized controlled trial of PK 10169 low molecular weight heparin for the prevention of deep vein thrombosis in patients undergoing elective hip surgery, N. Engl. J. Med. 315:925 (1987).
12. A. Planes, N. Vochelle, J. Ferry, D. Pryzrowski, J. Clerc, M. Fayola, M. Planes, Enoxaparine low molecular weight heparin: its use in the prevention of deep vein thrombosis following total hip replacement, Haemostasis 16:152 (1986).
13. R.P. Hull, T. Delorme, E. Genton, J. Hirsh, M. Gent, D. Sackett, P. McLaughlin, and P. Armstrong, Warfarin sodium versus low-dose heparin in the long term treatment of venous thrombosis, N. Engl. J. Med. 301:855 (1979).
14. J.R. Leclerc, W. Geerts, L. Desjardins, F. Jobin, F. Laroche, F. Delorme, S. Haviernick, S. Atkinson, and J. Bourgouin, Prevention of deep vein thrombosis after major knee surgery. A randomized, double-blind trial comparing a low molecular weight heparin fragment (Enoxaparin) with placebo, Thromb. Haemostas. 67:417 (1992).

15. C.I. Lagerstedt, C.G. Olsson, B.O. Fagher, B.W. Oquist, and U. Albrechtsson, Need for long-term anticoagulation in symptomatic calf vein thrombosis, Lancet ii 515 (1985).
16. W. Coon, and P. Willis, Recurrence of venous thromboembolism, Surgery 73:823 (1973).
17. J. Van Ryn-McKenna, F.A. Ofosu, E. Grey, J. Hirsh, and M.R. Buchanan, Effects of dermatan sulphate and heparin on inhibition of thrombus growth *in vivo*, Ann. N.Y. Acad. Sci. 556:304 (1989).
18. J.L. Okwusidi, M. Falcone, J. Van Ryn-McKenna, J. Hirsh, F.A. Ofosu, and M.R. Buchanan, Fibrin moderates the catalytic action of heparin but not that of dermatan sulphate on thrombin inhibition in human plasma, Thromb. Haemorrh. Dis. 1:77 (1990).
19. P.J. Hogg, and C.M. Jackson, Fibrin monomer protects thrombin from inactivation by heparin-antithrombin III: implications for heparin efficacy. Proc. Natl. Acad. Sci. (USA) 86:3619 (1989).
20. P.J. Hogg, and C.M. Jackson, Heparin promotes the binding of thrombin to fibrin polymer. Quantitative characterization of a thrombin-fibrin polymer - heparin ternary complex, J. Biol. Chem. 265:245 (1990).
21. J.I. Weitz, M. Huboda, D. Massel, J. Maraganore, and J. Hirsh, J. Clot-bound thrombin is protected from heparin-antithrombin III but is susceptible to inactivation by antithrombin III-independent inhibition, J. Clin. Invest. 86:385 (1990).
22. J.L. Okwusidi, N. Anvari, M. Kulcycky, M.A. Blajchman, M.R. Buchanan, and F.A. Ofosu, J. Lab. Clin. Med. 117:359 (1991).
23. D.E. Bagdy, E. Barabas, L. Graf, T.E. Peterson, and S. Magnusson, Hirudin, Methods Enzymol. 45:669 (1976).
24. R.P. Harvey, E. Dagryse, L. Stefani, L. Schamber, J.P. Cazeneve, M. Courtney, P. Tobstoskey, and J.P. Lecocg, Cloning and expression of cDNA coding for the anticoagulant hirudin from the bloodsucking leech, Hirudo medicinales, Proc. Nat. Acad. Sci. USA 83:1084 (1986).
25. E. Degryse, M. Acker, A. Bernt, J.P. Maffrand, C.R. Roitsch, and M. Courtney, Point mutation modifying the thrombin inhibition kinetics and antithrombotic activity *in vivo* of recombinant hirudin, Prot. Engin. 2:459 (1989).
26. J.L. Krstenansky, and S.J. Mao, C-terminus of hirudin using synthetic unsulfated Nα acetyl-hirudin 45-65, FEBS Lett. 211:10 (1987).
27. S.J.T. Mao, M.T. Yates, T.J. Owen, and J.L. Krstenansky, Interaction of hirudin with thrombin: identification of a minimal binding domain of hirudin that inhibits clotting activity, Biochemistry 27:8170 (1988).
28. J. DiMaio, B. Gibbs, D. Munn, J. Lefebvre, F. Ni, and Y. Konishi, Bifunctional thrombin inhibitors based on the sequence of hirudin$^{45-65}$, J. Biol. Chem. 265:21698 (1990).
29. P. Bourdon, J.W. Fenton II, and J.M. Maraganore, Affinity labelling of lysine-149 in the anion binding exosite of human α-thrombin with a Nα-dinitro fluorobenzyl - hirudin C-terminal peptide, Biochemistry 29:6379 (1990).
30. J.M. Maraganore, B. Chao, M.L. Joseph, J. Jablonski, K.L. Ramachandran, and J.W. Fenton II, Design and characterization of hirulogs: novel class of bivalent peptide inhibition of thrombin, Biochemistry 29:7095 (1990).
31. T. Kline, C. Hammond, P. Bourdon, and J.M. Maraganore, Hirulog peptides with scissile bond replacement resistent to thrombin cleavage, Biochem. Biophys. Res. Commun. 177:1049 (1991).

32. J. DiMaio, F. Ni, B. Gibbs, and Y. Konishi, A new class of potent thrombin inhibitor that incorporates a scissile pseudopeptide bond, FEBS Letts 282:47 (1991).
33. S. Bajusz, E. Barabas, P. Tolnag, E. Szell, and D. Bagdy, Inhibition of thrombin and trypsin by tripeptide aldehyde, Int. J. Pept. Prot. Res. 12:217 (1978).
34. C. Kettner, L. Merginger, and R. Knabb, The selective inhibition of thrombin by peptides of boroarginine, J. Biol. Chem. 265:18289 (1990).
35. R. Kikumoto, Y. Tanao, T. Tezuka, S. Tonomura, M. Mara, K. Ninomiya, A. Hijikata, and S. Okamoto, Selective inhibition of thrombin by (2R,4R)-4-methyl-1-(N2-((3-methyl-1,2,3,4-tetrahydro-8-quinolinyl)sulfonyl)-L-arginyl)-2-piperidinecarboxylic acid), Biochemistry 23:85 (1984).
36. D.M. Tollefsen, C.A. Pestka, M.J. Monafo, Activation of heparin cofactor II by dermatan sulfate, J. Biol. Chem. 258:6713 (1984).
37. F.A. Ofosu, G.J. Modi, L.M. Smith, A.L. Cerskus, J. Hirsh, and M.A. Blajchman, Heparan sulphate and dermatan sulphate inhibit the generation of thrombin activity in plasma by complementary pathways, Blood 64:742 (1984).
38. F.A. Ofosu, M.A. Blajchman, G.J. Modi, L.M. Smith, M.R. Buchanan, and J. Hirsh, The importance of thrombin inhibition for the expression of the anticoagulant activity of heparin, dermatan sulphate, low molecular weight heparin and pentosan polysulphate, Br. J. Haematol. 60:695 (1985).
39. W. Raake, R.J. Klausen, E. Meintsberger, P. Zeller, and H. Elling, Pharamcologic profile of the antithrombotic and bleeding actions of sulfated lactobionic acid amides, Sem. Thromb. Haemost. 17:Suppl 1. 129 (1991).
40. F.A. Ofosu, J. Fareed, L.M. Smith, N. Anvari, D. Hoppensteadt, and M.A. Blajchman, Eur. J. Biochem. 203:121 (1992).
41. J.W. Fenton II, J.I. Witting, C. Pouliott, and J. Fareen, Anion binding site exosite interactions with heparin and various polyanions, Ann. N.Y. Acad. Sci. 556:158 (1989).
42. J.I. Witting, P. Bourdon, D.X. Brezniak, J. Maraganore, and J.W. Fenton II, Thrombin specific inhibition by and slow cleavage of Hirulog-1, Biochem. J. 283:737 (1992).
43. R.E. Jordan, G.M. Oosta, W.T. Gardner, and R.D. Rosenberg, The kinetics of haemostatic enzyme antithrombin interactions in the presence of low molecular weight heparin, J. Biol. Chem. 255:10081 (1980).
44. D.M. Tollefsen, D.W. Majerus, and M.K. Blank, Heparin cofactor II. Purification and properties of a heparin-dependent inhibitor of thrombin in human plasma, J. Biol. Chem. 257:2162 (1982).
45. E.W. Davie, K. Fujikawa, and W. Kisiel, The coagulation cascade: initiation, maintenance and reguation, Biochemistry 30:10363 (1991).
46. F.A. Ofosu, J. Hirsh, C.T. Esmon, G.J. Modi, L.M. Smith, N. Anvari, M.R. Buchanan, J.W. Fenton II, and M.A. Blajchman, Unfractionated heparin inhibits the thrombin-catalyzed amplification reactions of coagulation more efficiently than those catalyzed by factor Xa, Biochem. J. 257:143 (1989).
47. F.A. Ofosu, J. Choay, N. Anvari, L.M. Smith, and M.A. Blajchman, Inhibition of factor X and factor V activation by dermatan sulphate and a pentasaccharide with high affinity for antithrombin III in human plasma, Eur. J. Biochem. 193:485 (1990).
48. X. Yang, M.A. Blajchman, S. Craven, L.M. Smith, N. Anvari, and F.A. Ofosu, Activation of factor V during intrinsic and extrinsic coagulation. Inhibition by heparin, hirudin and D-Phe-Pro-Arg CH_2CP, Biochem. J. 272:399 (1990).
49. D.D. Monkovic, and P.B. Tracy, Activation of human factor V by factor Xa and thrombin, Biochemistry 29:1118 (1990).

50. E. Marcianiak, Factor Xa inactivation by antithrombin III. Evidence for biological stabilization of factor Xa by factor V-phospholopid complexes, Br. J. Haematol. 24:391 (1973).
51. P.N. Walsh, R. Biggs, and G. Gagnatelli, Platelet antiheparin activity. Assay based on factor Xa inactivation by heparin and antifactor Xa, Br. J. Haematol. 26:405 (1974).
52. T.W. Barrowcliffe, F.J. Havercroft, G. Kemball-Cook, and U. Lindahl, The effect of Ca^{2+}, phospholipid and factor V on the antifactor Xa activity of heparin and its high affinity oligosaccharide, Biochem. J. 243:31 (1987).
53. F.A. Ofosu, J.W. Fenton II, J. Maraganore, M.A. Blajchman, X. Yang, L. Smith, N. Anvari, M.R. Buchanan, and J. Hirsh, Biochem. J. 283:893 (1992).
54. J.A. Koedam, R.I. Hamer, N.H. Beeser-Visser, B.N. Bouman, and J.J. Sixma, The effect of von Willebrand factor on activation of factor VIII by factor Xa, Eur.J. Biochem. 189:229 (1991).
55. C.T. Esmon, The regulation of natural anticoagulant pathways, Science 235:1348 (1987).
56. G.J. Broze, T.J. Girard, and W.F. Novotny, Regulation of coagulation by a multivalent Kunitz-type inhibitor, Biochemistry 29:7541 (1990).

PRE-CLINICAL AND CLINICAL STUDIES ON HIRULOG: A POTENT AND SPECIFIC DIRECT THROMBIN INHIBITOR

John M. Maraganore

Thrombosis and Hemostasis Research Group
Biogen, Inc.
14 Cambridge Center
Cambridge, MA 02142

Despite the use of heparin and aspirin, acute thrombotic events are frequent complications in a broad range of clinical settings including following surgery, during acute episodes of unstable angina, and during both mechanical and pharmacologic revascularization therapies. New antithrombotic drugs are required to interrupt thrombus formation in those clinical indications where heparin, aspirin, or their combination are of limited or insufficient benefit. The enzyme thrombin has emerged as a key target for the development of more effective antithrombotic drugs. Our studies have addressed the following questions:

1) Is direct antithrombin therapy more effective than heparin in prevention of thrombosis in animal models?

2) What is the basis for the improved actions of direct antithrombins?

3) Can a direct antithrombin be administered to humans at levels required to obtain optimal effects? Is it well-tolerated?

These questions have been answered using Hirulog[1], a specific and direct thrombin inhibitor. Hirulog is a peptide built by rational design using the protein hirudin as a model. It is a reversible, competitive inhibitor which binds to both thrombin's anion-binding exosite[2,3] and its catalytic site. The bivalent interactions of Hirulog with thrombin were established in kinetic[1] and X-ray crystallographic[4] studies. A space-filling representation of the Hirulog-thrombin complex is shown in Figure 1. Bivalent interactions assure potency and specificity: i) Hirulog inhibits thrombin-catalyzed hydrolysis of a chromogenic substrate with Ki=2.3 nM, while its constituent NH_2 and COOH-terminal domains inhibit thrombin with mM and μM affinities, respectively; and, 2) Hirulog fails to inhibit related proteinases, even the limit proteolytic derivative, gamma-thrombin, at concentrations > 10 μM. Both potency and specificity are required to address the questions raised above, and are critical predicates to development of new generation pharmacologic approaches.

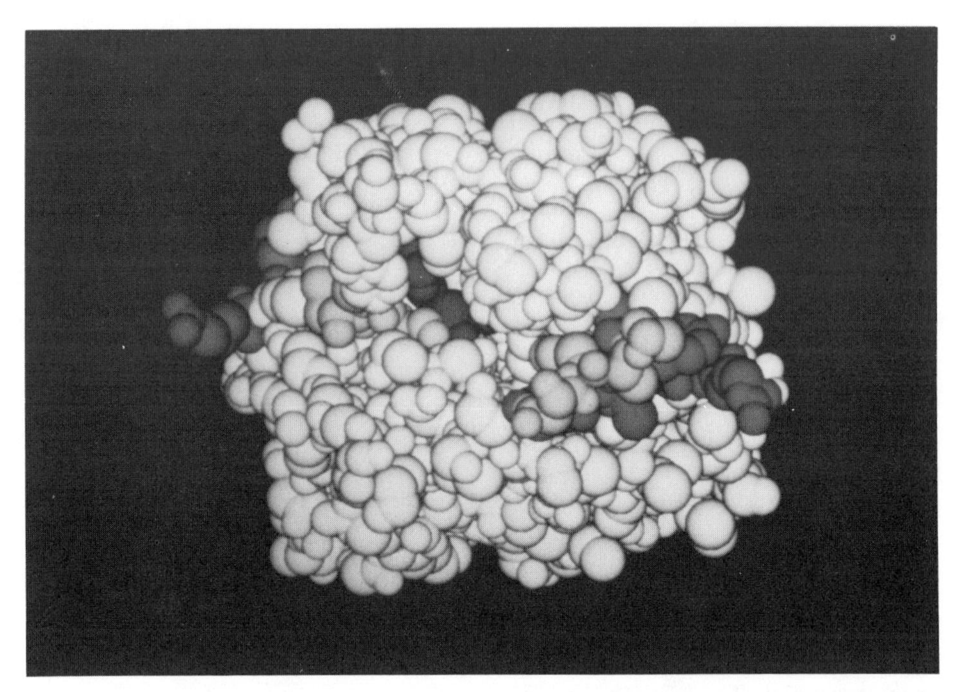

Figure 1 Space filling representation of the Hirulog-thrombin crystallographic structure - Hirulog binds to both thrombin's anion-binding exosite and catalytic site. Amino acids in thrombin implicated in heparin binding[29] are visible. The structure of the Hirulog-thrombin complex was solved to 2.2 Å resolution.

Evaluations of Hirulog in Animal Models of Thrombosis and Thrombolysis

The antithrombotic efficacy of Hirulog has been evaluated in animal models of venous and arterial thrombosis. In a rat model of venous thrombosis[5], injection of tissue thromboplastin combined with stasis led to formation of thrombus. When administered by intravenous infusion, Hirulog showed dose-dependent interruption of thrombus formation (Figure 2). The antithrombotic effects of Hirulog were correlated with prolongations in blood coagulation times. The maximal effects of the agent ($>90\%$ inhibition of thrombus weight) were observed at doses yielding prolongations in APTT to $\approx 275\%$ baseline.

In addition to demonstrating efficacy in models of venous thrombosis, Hirulog was also found to be effective in models of high-shear, platelet-dependent thrombus formation. In an exteriorized arterio-venous shunt model in baboons, IV Hirulog blocked recruitment of both ^{111}In-labelled platelets and ^{125}I-fibrin(ogen) in thrombus accumulated on segments of endarterectomized baboon aorta, collagen-coated Gortex, and Dacron vascular graft[6]. This model has been shown previously to be resistant to even supra-therapeutic doses of heparin, aspirin, or the combination of heparin and aspirin[7]. A direct comparison of the antithrombotic effects of Hirulog and heparin was performed in a porcine version of the Folts model[8]. Hirulog prevented recurrent cycles of platelet-dependent thrombus formation caused by the combination of critical-grade stenosis and deep arterial injury. Similar

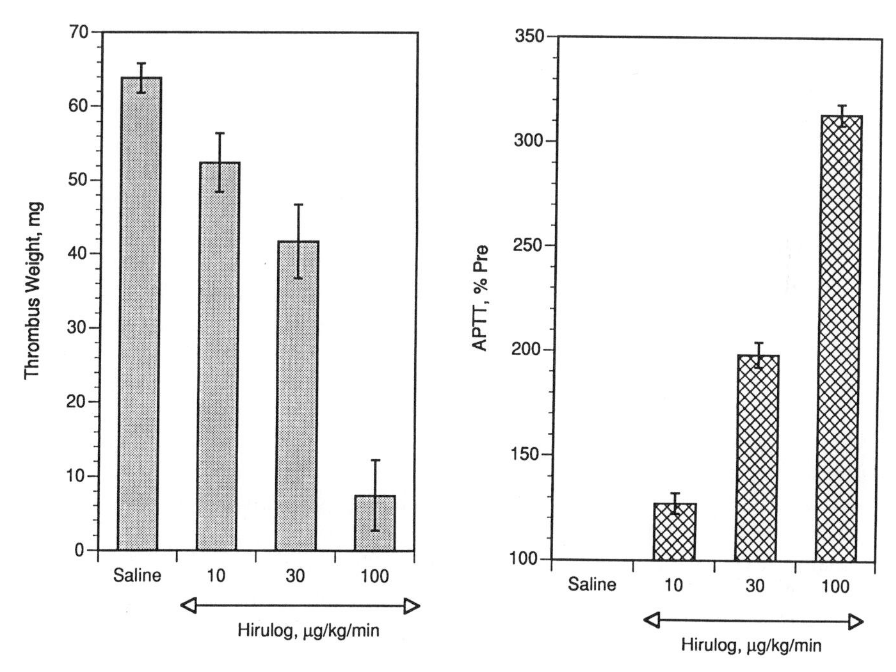

Figure 2 Antithrombotic and anticoagulant effects of Hirulog in a rat model of venous thrombosis.

effects were achieved by heparin, but only at a high dose of 500 U/kg (the equivalent of a 35,000 U bolus injection in a 70 kg human!). Importantly, interruption of thrombus formation by Hirulog was associated with modest prolongations of hemostatic parameters (APTT $<$ 300% baseline), whereas the effect of heparin was associated with marked changes in hemostasis (APTT $>$ 1000% baseline).

In addition to evaluating direct thrombin blockade in models of thrombus formation, Hirulog has also been tested in combination with thrombolytic agents for its effects on reperfusion of occlusive experimental thrombi and the prevention of reocclusion following successful lysis[9-10]. For example, in a canine copper-coil model of tPA-induced thrombolysis[11], the combination of Hirulog + tPA resulted in decreased times to reperfusion and either the delay or prevention of rethrombosis (Figure 3). The actions of Hirulog as an adjunct to tPA-induced lysis were clearly superior to those of placebo or IV heparin. Again, an important difference between the direct antithrombin and heparin was observed. Namely, even as a more effective antithrombotic, Hirulog administration was associated with modest changes in hemostatic parameters as compared to those obtained by heparin.

Animal studies on Hirulog support the following conclusions. First, direct antithrombins are superior to heparin in the prevention of thrombus formation and as adjunctive agents with thrombolytic therapy. Second, the relationship between systemic anticoagulant and antithrombotic effects for direct antithrombins differs substantially from that observed with heparin. This latter point bears particular relevance to the clinical development of antithrombin agents. Antithrombins may emerge as more potent agents for the management of thrombosis without increased risk for bleeding. Importantly, where the relationship between anticoagulant effect and clinical outcome has been established for heparin in both retrospective[11] and prospective[12] studies, this relationship must be abandoned in the clinical trials of direct thrombin inhibitors. This point is reinforced below.

Basis for Improved Antithrombotic Efficacy of Direct Thrombin Inhibitors

Notwithstanding the proven effectiveness of heparin, both clinical experience and studies highlighted above indicate limitations where direct thrombin inhibitors may prove beneficial. The antithrombotic actions of heparin are complex, in part due to the lack of a clear relationship between its pharmacokinetic and pharmacodynamic properties. Nevertheless, one might expect this agent to be highly efficient as it targets several points of the coagulation cascade and, further, it is a catalyst for numerous serpin-proteinase complexes. That is to say, one molecule of heparin can accelerate the serpin-dependent inhibition of many enzymes.

Why, then, might a direct thrombin inhibitor prove more effective than heparin. Of primary importance is the fact that thrombin plays a principle role in the regulation of blood coagulation[13], as it mediates amplification of its own zymogen activation through limited proteolysis of factor V, factor VIII and, as established quite recently[14] factor XI. Also, as the only member of the family of coagulation proteinases which can activate platelets[15] via a specific receptor[16], thrombin can modify the platelet surface for efficient assembly of the prothrombinase complex[17]. Altogether, this would indicate that the broad reactivity of the heparin-antithrombin III complex is not critical to the antithrombotic actions of heparin. Not surprisingly, then, the clinical efficacy of both heparin <u>and</u> low molecular weight heparinoids is more closely related to plasma levels of anti-IIa units, not anti-Xa units[18].

If, in fact, thrombin inhibition is fundamental to heparin's antithrombotic action, then its limited effects (as compared with direct thrombin inhibitors) must be explained by alternative factors. First, the dependence for heparin on antithrombin III may limit its actions in settings where levels of this plasma cofactor are quantitatively or qualitatively deficient. Such deficiencies are not limited to rare congenital abnormalities but, rather, broadly observed in many acquired clinical conditions[19]. Studies with Hirulog have confirmed that this agent is independent of a plasma cofactor[20].

Second, where heparin is neutralized by plasma and platelet-derived factors (e.g. platelet factor 4[21]), Hirulog escapes neutralization[20]. As shown in Figure 4, Hirulog shows equivalent activity in either platelet-poor plasma or collagen-activated platelet-rich plasma. In contrast, the effects of heparin are blocked to completion in collagen-activated platelet-rich plasma. These findings could explain the failure of heparin to prevent high-shear, platelet-dependent thrombus formation in numerous models. In particular, while <u>systemic</u>

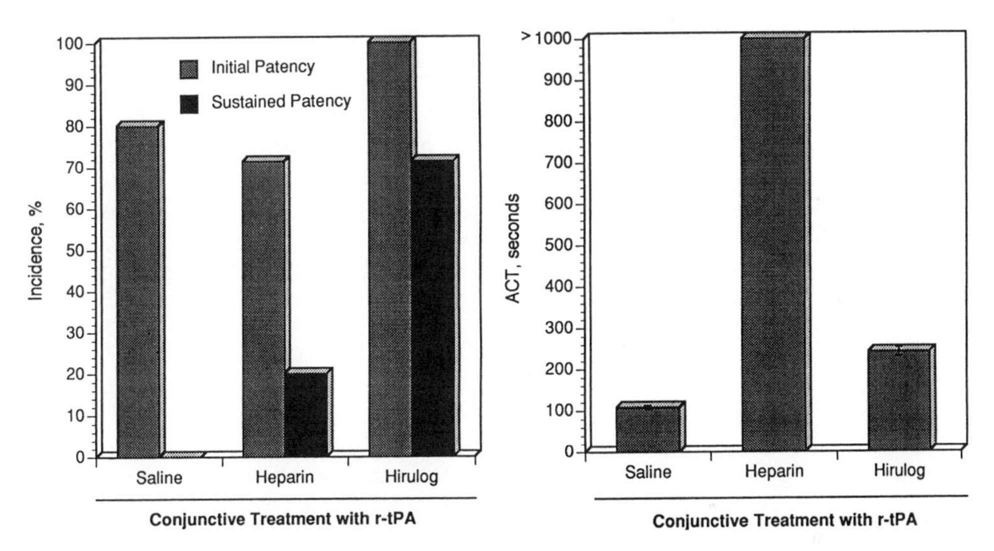

Figure 3 Effects of Hirulog, saline or heparin in a canine copper-coil model of tPA-induced thrombolysis - Hirulog administration resulted in improvements in both initial patency and sustained patency. As shown in the right panel administration was associated with modest prolongation of the activated clotting time (ACT).

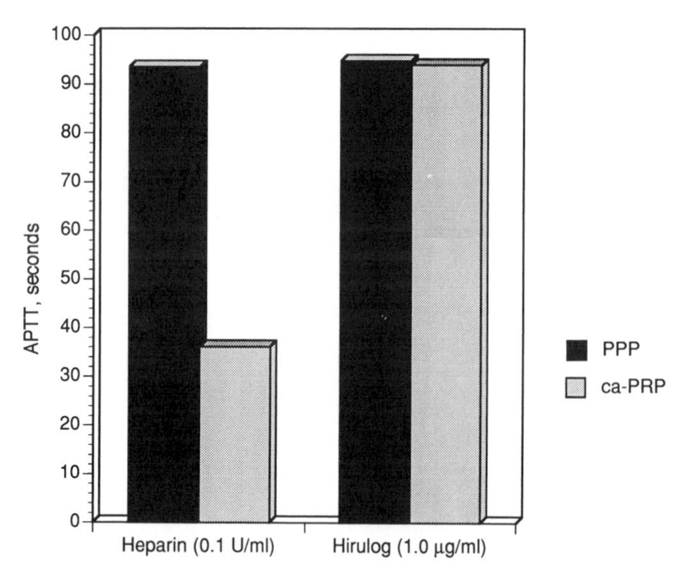

Figure 4 Heparin action is neutralized by platelet activation products: Hirulog escapes neutralization - Citrated platelet-poor plasma (PPP) and platelet-rich plasma (PRP) were obtained from 3 volunteers. Collagen was used to activate platelets in the PRP samples. Both PPP and collagen-activated PRP were supplemented with heparin (0.1 U/ml) or Hirulog (1.1 μg/ml) and then APTT was measured.

concentration of platelet factors are not sufficient to neutralize the systemic anticoagulant effects of heparin, higher, local concentrations may also block heparin action at sites of thrombus formation. Heparin-neutralizing factors may circulate at higher plasma levels during disease and infection. In contrast, due to its abilities to escape neutralization, Hirulog could inhibit arterial thrombus formation at relatively lower levels of systemic anticoagulation. As noted in the preceding section, this outcome has been observed, in fact, in animal models.

Finally, an important reason for improved antithrombotic efficacy of direct thrombin inhibitors has recently been identified through studies on clot-bound and matrix-bound thrombin[22-25]. Thrombin bound to the fibrin clot or to subendothelial matrix is resistant to neutralization by the heparin-antithrombin III complex and, yet, is still active as a procoagulant. In contrast to heparin, Hirulog is a highly efficient inhibitor of clot-bound thrombin. Clot-bound or matrix-bound thrombin can contribute to the thrombogenicity and thromboresistance of an existing clot or at a site of vessel injury and, thus, direct thrombin inhibitors may be particularly effective in the treatment of deep vein thrombosis, management of unstable angina, and prevention of reocclusion following thrombolytic therapy.

Three principle factors thus emerge as explanations for the improved antithrombotic actions of direct thrombin inhibitors in animal models: i) absence of a cofactor dependency; ii) escape from neutralization by platelet-derived substances; and iii) efficient

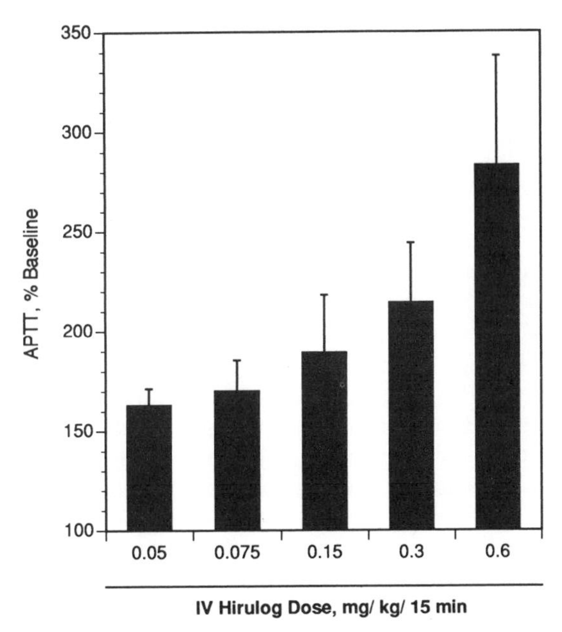

Figure 5 Dose-response for IV Hirulog administration in human volunteers - Hirulog was administered at the doses indicated as a 15 min infusion. Data are the mean ± SD peak anticoagulant effects obtained for four subjects in each group.

inhibition of clot-bound thrombin. Whether these factors are essential to the improvement of antithrombotic therapy in man remains to be seen. However, differences between heparin and direct antithrombins lead to some considerations pertinent to clinical studies. First, the absence of a cofactor dependence and of platelet neutralization may yield consistent antithrombotic effects and, if so, the potential lack of a need for laboratory monitoring of systemic anticoagulation. Second, since direct antithrombins are more efficient antithrombotic agents, the relationship now established for heparin between levels of anticoagulation and clinical outcome (both safety and efficacy) needs to be re-examined in applciations of these newer agents. The interrelationships between anticoagulant, antithrombotic, and antihemostatic effects of direct thrombin inhibitors will be understood only when optimal doses and regimens are established using clinical endpoints.

Clinical Studies of Direct Antithrombins

The actions of direct antithrombins *in vitro* and *in vivo* are promising in the context of the need for improved antithrombotic drugs, and clinical trials for these agents have been initiated. Hirulog was studied in human volunteers to examine its anticoagulant actions and tolerability[26]. In the first phase of the study, increasing doses of Hirulog were administered by IV infusion or subcutaneous injection. As shown in Figure 5, Hirulog administration in increasing IV doses was associated with dose-dependent prolongations in APTT values. Similar increases were observed in PT and TT measures. The administration of Hirulog was generally well-tolerated and, importantly, not associated with significant changes in bleeding times nor clinical evidence for bleeding. During prolonged infusions of the drug for 2-24 hrs in volunteers who had received aspirin, Hirulog showed immediate and stable anticoagulant effects. Again, the drug's administration was well-tolerated. Further, no interactions were observed between aspirin and the direct antithrombin. Altogether, these results are very encouraging, especially as doses administered to volunteers were shown previously to be effective in preventing thrombosis in animal models.

Studies of Hirulog in volunteers have already indicated important clinical differences with heparin. First, there was a close correlation between the antithrombins pharmacokinetics and pharmacodynamics. This offers a high degree of predictability when dosing patients with Hirulog. Second, the anticoagulant effects of Hirulog showed minimal intersubject variability. While not examined prospectively in this study, the consistent effects of Hirulog contrasts with historical experience with heparin, whose anticoagulant effects differ substantially amongst subjects. These two features of Hirulog suggest that optimal antithrombotic effects can be titrated to administered dose. With heparin, dose is titrated to patient anticoagulant responses. Since the establishment of effective heparin anticoagulation can take hours to days[28], wherein the patient remains at risk for thrombosis or is placed at risk for bleeding, direct antithrombin therapy with Hirulog may significantly improve patient care in the years to come.

CONCLUSION

Our studies with Hirulog demonstrate clearly that direct antithrombin therapy may solve the existing need for improved antithrombotic drugs. The next step is to extend this promise as fact by performing randomized, prospective clinical trials comparing Hirulog to existing therapies. Such studies are underway.

ACKNOWLEDGEMENTS

I wish to acknowledge the many collaborators whose work is reviewed in this chapter: John Fenton, Alexander Tulinsky, Laurence Harker, Andrew Kelly, Fumitoshi Asai, Thomas Muller, James Willerson, Jeffrey Weitz, Jack Hirsh, Tim Mant, and members of Biogen's Hirulog Team.

REFERENCES

1. J.M. Maraganore, P. Bourdon, J. Jablonski, K.L. Ramachandran, and J.W. Fenton II, Design and characterization of hirulogs : a novel class of bivalent peptide inhibitors of thrombin, Biochemistry 29:7095 (1990).
2. L.J. Berliner, Y. Sugawara, and J.W. Fenton II, Human α-thrombin binding to non-polymerized fibrin-Sepharose. Evidence for an anionic binding region, Biochemistry 24:7005 (1985).
3. D.H. Bing, M. Cory, and Fenton II, Exosite affinity labelling of human thrombins. Similar labelling on the A chain and B chain fragments of clotting α and non-clotting ß-thrombins, J. Biol. Chem. 252:8027 (1977).
4. E. Skrzypczak-Jankun, V.E. Carperos, K.G. Ravichandran, A. Tulinsky, M. Westbrook, and J.M. Maraganore, Structure of the hirugen and hirulog-1 complexes of α-thrombin, J. Mol. Biol. 221:1379 (1991).

5. J.M. Maraganore, T. Oshima, F. Asai, and A. Sugitachi, Comparison of anticoagulant and antithrombotic activities of hirulog-1 and argatroban (MD-805), Thromb. Haemostas. 65:651 (abstract) (1991).
6. A.B. Kelly, S.R. Hanson, B. Chao, J.M. Maraganore, and L.A.Harker, Bivalent antithrombin peptide interruption of thrombus formation *in vivo*, Thromb. Haemostas. 65:735 (abstract) (1991).
7. L.A. Harker, and S.R. Hanson, Experimental arterial thromboembolism in baboons. Mechanisms, quantitation and pharmacologic prevention, J. Clin. Invest. 64:559 (1979).
8. T.H. Muller, V. Koch, U. Gerster, and J.M. Maraganore, Hirulog, a synthetic hirudin-based peptide is superior to heparin in a porcine model of arterial thrombosis, Thromb. Haemostas. 65:1291 (abstract) (1991).
9. P. Klement, J. Hirsh, J.M. Maraganore, and J. Weitz, The effect of thrombin inhibitors on tissue plasminogen activator-induced thrombolysis in a rat model, Thromb. Haemostas. 65:735 (abstract) (1991).
10. J.M. Maraganore, S.K. Yao, J. McNatt, J. Edit. K. Cui, L.M. Buja, and J.T. Willerson, Hirudin-based peptidesaccelerate thrombolysis and delay reocclusions after treatment with recombinant tissue-type plasminogen activator, Thromb. Haemostas. 65:1188 (abstract) (1991).
11. R.D. Hull, G.E. Raskob, J. Hirsh, R.M. Jay, J.R.LeClerc, W.H. Geerts, D. Rosenbloom, D.L. Sackett, C. Anderson, and L. Harrison, Continuous intravenous heparin compared with intermittent subcutaneous heparin in the initial treatment of proximal vein thrombosis, N. Engl. J. Med. 315:1109 (1986).
12. D. Basu, A. Gallus, J. Hirsh, and J. Cade, A prospective study of the value of monitoring heparin treatment with the activated partial thromboplastin time, N. Engl. J. Med. 287:324 (1972).
13. C.T. Esmon, W.G. Owen, D.L. Duiguid, and C.M. Jackson, The action of thrombin on blood clotting factor V: conversion of factor V to a prothrombin binding protein, Biochim. Biophys. Acta 310:289 (1973).
14. D. Gailani, and G.J. Broze, Factor XI-activation in a revised model of blood coagulation, Science 253:909 (1991).
15. D.M. Tollefsen, J.R. Feagler, P.W. and Majerus, The binding of thrombin to the surface of platelets, J. Biol. Chem. 249:2646 (1974).
16. T.K.H. Vu, D.T. Hung, V.I. Wheaton, and S.R. Coughlin, Molecular cloning of a functional thrombin receptor reveals a novel proteolytic mechanism of receptor activation, Cell 64:1057 (1991).
17. K.G. Mann, R.J. Jenny, and S. Krishnaswamy, Cofactor proteins in the assembly and expression of blood clotting enzyme complexes, Annu. Rev. Biochem. 57:915 (1988).
18. M.R. Buchanan, B. Boneu, F.A. Ofosu, and J. Hirsh, The relative importance of thrombin inhibition and factor Xa inhibition to the antithrombotic effects of heparin, 65:98 (1985).
19. E. Thaler, and K. Lechner, Antithrombin III deficiency and thromboembolism, in Clinics in Haematology (Prentice CRM, ed.) vol. 10, p. 369, Saunders, Philadelphia, PA (1981).
20. J.M. Maraganore, B. Chao, J.I. Weitz, and J. Hirsh, Comparison of antithrombin activities of heparin and hirulog 1: Basis for improved antithrombotic properties of direct thrombin inhibitors, Thromb. Haemostas. 65:829 (abstract) (1991).
21. D.A. Lane, J. Denton, A.M. Flynn, L. Thunberg, and U. Lindahl, U. Anticoagulant activities of heparin oligosaccharides and their neutralization by platelet factor 4, Biochem. J. 218:725 (1984).

22. P.J. Hogg, and C.M. Jackson, Fibrin monomer protects thrombin from inactivation by heparin-antithrombin III: Implications for heparin efficacy, Proc. Natl. Acad. Sci. USA 86:3619 (1989).
23. J.I. Weitz, M. Hudoba, D. Massel, J.M. Maraganore, and J. Hirsh, Clot-bound thrombin is protected from heparin-antithrombin III but is susceptible to inactivation by antithrombin III-independent inhibitors, J. Clin. Invest. 86:385 (1990).
24. M. Mirshahi, J. Soria, C. Soria, F. Faivre, H. Lu, M. Courtney, C. Roitsch, D. Tripier, and J.P. Caen, Evaluation of the inhibitors of heparin and hirudin of coagulation activation during r-tPA-induced thrombolysis, Blood 74:1026 (1989).
25. R. Bar-Shavit, M. Benezra, A. Eldor, E. Hy-Am, J.W. Fenton II, G.D. Wilner, and I. Vlodavsky, Thrombin immobilized to extracellular matrix is a potent mitogen for vascular smooth muscle cells: non-enzymatic mode of action, Cell Regulation 1:453 (1990).
26. A. Dawson, P. Loynds, K. Findlen, E. Levin, T. Mant, J.M. Maraganore, D. Hanson, J. Wagner, and I. Fox, Hirulog 1: a bivalent thrombin inhibitor with potent anticoagulant properties in humans, Thromb. Haemostas. 65:830 (abstract) (1991).
27. M.K. Cruickshank, M.N. Levine, J. Hirsh, R. Roberts, and M. Siguenza, A standard heparin program for the management of heparin therapy, Arch. Intern. Med. 151:333 (1991).
28. F.C. Church, C.W. Pratt, C.M. Noyes, T. Kalayanamit, G.B. Sherrill, R.B. Tobin, and J.B. Meade, Structural and functional properties of human α-thrombin phosphorylated alpha-thrombin, and gamma thrombin.Identification of lysyl groups in alpha thrombin that are critical for heparin and fibrin(ogen) interaction, J. Biol. Chem. 264:18419 (1989).

THE EFFECT OF RECOMBINANT HIRUDIN ON ARTERIAL THROMBOSIS

R.B. Wallis

Research Centre
Ciba-Geigy Pharmaceuticals
Horsham
West Sussex

INTRODUCTION

"Hirudin" is the name given to what is now known to be a family of highly homologous polypeptide anticoagulants from the European medicinal leech, Hirudo medicinalis. It has been known since 1884 when an anticoagulant principal was extracted from leech heads but it was not until the 1950s that the activity was isolated and characterised as a polypeptide[1]. Several laboratories have now cloned and purified biologically active recombinant hirudins from yeast or bacterial systems. The recombinant molecules all lack a sulphate moeity on tyrosine 63 and are hence designated desulphatohirudins[2]. Differences in the amino acid sequences are reflected by the number of the hirudin variant (HV). Thus descriptors: HV1, HV2, etc. are used.

Recombinant desulphatohirudin variant 1 (rHV1) is a highly selective inhibitor of thrombin. No direct inhibitory effect has yet been described on any other protease even though many, including those in the coagulation, fibrinolytic, complement and digestive systems, have been investigated and high concentrations of hirudin have been used[3]. rHV1 inhibits all of the effects of thrombin where catalytic activity is required. Thus, not surprisingly, it prolongs the clotting time of plasma whether this is induced directly with thrombin or through the intrinsic (activated partial thromboplastin time (APTT)) or the extrinsic pathways (prothrombin time)[3]. In this article I shall concentrate on the effects of hirudins on the function of platelets as they seem to have a more important role in arterial thrombosis.

Microscopical examination of thrombi taken from rats where the thrombus was induced in the aorta in high blood flow demonstrates the large enrichment of platelets compared to their concentration in the normal circulation. A major part of the thrombus comprises platelets in this situation. This contrasts with venous thrombi, formed in static blood *in vivo*, where there is little change in the concentration of any of the blood components[4]. rHV1 inhibits thrombus formation in each of these situations but approximately ten times the dose is required to inhibit the platelet-rich arterial thrombosis[5]. This observation is attributable to the difference in potency of thrombin on platelets and on fibrinogen. The dissociation constant for the interaction of thrombin with platelets is in the nmolar range whilst the Km for fibrinogen cleavage is in the μmolar range. Hence low, nmolar, concentrations of thrombin can activate platelets without affecting fibrinogen. This means that when a large amount of thrombin is formed *in vivo*, increasing amounts of hirudin will initially prevent fibrin formation but higher concentrations will be required to titrate the thrombin to below the concentration required to activate platelets.

The effect of hirudins on platelets is very specific as there is no influence on platelet aggregation induced by any of the other commonly used agonists except thrombin. The effect on thrombin-induced responses is attributable to inhibition of the thrombin as catalytically active thrombin is required in order to activate platelets. All the effects of thrombin on platelets are affected equally by hirudins thus, aggregation, ATP secretion, thromboxane biosynthesis and expression of the procoagulant are all inhibited[6].

The inhibitory effect of hirudins on arterial thrombosis has been extensively studied in many different types of model ranging from arterial injury to lysis of existing thrombi and in a number of different species.

Arterial Injury in the Rat

In animals it is difficult to induce a reproducible injury to an arterial wall without recourse to quite severe electrical or chemical injury and then it is by no means certain that the injury caused resembles that which occurs in man. Probably of greater relevance to the clinic are those models where vascular damage is caused by mechanical means. In the rat, a model of arterial thrombosis where mechanical damage is elicited to the dorsal aorta by application of a pressure clamp for 1 min has been developed[6]. Restoration of blood flow for 45 min. after clamping allows a large thrombus to develop. The resultant morphology shows large deposits of platelets, attached to deep tears in the vessel wall, partially obstructing the lumen. Incorporation of both fibrin(ogen) and platelets at the site of injury and insignificant increases in albumen or erythrocytes are demonstrated by radiometric analysis. Thrombosis was evaluated by the incorporation of radiolabelled platelets and fibrinogen on to the injured segment. rHV1, dose dependently, inhibited both platelet and fibrin deposition up to about 75%[6]. The doses (3-10 mg/kg s.c.) required to inhibit caused a prolongation of the APTT to between 2 and 4 times the control value. Comparison of the effects of rHV1 with those of heparin indicate that, for equivalent levels of anticoagulation measured by APTT, rHV1 is a more effective antithrombotic. In fact, prolongation of APTT to the same (4 times the control value) extent resulted in no inhibition at all by either unfractionated or a low molecular weight heparin[6].

Angioplasty in the Pig

When an angioplasty balloon is inflated in the carotid artery of the pig, very similar morphological changes to those observed after similar procedures in man can be demonstrated[7]. There is considerable damage to the endothelial cells and in about 50% of the animals there are deep tears in the intima that penetrate the internal elastic lamina into the media. A large thrombus forms at the site of damage and it comprises mainly platelets. Thrombus formation was monitored by the incorporation of radiolabelled platelets and fibrinogen and by morphometric analysis of the thrombi formed. Intravenous infusions of rHV1 (0.7 or 1 mg/kg/h) abolished thrombus formation and reduced the platelet incorporation by about 85% such that only a single layer or less of platelets remained[8,9]. The APTT was maintained at 2-3 times the control at these doses. Heparin infusions that caused a similar effect on the APTT were much less effective, once again demonstrating the superiority of rHV1 at equivalent levels of anticoagulation.

Thrombolysis

The lysis of arterial thrombi in man by both streptokinase and by tissue plasminogen activator (t-PA) causes an increase in both circulating thrombin-antithrombin complexes[10] and the systemic production of thromboxane A2[11,12] indicating the highly prothrombotic nature of this procedure and illustrating two possible reasons, thrombin generation and platelet activation, why there is such a high incidence of acute reocclusion in the clinic. The rate of lysis and the susceptibility to reocclusion seems to depend on the balance between thrombolysis and thrombogenesis. The high thrombogenicity of existing thrombi is probably caused by the capacity of fibrin to bind thrombin during its formation and the fact that this thrombin retains its activity. This activity of clot-bound thrombin is expressed during lysis and for example can cleave further fibrinogen, releasing FpA[13] and can therefore presumably cause the recruitment of further fibrin and platelets on to the clot surface. Recombinant hirudins inhibit FpA formation caused by plasma clots whereas heparin is much less effective[13,14]. In order to reduce the problem of reocclusion, thrombolysis is normally carried out with conjunctive therapy with an anticoagulant such as heparin. In spite of this, the reocclusion rate is very high and clearly there would be a significant advantage in reducing it. Data from animal models suggest that rHVl may be a very effective alternative to heparin. In pigs, coadministration of rHV1 (1 mg/kg bolus i.v. plus 0.7 mg/kg/h infusion) with t-PA resulted in much faster and a much higher rate of recanalisation of carotid arteries which were occluded following balloon angioplasty, than t-PA alone (Table 1)[15].

In dogs, coronary artery thrombi were lysed much more rapidly, resulting in reestablishment of blood flow, when rHV1 (1.5 mg/kg bolus i.v. plus 1.5 mg/kg/h infusion) was given with the t-PA than if t-PA was used alone (Table 2)[16]. Heparin was a much less effective alternative. In addition there were no reocclusions in the rHV1 plus t-PA group whereas all animals in the group given t-PA alone and 5 of 6 animals in the group given heparin plus t-PA reoccluded.

Table 1 Effect of adjunctive therapy on thrombolysis with t-PA in the pig (ref[15]).

Treatment	Number recanalised	Time to recanalisation
saline	0/3	-
t-PA	1/6	120
t-PA + rHV1	7/7	43 ± 28

Table 2 Effect of adjunctive therapy on thrombolysis and re-occlusion with t-PA in the dog (ref[16]).

Treatment	Time to recanalisation(min)	Number of reoclusions
t-PA	43 ± 16	7/7
t-PA + heparin	35 ± 22	5/6
t-PA+ rHV1	19 ± 10	0/6

Specific inhibition of thrombin with rHV1 is clearly a very effective antithrombotic strategy in the arterial circulation where the thrombi are dominated by platelet incorporation. Results from experiments in animals clearly indicate the great promise of such an agent as adjunctive therapy in angioplasty and thrombolysis. We await with great interest results from studies in human patients.

REFERENCES

1. F. Markwardt, The comeback of hirudin - an old established anticoagulant agent, Folia Haematol. 115: 0 (1988).
2. W.E. Märki, and R.B. Wallis, The anticoagulant and antithrombotic properties of hirudins, Thromb. Haemostas. 64:344 (1990).
3. M.D. Talbot, Biology of recombinant hirudin (CGP 39393): A new prospect in the treatment of thrombosis, Sem. Thromb. Haemostas. 15:293 (1989).
4. K.D. Butler, C.M. Lees, K.A. Mitchell, R.F. Peters, M.D. Talbot, M.F. Tweed, and R.B. Wallis, A comparison of three experimental models of thrombosis using morphometric and radiometric analysis, Thromb. Haemostas. 62:Abs 373 (1989).
5. F. Markwardt, Development of hirudin as an antithrombotic agent, Sem. Thromb. Haemostas. 15:269 (1989).
6. M.D. Talbot, J. Ambler, K.D. Butler, C.M. Lees, K.A.Mitchell, R.F. Peters, M.F. Tweed, and R.B. Wallis, The effects of recombinant desulphatohirudin on arterial thrombosis in rats, Haemostasis 21(suppl):73 (1991).
7. P.M. Steele, J.H. Chesebro, A.W. Stanson, D.R. Holmes, M.K. Dewanjee, L. Badimon, and V. Fuster, Balloon angioplasty, Natural history of the pathophysiological response to injury in a pig model, Circ. Res. 57:105 (1985).
8. M. Heras, J.H. Chesebro, W.J. Penny, K.R. Bailey, L. Badimon, and V. Fuster, Effects of thrombin inhibition on the development of acute platelet thrombus deposition during angioplasty in pigs: heparin versus recombinant hirudin a direct thrombin inhibitor, Circulation 79:657 (1989).
9. M. Heras, J.H. Chesebro, M.W.I. Webster, J.S. Mruk, D.E. Grill, W.J. Penny, E.J.W. Bowie, L. Badimon, and V. Fuster, Hirudin, heparin and placebo during deep arterial injury in the pig: the *in vivo* role of thrombin in platelet mediated thrombosis, Circulation 82:1476 (1990).
10. D.C. Gulba, M. Barthels, G-H. Reil, and P.R. Lichtlen, Thrombin/antithrombin III complex level as early predictor of reocclusion after successful thrombolysis, Lancet ii:97 (1988).
11. D.M. Kerins, L. Roy, G.A. Fitzgerald, and D.J. Fitzgerald, Platelet and vascular function during coronary thrombolysis with tissue type plasminogen activator, Circulation 80:1718 (1989).
12. D.J. Fitzgerald, F. Catella, L. Roy, and G.A. FitzGerald, Marked platelet activation *in vivo* after intravenous streptokinase in patients with acute myocardial infarction, Circulation 77:142 (1989).
13. M. Mirshahi, J. Soria, C. Soria, R. Faivre, H. Lu, M. Courtney, C. Roitsch, D. Tripier, and J.P. Caen, Evaluation of the inhibition by heparin and hirudin of coagulation activation during r-tPA-induced thrombolysis, Blood 74:1025 (1989).
14. J.I. Weitz, M. Hudoba, D. Massel, J. Maraganore, and J.Hirsh, Clot-bound thrombin is protected from inhibition by heparin-antithrombin III but is susceptible to inactivation by antithrombin III-independent inhibitors, J. Clin. Invest. 86:385 (1990).

15. J.S. Mruk, J.H. Chesebro, M.W.I. Webster, M. Heras, D.E. Grill, and V. Fuster, Hirudin markedly enhances thrombolysis with rt-PA, Circulation Abs 533 (1990).
16. E.J. Haskel, N.A. Nager, B.E. Sobel, and D.R. Abendschein, Relative efficacy of antithrombin compared with antiplatelet agents in accelerating coronary thrombolysis and preventing early reocclusion, Circulation 83:1048 (1991).

INDEX